T0298681

Surveying with Geomatics and R

Surveying with Geomatics and R

Marcelo de Carvalho Alves and Luciana Sanches

CRC Press
Taylor & Francis Group
Boca Raton London New York

CRC Press is an imprint of the
Taylor & Francis Group, an **informa** business

First edition published 2022
by CRC Press
6000 Broken Sound Parkway NW, Suite 300, Boca Raton, FL 33487-2742

and by CRC Press
4 Park Square, Milton Park, Abingdon, Oxon, OX14 4RN

Library of Congress Cataloging-in-Publication Data

Names: Alves, Marcelo de Carvalho, author. | Sanches, Luciana, author.
Title: Surveying with geomatics and R / Marcelo de Carvalho Alves and Luciana Sanches.
Description: Boca Raton : CRC Press, [2022] | Includes bibliographical references and index.
Identifiers: LCCN 2021036434 (print) | LCCN 2021036435 (ebook) | ISBN 9781032015033 (hardback) | ISBN 9781032026213 (paperback) | ISBN 9781003184263 (ebook)
Subjects: LCSH: Surveying--Data processing. | Geomatics. | R (Computer program language)
Classification: LCC TA556.M38 A48 2022 (print) | LCC TA556.M38 (ebook) | DDC 526.90285--dc23/eng/20211006
LC record available at https://lccn.loc.gov/2021036434
LC ebook record available at https://lccn.loc.gov/2021036435

ISBN: 978-1-032-01503-3 (hbk)
ISBN: 978-1-032-02621-3 (pbk)
ISBN: 978-1-003-18426-3 (ebk)

DOI: 10.1201/9781003184263

Typeset in Latin Modern
by KnowledgeWorks Global Ltd.

Publisher's note: This book has been prepared from camera-ready copy provided by the authors.

Contents

About the Authors

Marcelo de Carvalho Alves

Dr. Alves is associate professor at the Federal University de Lavras, Brazil. His education includes master's, doctoral, and post-doctoral degrees in Agricultural Engineering at Federal University of Lavras, Brazil. He has varied research interests and has published on surveying, remote sensing, geocomputation and agriculture applications. He has over 20 years of extensive experience in data science, digital image processing and modeling using multiscale, multidisciplinary, multispectral and multitemporal concepts applied to different environments. Experimental field-sites included a tropical forest, savanna, wetland and agricultural fields in Brazil. His research has been predominantly funded by CNPq, CAPES, FAPEMIG and FAPEMAT. Over the years, he has built up a large portfolio of research grants mostly relating to applied and theoretical remote sensing, broadly in the context of vegetation cover, plant diseases and related impacts of climate changes.

Luciana Sanches

Dr. Sanches graduated with a degree in Sanitary Engineering from the Federal University of Mato Grosso, Brazil, a master's degree in Sanitation, Environment and Water Resources from the Federal University of Minas Gerais, a PhD in Road Engineering, Hydraulic Channels and Ports from Universidad de Cantabria, Spain, a post-doctorate degree in Environmental Physics, Brazil, and a post-doctorate degree in Environmental Sciences from the University of Reading, United Kingdom. She specialized in workplace safety engineering and in project development and management for the Municipal Water Resources Management by the National Water Agency. She is currently associate professor at the Federal University of Mato Grosso, and worked for more than 20 years in research on atmosphere-biosphere interaction, hydrometeorology in meant temporal-spatial scales with interpretation based in environmental modeling and remote sensing. She has been applying geomatics in teaching and research activities to support the interpretation of environmental dynamics.

Preface

This book is the result of experience in teaching, research and extension activities developed by the authors over their 20-year professional careers. With the multidisciplinary association of areas such as geomatics, electronics and robotics, science paradigms can be better understood, such as the extension of the space universe in which we live and the origin of life on Earth. The science, art and technology of surveying has advanced rapidly with the need to obtain maps with locations of borders, realization of practices and constructions in rural areas, urban areas, preservation of natural resources and knowledge about ecological aspects of the inhabitants of the planet Earth. Scientific principles are used to encompass a broad range of more specialized fields of geomatic engineering, each with a more specific emphasis on particular areas of land surveying, photogrammetry, remote sensing, cartography, geographic information systems and science. The traditional instruments used in the 1960s and 1970s, such as theodolites, surveyor levels and surveying chains, have been complemented by a set of new high-tech instruments. Currently, depending on the quality of the topographic survey, measuring the environment with geomatics requires the use of electronic instruments, such as electronic total station, global navigation satellite systems (GNSS), cell phones connected to geographic information systems and with geographic applications, 3D mobile mapping systems, laser scanners, digital cameras, remote sensors and digital printers. The scope of the chapters covered was chosen for applications in agricultural engineering science problems using computer programming routines, through the R software and R packages. Exercises are solved at the end of each chapter to help understand the theories developed in the 14 chapters. This is the first book ever with a theoretical and practical approach on the use of geomatics with science and R software. As a study guide, slides and illustrative videos are presented about the subject covered in each chapter. The development of scientific research on geomatics is stimulated through activity proposals that can be used or adapted by the student to assess the applicability of the subject matter covered in each chapter. Learning outcome assessment strategies are also included in order to expand the possibilities of solving geomatics problems with R and science by students, teachers, researchers and users of geomatics.

Surveying with Geomatics and R fills a gap in the literature presenting basic concepts and practical material on surveying with geomatics using R and R packages. Thereby, we hope to engage students in the learning process of surveying with real field examples and different degrees of complexity along the book development and improved results based on existing literature. We explore surveying problems based on field observations and geospatial advanced technology. Thus, the book can be applied in geocomputation, remote sensing, geography and cartography courses focusing on surveying tasks. We include a wide range of case studies as motivational self-paced tutorials. Computer programming routines are detailed and linked with theories and applications of each chapter. The academic and professional community can use a free software to develop complex surveying problems. Certainly, after this book, a new way of teaching surveying courses will arise with didactic motivating examples and development possibilities of R free software user-defined functions.

With the advancement of surveying science and techniques, there was a need to create a broader term to encompass methods of mathematical modeling, georeferencing, cartographic representation, geocomputation and electronics applied to surveying tasks, giving rise to the term "geomatics". With this, there was a need to produce a book using theoretical and practical surveying fundamentals associated with geospatial information, R software and R packages. The approach of covering surveying topics associated with scientific projects and examples of manuscript preparation in the surveying area give this work unprecedented prominence. The mathematical modeling, geocomputation, mapping and image processing techniques used in examples in the book expand the application potential of this work for use in a wider range of didactic courses in which geomatics can be applied. We also present relevant topics updated with scientific references and technological description of instruments and methods for surveying the Earth's environment. All chapters are structured with learning questions, learning outcomes, computation, solved exercises, homework, resources on the Internet, research suggestion, and learning outcome assessment strategies in order to expand the scientific capability of each topic covered on the use of surveying with geomatics and R.

The basic idea of the book is to advance slowly in surveying topics, with increasing order of complexity from the first to the last chapter. The book covers 14 chapters about surveying applications with geomatics and R. Some techniques presented at the beginning of the book, with direct measurements of distance and stadia may be considered obsolete; however, they are important for understanding surveying in practice and can be used to create sophisticated technological instruments using the same basic principle. In Chapter 1, scientific applications of surveying are introduced. In Chapter 2, we present introduction to geomatics and measurement units. In Chapter 3, theory on measurement errors is presented. Chapter 4 elucidates questions about angle and direction observations in geomatics and R. Chapter 5 presentes direct distance and angle measurements as a practical and low-cost measurement option. In Chapter 6, information on stadia indirect measurements with mechanical optical instruments, theodolites and optical levels is covered. In Chapter 7, aspects involved with electronic distance and level measurements are presented. In Chapter 8, radial traverse survey examples and calculations are addressed. In Chapter 9, closed-path traverse surveying is detailed with traversing adjustment examples and calculations. In Chapter 10, questions are raised about intersection surveying for determining coordinates of points in inaccessible locations. In Chapter 11, a practical way to elaborate a surveying descriptive memorial, with inverse, area and perimeter calculation of closed-path polygon surveying is addressed. In Chapter 12, we present coordinate reference systems for geodetic surveying with geomatics and R. In Chapter 13, cartographic coordinate projection systems are evaluated in theory and practice with geomatics and R. Finally, in Chapter 14, we address questions on how global navigation satellite systems have been used for dynamic surveying, speed calculation and mapping.

The practical examples solved cover the topographic survey with field data measurements and different survey equipments. We started the book exploring basic geographic data and tools with increasing evolving techniques and complexity of analyses until the end of the book. The number of used packages and calculation complexity increased along the book with applied geographical examples. At the end of the book, we present how to use international reference for geographic information representation and reprodution worldwide. In each chapter we show study cases with computation practices related to the chapter subject in geomatics. The book is innovative, presenting a large number of surveying example applications with field equipment and R packages, exploring scientific possibilities of using geomatics for hypothesis test answers.

Marcelo de Carvalho Alves and Luciana Sanches, Lavras, June 2021

1

Scientific Applications of Surveying

1.1 Learning Questions

The emergent learning questions answered through reading the chapter are as follows:

- How to perform science with geomatics.
- What the relationship is between applications of geomatics to scientific practice.
- How to solve geographic problems objectively.
- How to carry out scientific research planning in geomatics.

1.2 Learning Outcomes

The learning outcomes expected from reading the chapter are as follows:

- Understand the use of geomatics in everyday life.
- Appreciate the variety and diversity of applications of geomatics as a science.
- Identify hypotheses in scientific studies of geographic problems.
- Check an example of writing a scientific project and article manuscript in geomatics with R.
- Check an example of scientific dissemination of geomatics with R.

1.3 Introduction

Geospatial data and analysis have become so commonplace that society often takes them for granted. For example, most people are not concerned about how the fastest route to a local is determined when an Internet mapping site (Bing Maps, Google Maps, Yahoo! Maps) or personal cell phone navigation device is used. The only concern is with the correct functioning of the device. Therefore, it became essential to worry about improving the scientific applications of geomatics in order to have the highest possible accuracy and efficiency in the decision making process (Jensen and Jensen, 2012).

Geomatics has been used in everyday activities of working and living on Earth. With this, the following everyday uses of geomatics are highlighted (Longley et al., 2001):

- Efficiency in decision making;
- Application in various areas related to socioeconomics and environment;

- Use in mapping, measuring, managing, monitoring and modeling operations;
- Measurable economic benefits;
- Combinations with other technologies.

The factors that determined the everyday use of geomatics in modern society are (Longley et al., 2001):

- Increased availability of geographic information systems on the Internet and local enterprise networks;
- Relative price reductions of geomatics hardware and software;
- Increased awareness of the population to include geographic dimension in decision making;
- Possibility of user interaction with windowed environments;
- Technology supporting applications to visualize, manage and analyze data, as well as linking to other software;
- Availability of satellite positioning system;
- Availability of user-friendly applications;
- Accumulated experience in using geomatics applications.

1.4   Context of Information in Geomatics

The geospatial information used occurs in a context consisting of the set of knowledge acquired in scientific research and geospatial applications, technologies applied to develop and manage computer systems and the institutions created to facilitate the acquisition and development of geospatial information. The context of geospatial information affected different sectors of the economy (Figure 1.1) (Lo and Yeung, 2007).

The main sources of geospatial information production are (Burrough and McDonnell, 1998):

- Topographic mapping;
- Property registration and cadastre;
- Hydrographic mapping;
- Military organizations;
- Remote sensing and satellite agencies and companies;
- Survey of natural resources, such as geology, hydrology, soil, ecology, biogeography, meteorology, climatology, and oceonography.

The main types of geographic data available are (Burrough and McDonnell, 1998):

- Topographic maps at different scales;
- Imagery at different data collection altitudes and resolutions;
- Administrative boundaries, census data, zip code, statistics, people, land cover and land use at different resolutions;
- Marketing survey data;
- Utility data, such as gas, water, sewage, power lines, and Internet network and their location;
- Data on rocks, water, soil, atmosphere, biological activity, natural disasters, and other types.

The main applications of geographic data are (Burrough and McDonnell, 1998; Neves et al., 1998; Silva and Assad, 1998):

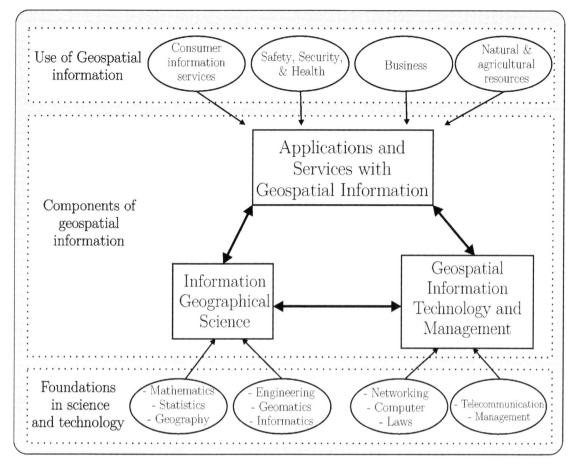

FIGURE 1.1: Geospatial information context.

- Agricultural sciences, for monitoring, management, description and scenario building;
- Archaeology, for description and study of the past;
- Environmental monitoring and management;
- Health and epidemiology, for locating diseases in relation to the environment;
- Emergency services, for optimizing routes for ambulance, police, fire escape, investigation and location of crimes;
- Navigation, for air, sea, and land;
- Marketing, for locating places and target groups, delivery optimization;
- Regional and local cost planning, maintenance, and site management;
- Planning and management of highways, railroads, and airways;
- Property and inventory evaluation, calculation of cut, fill, and volume of materials;
- Social studies to analyze population movement, local and regional development;
- Tourism, for locating and managing attractions and facilities;
- Everyday utilities for locating, managing, and planning water, drainage, gas, electricity, telephone, and cabling services.

With the need to improve the quality of products, services, and processes, geomatics has been used at different structural levels of government organizations in the following applications (Figure 1.2) (Longley et al., 2001):

- Inventory resources and infrastructure, planning transportation routing, improving public service delivery, managing land development, and generating revenue by increased economic activity;

- Use of geomatics in long-term geographic problems with health, safety and welfare of citizens, incorporating public values in decision making, providing services in a fair and equitable manner, and representing citizens' opinions by democratic work;
- Applications of geomatics in public health risk monitoring, housing stock management, social welfare fund allocation, crime tracking, geodemography analysis, operational, tactical and strategic decision making in enforcement, health planning and education management;
- Asset inventory, policy analysis, modeling and strategic planning.

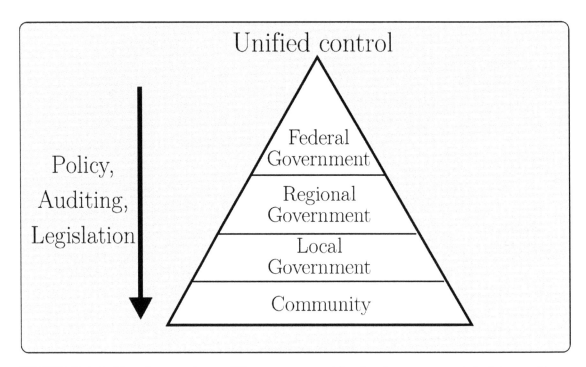

FIGURE 1.2: Use of geomatics at different structural levels of governmental decision making.

1.5   Scientific Applications of Geomatics

In the development of scientific applications of geomatics, some fundamental concepts are presented in order to clarify any conceptual doubts about science, scientist, research, experiment, and hypothesis (Table 1.1).

In the deductive scientific method, a chain of descending reasoning of analysis from the general to the particular is used to reach a conclusion. In the inductive scientific method, inferences about a general or universal truth not contained in the observations are made from particular observed data.

In basic research, useful knowledge is generated for the advancement of science with no foreseen practical application. In applied research, the generation of knowledge for practical application is directed to the solution of specific problems.

Since in science the goal of knowing and solving geographic problems varied according to the complexity of available scientific principles and techniques, some project goals used in geomatics are (Longley et al., 2001):

- Rational, effective and efficient allocation of resources according to criteria;
- Monitoring and understanding the geospatial context;
- Understanding regional differences of an object or process;
- Understanding processes in natural and anthropic environments;
- Prescribing strategies for environmental maintenance, air quality, soil and water conservation.

Applications of geomatics should be based on sound concepts and theory to solve different types of geographic problems (Longley et al., 2001).

1.6 Elaboration of Scientific Projects in Geomatics

In the process of developing topography applications, the following phases are addressed to obtain the knowledge needed to make decisions using geomatics:

- Hypothesis testing;
- Data collection *in situ*;
- Geospatial infrastructure data collection;
- Definition of systematic methodology to evaluate and interpret the results obtained (Figure 1.3).

The full content of known digital libraries can be used to perform keyword searches for geomatics related subjects of interest (Arvanitou et al., 2021):

- Web of Science[1];
- ScienceDirect[2];
- IEEExplore[3];
- ACM[4];
- Scopus[5];
- Google Scholar[6].

In the preparation of the scientific project in the area of geomatics some terms can be written with the verb form in the future tense, because the text is prepared in the context of presenting a proposition to study and generate relevant scientific conclusions about a particular geographic problem. In the elaboration of the research project, fundamental concepts are important in scientific surveying applications used in the elaboration of a scientific research proposal with geomatics and R (Table 1.2).

---

[1] https://www.periodicos.capes.gov.br/?option=com_pcollection&mn=70&smn=79&cid=81
[2] https://www.sciencedirect.com/
[3] https://ieeexplore.ieee.org/Xplore/home.jsp
[4] https://dl.acm.org/
[5] https://www.scopus.com/home.uri
[6] https://scholar.google.com/

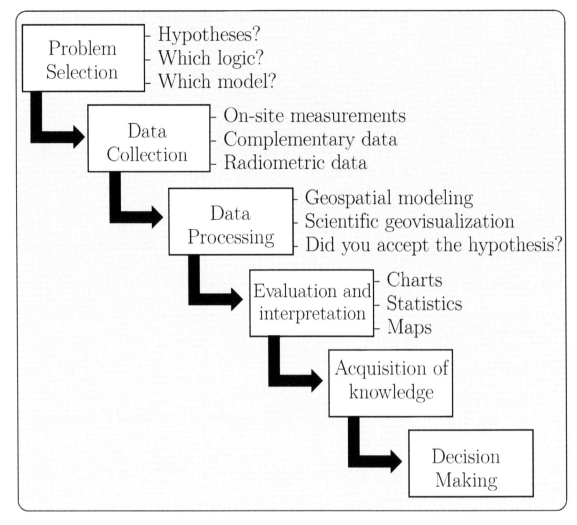

FIGURE 1.3: Stages of decision making in application science with geomatics.

TABLE 1.1: Fundamental concepts used in scientific surveying applications.

Term	Meaning
Science	Knowing and solving geographical problems
Scientist	Individual who generates knowledge
Research	Set of activities oriented toward the search for knowledge
Experiment	Planned activity designed to obtain new facts, confirm the results of previous experiments, or generate or validate technologies
Hypothesis	Testable proposition in the experiment involved in solving problems
Methodology	Use of experimental procedures

1.7 Implementation of Scientific Projects in Geomatics

In the implementation cycle of scientific projects in geomatics and R, the development occurred as a continuous and recursive process in which modifications and improvements can be made according to necessary adaptations in the face of new scientific and technological advances that exist. The model is divided into different working phases of the development process involving: planning, analysis, design, implementation and support (Figure 1.4) (Lo and Yeung, 2007).

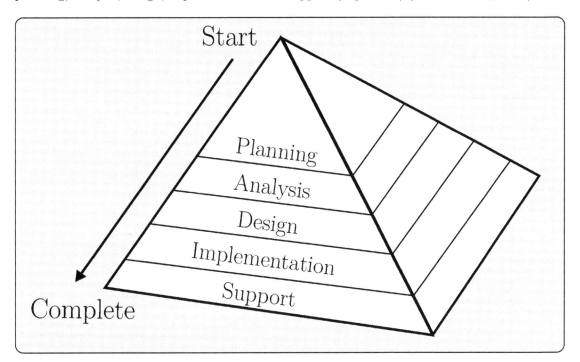

FIGURE 1.4: Pyramid model of the project implementation cycle in geomatics.

The implementation of a geomatics project with R can encompass different project management and support activities, such as (Lo and Yeung, 2007):

- Planning, localization and control;
- Formal review of business cases and project proposals;
- Availability of studies by simulation or prototypes;
- Data and software quality control;
- Software evaluation and establishment of standards;
- Software acquisition, installation and version control;
- Preparation and production of technical documents and user guides;
- Risk analysis with contingency plan to recover data and system failure.

1.8   Scientific Dissemination in Geomatics

In scientific dissemination, considerations of scientific writing in the form of a scientific article and scientific presentation in the format of a slide show are presented.

TABLE 1.2: Description of topics used in the development of scientific projects in geomatics.

Topic	Description
Title	Should be succinct and contain in a few words what the project is intended to accomplish
Autor	Project proponent, usually the project coordinator. Provides information on the call for proposals to which the project will be submitted
Abstract and Keywords	Include index words in the abstract that are not inserted in the title
Introduction	Describes the issue directly, pointing out the problem and the generating demand. Mentions the diagnosis of the problem and how you will solve it. Obtain information about previous studies on similar issues
Hypothesis	Should be written in an affirmative way in order to elucidate the tested proposition of the project
Objectives	Should be written in a way that leaves no doubt as to what is intended to be achieved in the project
Literature review	Recent scientific articles published in high impact journals on the researched subject
Methods	Present methodology and information that will be used in the project
Expected results	Present the expected results based on data analysis
Execution schedule	Time schedule in which the project will be carried out
Technology diffusion	Teaching, research and extension activities associated with the project
Members	Researchers' team, with information about branches, institution and how they contributed to the project
References	Include the bibliographical references used

1.8.1   Scientific writing

Scientific writing, although an indispensable step in the scientific process, has often been over-looked in some undergraduate courses in favor of maximizing class time devoted to scientific concepts. However, the ability to effectively communicate research results is crucial to success in science. Students and professional scientists are judged by the amount of papers published and the number of citations those papers received. Therefore, a solid foundation in scientific writing can better prepare undergraduate and graduate students for productive academic careers (Turbek et al., 2016).

When writing a scientific paper, the structure of the manuscript should be similar to that of an hourglass, with an open and broad beginning, tapering off as it narrows down to the relevant literature that determines the relevance of the paper as unpublished, definition of the hypothesis, material and methods used. The knowledge gap that will be evaluated should be clear by this point. In the results phase, corresponding to the middle of the hourglass, the objectives are achieved through of geomatics and statistical techniques used to evaluate the acceptance of the hypothesis. In the discussion phase of the results obtained, the hourglass is again extended to interpret the results based on the existing literature, in order to make it clear that the knowledge gap previously detected is filled, culminating in conclusions and implication of the work (Figure 1.5).

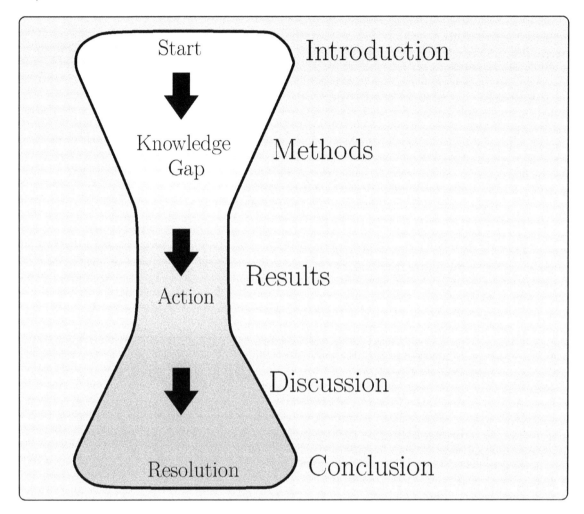

FIGURE 1.5: Structure of a scientific paper compared to the outflow of an hourglass.

Therefore, in the Introduction one should start from broad ideas to specific questions, using a structure that resembles the image of a funnel or an inverted pyramid. The operation of the Discussion and Introduction sections are opposite and mirrored. In the Discussion, the pyramid presents the conventional format, starting from specific questions (study findings) to more comprehensive elaborations (Figure 1.6) (Cáceres et al., 2011).

The immutable characteristics of good scientific writing distinct among other literature are (Lindsay, 2011):

- Accuracy;
- Clarity;
- Brevity.

A vague text cannot be considered a scientific text because of its lack of clarity or ambiguity. A prolix and unnecessarily discursive scientific manuscript is considered poor in terms of scientific writing. Therefore, a clear, precise and brief text can be read and understood by a larger number of readers (Lindsay, 2011).

In the preparation of the scientific article in the area of geomatics, the topics should be written with the verb form in the past tense, because the text was prepared in the context of presenting the results of a work in which relevant scientific conclusions are generated about a particular geographic problem.

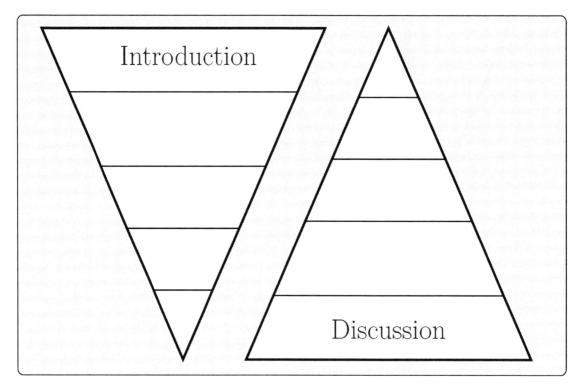

FIGURE 1.6: Mirrored structure of the Introduction and Discussion sections.

1.8.2 Scientific dissemination

Formal scientific dissemination has been disseminated through presentations in different types of media. The use of short video presentations has been a form of presentation used by researchers and in scientific journals and books. However, transforming a scientific publication into a short video with accessible language can be a complex mission for researchers with little affinity and availability of these resources.

Dissemination of science means allowing other people to learn about the research conducted. Scientists working in academia (universities, research centers) and in research and development in large corporations and small companies need to disseminate research results to gain respect and credibility in society and boost careers. This can create a market for products and attract talented employees, as well as build a network of collaborations with other research groups. This can increase the chances of obtaining funding for projects and research with public and private investors (CAPES, 2020).

Scientists should inform contributors and investors about accountability and if possible the positive impact of research on health and socio-economic progress of society. An effective communication strategy can increase the likelihood that it will attract the attention of decision makers and that science will be used to support decisions about strategic evidence-based policy priorities to meet the needs of the population (CAPES, 2020).

Researchers can disseminate scientific work through articles, review papers, workshops, posters, talks at conferences and seminars and in reports. In addition, written and visual materials (video and infographics) can be used for flyers, brochures, press releases, websites, newsletters, blogs, and the entire broad spectrum of social media (CAPES, 2020).

In preparing the presentation of the scientific article through slides, the following aspects on each topic proposed in the disclosure of the research are considered (Table 1.3).

TABLE 1.3: Aspects considered in developing presentation topics for scientific papers in geomatics with slides.

Topic	Aspect Considered
Title	Should be brief and contain in a few words what is accomplished with the work
Authors	Include authors, advisors and affiliation according to the norms. Authors' names should appear on the front cover, below the title
Introduction	Should be straight to the point, pointing out the problem and the demand that generated it. Mentions the diagnosis of the problem and how it was solved. Gets information about previous studies on similar issues. The text should be presented in a summarized form in topics that facilitate the sequence of ideas
Objectives	Should be written in a way that leaves no doubt as to what the work intended to achieve
Methods	Materials and methods should be presented in detail, but without exaggeration, as variable studied, metadata information, analysis
Results	Should present the results obtained based on data analysis. Images, graphs, tables should be legible
Discussion	Should discuss the results obtained based on the available literature
Conclusion	Should be addressed with the execution of the work within what was proposed in the objective
References	Include the references used

1.8.3 Predatory journals

With the current technological development, one can publish a book or scientific article on a website with all the necessary features to simulate the image and procedures of a major publisher. In predatory journals, classified as open access, the scientific quality of the publications is highly questionable. In these journals, peer review of low quality academic manuscripts is requested from the authors. The amount of articles in these journals that are actually read, cited, or had any significant research impact in the same area of knowledge is low. Citation statistics over a five-year period in Google Scholar for 250 random articles published in predatory journals in 2014 averaged 2.6 citations per article, with 56% of the articles having no citations at all. Based on a comparative random sample of articles published in approximately 25000 peer-reviewed journals included in the Scopus index, an average of 18.1 article citations was observed over the same period, with only 9% of articles having no citations at all. Therefore, we concluded that articles published in predatory journals presented little scientific impact (Björk et al., 2020).

With the subversion of the scientific publication process, not validated by peers, the basic foundation of communication in science can be discredited, ultimately turning it into mere opinion pieces disguised as scientific articles, without any validation of the published content. Therefore, the choice of the journal in which to publish a paper can be a challenge full of pitfalls for the less and more experienced authors, and it is important to evaluate if some knowledge about the credibility of the journal and the publisher has been made available, besides exercising the authors' critical sense in choosing a journal with credibility (Penedo and Borges, 2017).

1.9   Computation

As a computation practice, a scientific project is carried out with the goal of mapping the COVID-19 pandemic in South America in the year 2020 and making a scientific summary about the project and the manuscript paper. The COVID-19 data are obtained from online national-level government sources using the R package, `COVID19`. The R package `rnaturalearth` is used to obtain the country database. Other topography functions are performed with the Rrnaturalearth packages, `sf`, `dplyr`, `tidyr`, `rgdal`, `tmap` and `ggplot2`.

1.9.1   Preparation of the project abstract

The title, abstract and keywords of the project are prepared as follows:

Geospatial and Temporal Progress of COVID-19 in South America in 2020

Marcelo de Carvalho Alves - Federal University of Lavras, Agricultural Engineering Department, email: *marcelo.alves@ufla.br*

Luciana Sanches - Federal University of Mato Grosso, Department of Sanitary and Environmental Engineering, email: *lsanches@hotmail.com*

Abstract

With the advent of the new SARS-CoV-2 coronavirus, causing COVID-19, there have been unprecedented socioeconomic changes on a global scale. On March 11, 2020, with more than 100,000 cases of COVID-19 on Earth, the World Health Organization (WHO) declared a pandemic scenario. We aimed to evaluate a topographic analysis methodology to assess the progress of the COVID-19 pandemic in 13 South American countries. We hypothesized that through graphs of temporal variation of the disease and quantile choropleth maps, it is possible to comparatively evaluate the progress of the disease in South American countries. Data on the incidence of COVID-19 in the South American population will be obtained from the R package `COVID19` referring to government sources at the national level until 12/31/2020. Time-varying curves will be performed on deaths, confirmed cases, recovery cases and performed tests of COVID-19 in South American countries.

Keywords: confirmation, deaths, epidemiology, geomatics, recovery, testing.

1.9.2   Elaboration of the scientific article

Abstract

With the advent of the new SARS-CoV-2 coronavirus, causing COVID-19, there have been unprecedented socioeconomic changes on a global scale. On March 11, 2020, with more than 100,000 cases of COVID-19 on Earth, the World Health Organization (WHO) declared a pandemic scenario. We aimed to evaluate a geomatics analysis methodology to assess the progress of the COVID-19 pandemic in 13 South American countries. Data on the incidence of COVID-19 in the South American population were obtained by the R package `COVID19` referring to government sources at the national level until 12/31/2020. Time variation curves were performed on deaths, confirmed cases, recovery cases, and tests performed of COVID-19 in South American countries. Disease time-variation plots and choropleth quantile maps enabled to comparatively evaluate the progress of the disease in South American countries in the periods 4/19/2020, 7/28/2020, 9/16/2020 and 11/4/2020. COVID-19 temporal variation plots and quantile choropleth maps enabled comparatively assessing disease spatial progress in South American countries.

An abridged version of the topics required in a scientific paper with the codes used in the computation practice are briefly presented below.

1.9.3 Introduction

The SARS-CoV-2 coronavirus, causing COVID-19 disease, has generated unprecedented socioeconomic disruption on a global scale. On March 11, 2020, the World Health Organization (WHO) declared the progress of the disease as pandemic. An increasing number of patients required intensive care unit (ICU) beds, and the rate of spread of the disease may peak with demand on ICU bed capacity in many countries (Sun et al., 2020; González-Bustamante, 2021).

The first cases of COVID-19 in South American countries occurred between late February and early March. The Argentine president, Alberto Fernández, considered the pandemic a severe threat to his country (González-Bustamante, 2021). Starting in the third week of March, a large number of South American countries implemented a series of measures to prevent the pandemic. In some countries, such as Uruguay and Paraguay, the pandemic progress was low, while others, such as Brazil and Peru, were more affected by COVID-19. In addition, it deserves consideration that the crisis caused by the virus was combined with country management problems. In countries like Bolivia and Chile, there was a context of superficial trust in institutions, massive protests, and social unrest in the months before the pandemic. Between April and May 2020, several countries reduced intensity of the measures initially implemented (González-Bustamante, 2021).

Thus, based on the hypothesis that it is possible to comparatively assess the progress of the disease in South American countries using time-varying plots and choropleth quantile maps, we evaluated the applicability of geomatics techniques to assess the progress of the COVID-19 pandemic in 13 South American countries.

1.9.4 Methods

Latin America's development during the 20th century has been associated with historical socioeconomic processes of inequality and lack of trust in institutions (Grassi and Memoli, 2016). In addition, there has been an inability to adequately provide public goods and services to the population (González-Bustamante, 2021).

COVID-19 data were obtained from government sources at the national level obtained from the Internet through the R package `COVID19` (Guidotti, 2021), to evaluate the temporal progress and map the COVID-19 pandemic in South America in the year 2020. Temporal variation on deaths, confirmed cases, recovery cases, and tests performed of COVID-19 in South American countries were determined between 01/22/2020 to 12/31/2020. Four reference dates for mapping COVID-19 in South America were set on 4/19/2020, 7/28/2020, 9/16/2020, and 11/4/2020, respectively. Summary statistics were determined on the data of deaths, confirmed cases, recovery cases, and tests performed of COVID-19 in South America in the same periods when the mapping was performed.

The R package `rnaturalearth` (South, 2021) was used to obtain the country database. Geocomputation functions were used to perform operations to obtain a subset and joins in a geographic database. Mappings were performed with the R packages sf (Pebesma, 2021, 2018), dplyr (Wickham et al., 2021), tidyr (Wickham, 2021), rgdal (Bivand et al., 2019), tmap (Tennekes, 2021, 2018), and ggplot2 (Wickham et al., 2020).

1.9.4.1   Installing R packages

The install.packages function was used to install the sf, tmap, dplyr, COVID19, readr, rnaturalearth, rgdal, gridExtra and ggplot2 packages through R console.

```
install.packages("sf")
install.packages("tmap")
install.packages("dplyr")
install.packages("COVID19")
install.packages("rnaturalearth")
install.packages("rgdal")
install.packages("ggplot2")
install.packages("tidyr")
install.packages("readr", repos=c("http://rstudio.org/_packages",
                          "http://cran.rstudio.com"))
install.packages("gridExtra")
```

1.9.4.2   Enabling R packages

The library function was used to enable the sf, tmap, dplyr, COVID19, readr, rnaturalearth, rgdal, raster, gridExtra and ggplot2 packages through R console.

```
library(sf)
library(tmap)
library(dplyr)
library(COVID19)
library(rnaturalearth)
library(rgdal)
library(ggplot2)
library(tidyr)
library(readr)
library(raster)
library(gridExtra)
```

1.9.4.3   Obtaining updated COVID-19 data

The COVID19 function was used to get up-to-date COVID-19 data at the country level by 12/31/2020.

```
d <- COVID19::covid19(end = "2020-12-31")
```

The data can be exported to a directory on your computer using the write.csv function for further use.

```
write.csv(d, "files/d.csv")
```

The data can then be imported into the computer to perform temporal and geospatial analysis.

```
d <- readr::read_csv("files/d.csv")
```

1.9.4.4 Obtaining geospatial polygons with country borders

Geospatial polygons with country borders can be obtained with the ne_download function. For a more detailed analysis of this function, a review of a vignette[7] on the subject is recommended.

```
# Get geographic data
world_rnatural <- rnaturalearth::ne_download(returnclass = "sf")
names(world_iso) # Evaluate available variables in database
```

Geospatial polygons can be exported into a particular directory for posterior use through of the st_write function.

```
st_write(world_iso, 'G:/covid/world_iso.shp', "world_iso.shp")
```

The read_sf function was used to import the polygons back into R.

```
world_iso <- sf::read_sf("files/world_iso.shp")
```

1.9.4.5 Obtaining subset of polygons with country borders of South America

A subset of polygons with country borders was taken in South America and the countries of South America: Argentina, Chile, Falkland Islands, Uruguay, Brazil, Bolivia, Peru, Colombia, Venezuela, Guyana, Suriname, Ecuador, and Paraguay.

```
# Subset in South America
world_america <- world_iso[world_iso$CONTINENT == "South America",]

# Subset in South American countries
ARG <- world_iso[world_iso$ISO_A3_EH == "ARG",] # Argentina
CHL <- world_iso[world_iso$ISO_A3_EH == "CHL",] # Chile
FLK <- world_iso[world_iso$ISO_A3_EH == "FLK",] # I. Falkland
URY <- world_iso[world_iso$ISO_A3_EH == "URY",] # Uruguay
BRA <- world_iso[world_iso$ISO_A3_EH == "BRA",] # Brazil
BOL <- world_iso[world_iso$ISO_A3_EH == "BOL",] # Bolivia
PER <- world_iso[world_iso$ISO_A3_EH == "PER",] # Peru
```

---

[7]https://rdrr.io/cran/rnaturalearth/f/vignettes/what-is-a-country.Rmd

```
COL <- world_iso[world_iso$ISO_A3_EH == "COL",] # Colombia
VEN <- world_iso[world_iso$ISO_A3_EH == "VEN",] # Venezuela
GUY <- world_iso[world_iso$ISO_A3_EH == "GUY",] # Guyana
SUR <- world_iso[world_iso$ISO_A3_EH == "SUR",] # Suriname
ECU <- world_iso[world_iso$ISO_A3_EH == "ECU",] # Ecuador
PRY <- world_iso[world_iso$ISO_A3_EH == "PRY",] # Paraguay
```

1.9.4.6 Merging global COVID-19 data with country polygons in South America

Global COVID-19 data was merged with country polygons in South America and for each of the 13 countries evaluated.

```
# Merge COVID-19 data in South America
w <- dplyr::left_join(world_america, d, by = c("ISO_A3_EH"= "id"))
# Merging of COVID-19 data in South American countries
ar = dplyr::left_join(ARG, d, by = c("ISO_A3_EH"= "id")) # Argentina
ch = dplyr::left_join(CHL, d, by = c("ISO_A3_EH"= "id")) # Chile
fl = dplyr::left_join(FLK, d, by = c("ISO_A3_EH"= "id")) # I. Falkland
ur = dplyr::left_join(URY, d, by = c("ISO_A3_EH"= "id")) # Uruguay
br = dplyr::left_join(BRA, d, by = c("ISO_A3_EH"= "id")) # Brazil
bo = dplyr::left_join(BOL, d, by = c("ISO_A3_EH"= "id")) # Bolivia
pe = dplyr::left_join(PER, d, by = c("ISO_A3_EH"= "id")) # Peru
co = dplyr::left_join(COL, d, by = c("ISO_A3_EH"= "id")) # Colombia
ve = dplyr::left_join(VEN, d, by = c("ISO_A3_EH"= "id")) # Venezuela
gu = dplyr::left_join(GUY, d, by = c("ISO_A3_EH"= "id")) # Guyana
su = dplyr::left_join(SUR, d, by = c("ISO_A3_EH"= "id")) # Suriname
ec = dplyr::left_join(ECU, d, by = c("ISO_A3_EH"= "id")) # Ecuador
pr = dplyr::left_join(PRY, d, by = c("ISO_A3_EH"= "id")) # Paraguay
# Group the results
all <- rbind(ar, ch, fl, ur, br, bo, pe, co, ve, gu, su, ec, pr)
```

The spatial polygons of South America with the COVID-19 data were exported for further use.

```
st_write(w, 'G:/covid/w.shp', "w.shp")
```

1.9.4.7 Setting reference dates for mapping COVID-19 in South America

Four reference dates for mapping COVID-19 in South America were set on 4/19/2020, 7/28/2020, 9/16/2020, and 11/4/2020. The `filter` function was used to obtain the data at the periods of interest.

```
# COVID-19 data on 4/19/2020
w_200 <- w %>%
  filter(date == as.Date("2020-04-19", na.rm = T))
# COVID-19 data on 7/28/2020
w_100 <- w %>%
```

```
   filter(date == as.Date("2020-07-28", na.rm = T))
# COVID-19 data on 9/16/2020
w_50 <- w %>%
   filter(date == as.Date("2020-09-16", na.rm = T))
# COVID-19 data on 11/4/2020
w_1 <- w %>%
   filter(date == as.Date("2020-11-4", na.rm = T))
```

1.9.4.8 Area description

The population of South America has not been evenly distributed, with sparse areas alongside others of relatively high density due to physical and human factors. Among the causes of population distribution in South America are desert regions, such as Patagonia, the dry pampa, Atacama, and Sechura; equatorial forest zones, such as the Amazon; and grassland areas, with extensive cattle raising and lower population density. Based on mapping of the population in South America, Brazil had the largest population followed by Colombia, Argentina, Peru, Venezuela, Chile, Ecuador, Bolivia, Paraguay, Uruguay, Guyana, Suriname, and the Falkland Islands (Figure 1.7).

```
qtm(w,fill="population", fill.style="cat", text="ISO_A3_EH",
text.col="black", fill.palette="Spectral")
```

The same mapped results were observed in graphical format for better comparison (Figure 1.8).

```
ggplot(all, aes(x = ISO_A3_EH, y = population)) +
  geom_bar(stat="identity", fill="grey50") +
  xlab("Country abbreviation")+
  ylab("Population")
```

1.9.5 Results

The temporal variation on deaths, confirmed cases, recovery cases and tests performed (Figure 1.9) of COVID-19 in South American countries were determined using the ggplot2 function.

```
# Deaths
a<-ggplot(all, aes(x = date, y = deaths)) +
  geom_line(aes(color = NAME_LONG), size = 1) +
  scale_color_manual(values = c("blue", "red", "yellow", "green",
    "black", "gray", "tan3","purple", "brown","orange","cyan",
    "darkorchid","coral"))+
  labs(color = "Country")
# Confirmed cases
b<-ggplot(all, aes(x = date, y = confirmed)) +
  geom_line(aes(color = NAME_LONG), size = 1) +
  scale_color_manual(values = c("blue", "red", "yellow", "green",
    "black", "gray", "tan3","purple", "brown","orange","cyan",
```

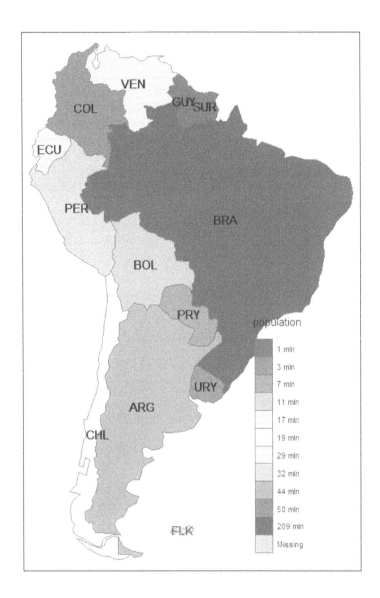

FIGURE 1.7: Population mapping of South American countries.

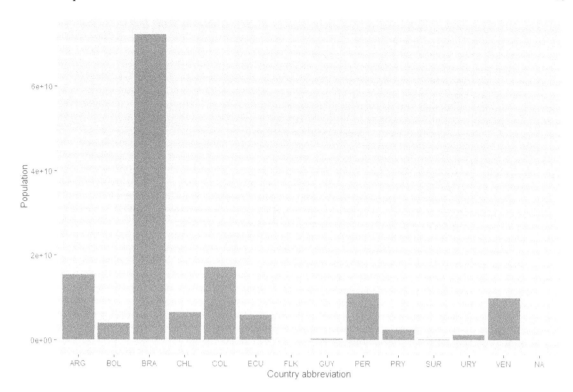

FIGURE 1.8: Population of South American countries in a barplot.

```
  "darkorchid","coral"))+
  labs(color = "Country")
# Recovery cases
c<-ggplot(all, aes(x = date, y = recovered)) +
  geom_line(aes(color = NAME_LONG), size = 1) +
  scale_color_manual(values = c("blue", "red", "yellow", "green",
  "black", "gray", "tan3","purple", "brown","orange","cyan",
  "darkorchid","coral"))+
  labs(color = "Country")
# Tests performed
d<-ggplot(all, aes(x = date, y = tests)) +
  geom_line(aes(color = NAME_LONG), size = 1) +
  scale_color_manual(values = c("blue", "red", "yellow", "green",
  "black", "gray", "tan3","purple", "brown","orange","cyan",
  "darkorchid","coral"))+
  labs(color = "Country")
grid.arrange(a, b,c,d, ncol=2)
```

Based on the disease progress curves, the highest increasing number of deaths was observed in Brazil, followed by Peru, Colombia, Argentina, Chile, Ecuador, and Bolivia. In the other countries, no significant trend of increasing deaths over time was observed. The death curve seemed to reach a point near the climax in Brazil, Peru and Colombia after October, with values of approximately 150,000 and 30,000 and 25,000 deaths, respectively. Similar patterns of temporal variation were observed with respect to the number of confirmations and cases of recovery from the disease, so that after October more than 5 million confirmed cases and 4 million cases of recovery were

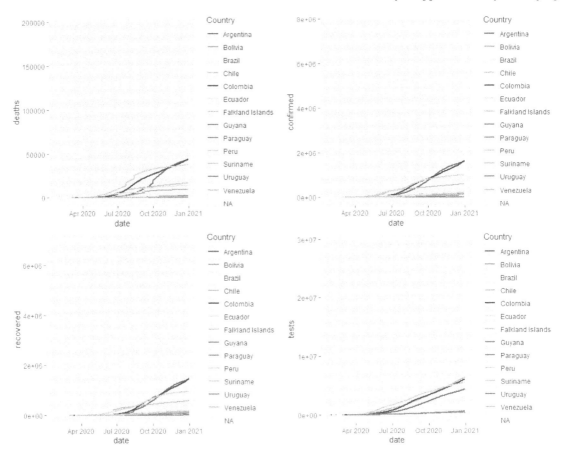

FIGURE 1.9: Time variation of COVID-19 deaths, confirmed cases, recovery cases and tests performed in South American countries in 2020.

observed in Brazil, respectively. In the case of testing to assess the incidence of the disease, the highest amount of testing was done in Peru, followed by Chile and Ecuador, with similar distribution of the number of tests over time, reaching values above 4 million tests after October. In the other countries, the number of tests performed was approximately 500,000 after October.

Summary statistics were determined on deaths, confirmed cases, recovery cases, and tests performed for COVID-19 in South America on 4/19/2020, 7/28/2020, 9/16/2020, and 11/4/2020, through the summary function (Table 1.4).

```
# Deaths
summary(w_1[12])   # 11/4/2020
summary(w_50[12])  # 9/16/2020
summary(w_100[12]) # 7/28/2020
summary(w_200[12]) # 4/19/2020
# Confirmed cases
summary(w_1[10])   # 11/4/2020
summary(w_50[10])  # 9/16/2020
summary(w_100[10]) # 7/28/2020
summary(w_200[10]) # 4/19/2020
# Recovery cases
summary(w_1[11])   # 11/4/2020
```

```
summary(w_50[11]) # 9/16/2020
summary(w_100[11]) # 7/28/2020
summary(w_200[11]) # 4/19/2020
# Tests performed
summary(w_1[9]) # 11/4/2020
summary(w_50[9]) # 9/16/2020
summary(w_100[9]) # 7/28/2020
summary(w_200[9]) # 4/19/2020
```

Based on the summary statistics, regarding the number of deaths in South America, there was an increase in the minimum, average and maximum value of deaths over time in the four time periods evaluated, with the minimum value of one death being observed on 4/19/2020 and the maximum of 161246 deaths on 11/4/2020. Regarding the number of confirmations, recoveries and tests, there was also an increase in minimum, average and maximum values over the four evaluation periods.

TABLE 1.4: Summary statistics of deaths, confirmed cases, recovery cases and tests of COVID-19 performed in South America, on 4/19/2020, 7/28/2020, 9/16/2020 and 11/4/2020.

Variable	Statistic	Time Period			
		4/19/2020	7/28/2020	9/16/2020	11/4/2020
Deaths	Minimum	1.00	20.0	45	61
Deaths	First Quartile	8.75	42.5	407	643
Deaths	Median	82.00	2949.5	9237	10731
Deaths	Mean	326.17	11639.8	19463	24970
Deaths	Third Quartile	262.75	9682.0	15170	32599
Deaths	Maximum	2491.00	89060.0	134248	161246
Confirmed	Minimum	10.0	396	1856	3245
Confirmed	First Quartile	243.5	3883	23976	50248
Confirmed	Median	1683.5	77303	125198	156748
Confirmed	Mean	6883.2	322351	607269	818330
Confirmed	Third Quartile	9623.0	287995	625856	960862
Confirmed	Maximum	39197.0	2503681	4425451	5594098
Recovery	Minimum	6.0	181	1302	2770
Recovery	First Quartile	38.5	2520	12827	35115
Recovery	Median	503.5	28627	92047	130945
Recovery	Mean	1687.8	230339	511714	732757
Recovery	Third Quartile	1995.8	168543	481411	867410
Recovery	Maximum	6811.0	1885035	3853829	5078162
Tests	Minimum	4135	107321	205278	330875
Tests	First Quartile	10995	142118	277645	374486
Tests	Median	28828	827583	1967410	3217537
Tests	Mean	50023	1448498	2962591	4402835
Tests	Third Quartile	75418	1563693	3018946	44278558
Tests	Maximum	143745	6359398	14149045	21518065

The mapping of deaths (Figure 1.10), confirmed cases (Figure 1.11), recovery cases (Figure 1.12) and tests (Figure 1.13) of COVID-19 in South America were performed on 4/19/2020, 7/28/2020, 9/16/2020, and 11/4/2020, through of the `qtm` and `tmap_arrange` functions.

```
# Deaths
m1<-qtm(w_1,fill="deaths",fill.style="quantile", text="ISO_A3_EH",
text.col="black", fill.palette="OrRd", title="11/4/2020")
m2<-qtm(w_50,fill="deaths",fill.style="quantile", text="ISO_A3_EH",
text.col="black", fill.palette="OrRd", title="9/16/2020")
m3<-qtm(w_100,fill="deaths",fill.style="quantile", text="ISO_A3_EH",
text.col="black", fill.palette="OrRd", title="7/28/2020")
m4<-qtm(w_200,fill="deaths",fill.style="quantile", text="ISO_A3_EH",
text.col="black", fill.palette="OrRd", title="4/19/2020")
current.mode <- tmap_mode("plot")
tmap_arrange(m1, m2, m3, m4, nrow=2)
```

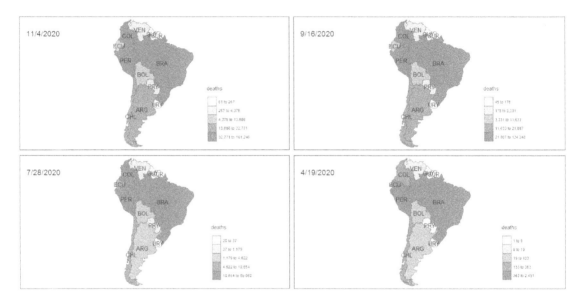

FIGURE 1.10: Quantile mapping of the number of deaths by COVID-19 on 4/19/2020, 7/28/2020, 9/16/2020, and 11/4/2020.

```
# Confirmed cases
c1<-qtm(w_1,fill="confirmed",fill.style="quantile", text="ISO_A3_EH",
text.col="black", fill.palette="OrRd", title="11/4/2020")
c2<-qtm(w_50,fill="confirmed",fill.style="quantile", text="ISO_A3_EH",
text.col="black", fill.palette="OrRd", title="9/16/2020")
c3<-qtm(w_100,fill="confirmed",fill.style="quantile", text="ISO_A3_EH",
text.col="black", fill.palette="OrRd", title="7/28/2020")
c4<-qtm(w_200,fill="confirmed",fill.style="quantile", text="ISO_A3_EH",
text.col="black", fill.palette="OrRd", title="4/19/2020")
current.mode <- tmap_mode("plot")
tmap_arrange(c1, c2, c3, c4, nrow=2)


# Recovery cases
r1<-qtm(w_1,fill="recovered",fill.style="quantile", text="ISO_A3_EH",
```

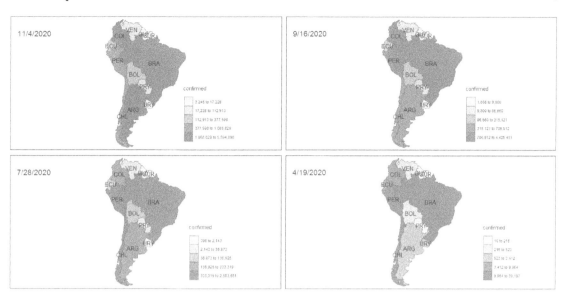

FIGURE 1.11: Quantile mapping of the number of confirmed cases of COVID-19 on 4/19/2020, 7/28/2020, 9/16/2020, and 11/4/2020.

```
text.col="black", fill.palette="OrRd", title="11/4/2020")
r2<-qtm(w_50,fill="recovered",fill.style="quantile", text="ISO_A3_EH",
text.col="black", fill.palette="OrRd", title="9/16/2020")
r3<-qtm(w_100,fill="recovered",fill.style="quantile", text="ISO_A3_EH",
text.col="black", fill.palette="OrRd", title="7/28/2020")
r4<-qtm(w_200,fill="recovered",fill.style="quantile", text="ISO_A3_EH",
text.col="black", fill.palette="OrRd", title="4/19/2020")
current.mode <- tmap_mode("plot")
tmap_arrange(r1, r2, r3, r4, nrow=2)



# Tests performed
t1<-qtm(w_1,fill="tests",fill.style="quantile", text="ISO_A3_EH",
text.col="black", fill.palette="OrRd", title="11/4/2020")
t2<-qtm(w_50,fill="tests",fill.style="quantile", text="ISO_A3_EH",
text.col="black", fill.palette="OrRd", title="9/16/2020")
t3<-qtm(w_100,fill="tests",fill.style="quantile", text="ISO_A3_EH",
text.col="black", fill.palette="OrRd", title="7/28/2020")
t4<-qtm(w_200,fill="tests",fill.style="quantile", text="ISO_A3_EH",
text.col="black", fill.palette="OrRd", title="4/19/2020")
current.mode <- tmap_mode("plot")
tmap_arrange(t1, t2, t3, t4, nrow=2)
```

In relation to the number of deaths, in the first evaluation period, on 04/19/2020, in Brazil, Peru and Ecuador there were more deaths, followed by Colombia and Ecuador. In the same period, in Paraguay, Guyana and Suriname there were fewer deaths. In Peru and Chile there was a reduction in the number of deaths in the last period. In Argentina, there was increase in confirmed cases and recovery of the disease in the last evaluated period. On 07/28/2020, there were more deaths in Brazil, Peru and Chile, followed by Colombia and Argentina. On 09/16/2020, there were more

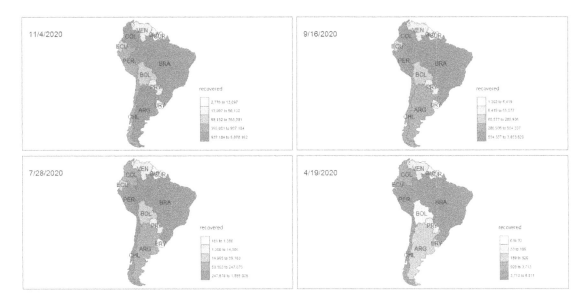

FIGURE 1.12: Quantile mapping of the number of COVID-19 recovery cases on 4/19/2020, 7/28/2020, 9/16/2020, and 11/4/2020.

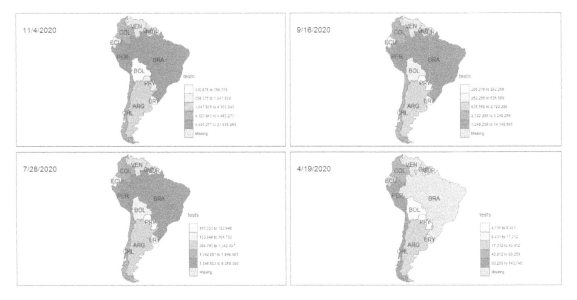

FIGURE 1.13: Quantile mapping of the number of COVID-19 tests performed on 4/19/2020, 7/28/2020, 9/16/2020, and 11/4/2020.

deaths in Brazil, Peru and Colombia. On 11/04/2020, there were more deaths in Brazil, Colombia and Argentina. Similar patterns of confirmed cases mapped by quantiles were observed in relation to the number of deaths. The highest number of tests was performed in Chile and Peru in all evaluation periods.

1.9.6 Discussion

The pandemic is still ongoing in the South American countries as of the writing of this book. In Brazil, the number of deaths was much higher than in other countries of the continent, which may cause political and economic repercussions.

Furthermore, as information about tests performed for COVID-19 was not available, it can be inferred that there may be problems in overcoming the pandemic related to the state capacity to overcome the destructive progress of the disease. Nevertheless, we observed a reduction in the number of deaths in Peru and Chile in the last period evaluated together with an increase in recovery cases. According to González-Bustamante (2021), even with socioeconomic crisis, suppression strategies associated with consistent testing of COVID-19 were instrumental in addressing the epidemic. As long as no vaccine was available, a return to normal activities as in the pre-pandemic period may be risky. Maintaining the balance between the adverse economic effects associated with paralyzing countries in the face of rigorous intervention may be a complex and challenging task for South American governments in the coming months.

Studies could still be conducted to assess the pattern of spatial dependence of the disease in South America and its relationship with climatic factors, population data and more detailed disease data granularity by regions within countries, in order to better elucidate strategies for pandemic control and mitigation.

1.9.7 Conclusions

Temporal variation plots of COVID-19 progress and quantile choropleth maps enabled comparative assessment of the disease spatio-temporal progress in South American countries.

1.10 Solved Exercises

1.10.1 List the topics used in the preparation of a scientific paper in the area of geomatics.

A: Title; Authors; Introduction; Objectives; Methodology; Results; Conclusions; References.

1.10.2 List the topics used in the elaboration of a scientific project in the field of geomatics.

A: Title; Authors; Introduction; Hypothesis; Objectives; Methodology; Expected Results; Technology Dissemination; Team; References.

1.10.3   List the main applications of geomatics.

A: Agricultural sciences, archaeology, environmental monitoring and management, health and epidemiology, emergency services, navigation, marketing, regional and local cost planning, planning and management of highways, railroads, and airways, property valuation and inventory, social studies, tourism, everyday utilities for locating, managing, and planning water, drainage, gas, electricity, telephone, and cabling services.

1.10.4   The scientific question asked in a research project in geomatics is:

    a. Methodology.
    b. Conclusions.
    c. Bibliographical references.
    d. Objectives.
    e. Hypotheses. [X]

1.11   Homework

Based on the first two solved exercises presented by the teacher, specify the content of each answer topic according to what is presented in the theoretical approach of the chapter.

1.12   Resources on the Internet

As a study guide, slides and illustrative videos are presented about the subject covered in the chapter (Table 1.5).

TABLE 1.5: Slide show and video presentation on scientific applications of geomatics with R.

Guide	Address for Access
1	Slides on project preparation[8]
2	Surveying examples[9]
3	Collection and analysis of geospatial data from the NASA AppEEARS platform[10]
4	How to Write an Effective Research Paper[11]

---

[8] http://www.sergeo.deg.ufla.br/geomatica/book/c1/presentation.html#/
[9] https://youtu.be/SPCewaAfqPA
[10] https://youtu.be/Gb9E4TkTdrc
[11] https://youtu.be/cMJWtNDqGzI

1.13 Research Suggestion

The development of scientific research on geomatics is stimulated through activity proposals that can be used or adapted by the student to assess the applicability of the subject matter covered in the chapter (Table 1.6).

TABLE 1.6: Practical and research activities used or adapted by students using scientific applications of geomatics with R.

Activity	Description
1	Collecting geospatial data infrastructure through the NASA AppEEARS platform[12]
2	Collecting of geospatial data infrastructure from the Sidra IBGE system[13]
3	Performing spatio-temporal analysis of COVID-19 in other regions, seasons, and surveying methodologies

1.14 Learning Outcome Assessment Strategy

Perform a summary of the chapter, "Scientific Applications of Geomatics with R", on a single A4 page in order to show the student's abilities to summarize a subject presenting key points considered of greater importance today.

---

[12]https://lpdaac.usgs.gov/tools/appeears/
[13]https://sidra.ibge.gov.br/pesquisa/pam/tabelas

2

Introduction to Measurement Units

2.1 Learning Questions

The emergent learning questions answered through reading the chapter are as follows:

- What is the difference between geomatics and other sciences?
- How did the history of topographic measurements on the Earth's surface come about?
- What is the difference between geodetic and plane surveys?
- What are the applications of surveying?
- What are the legal requirements for the application of geomatics?
- What are the main federal survey and mapping agencies, professional organizations and journals on geomatics?
- How can the future of geomatics be?
- How can R software and R packages be used in geomatics?
- How to perform calculations involving units of measure in R and `measurements` package.

2.2 Learning Outcomes

The learning outcomes expected from reading the chapter are as follows:

- Understand the meaning of geomatics and its difference from other sciences.
- Briefly know the history of the evolution of topographic measurements on the Earth's surface.
- Understand the difference between geodesic and plane surveys in the analysis of positional accuracy of measurement systems.
- Understand surface, aerial and satellite surveys and their applications.
- Understand the legal requirements for the application of geomatics and the main federal agencies of survey and mapping, professional organizations and periodicals.
- Know the future of geomatics and its function with the use of the R Program with geomatics packages.
- Apply calculations involving units of measure using R software and the `measurements` package.

2.3 Introduction

Topography, from the Greek topos (place) and graphein (describe), is traditionally defined as science applied to measuring and representing the configuration of position and altitude of a part

DOI: 10.1201/9781003184263-2

of the environment with improvements on its surface (Borges, 2013). Besides being science and technology, topography is defined as the art and technology to determine relative positions of points on the Earth's surface, with the instrument appropriate to the accuracy and precision required. With the advent of geocomputation and electronics associated with topographic surveys, the term "topography" has been replaced by geomatics (Alves and Silva, 2016), encompassing methods of mathematical modeling, georeferencing, cartographic representation and geoinformation for surveying terrestrial environments (Silva and Segantine, 2015).

In general, topography and geomatics have been used in disciplines with methods to measure and collect information about the physical environment, and process and disseminate a variety of multidisciplinary products (Alves and Silva, 2016). Some terms such as "geomatics", "geomatics engineering", and "geoinformatics" are applied in the science of geographic information as an approach to obtain, manage and apply the disciplines of topography, photogrammetry, remote sensing, cartography and geocomputation as components of geomatics to solve spatial and temporal problems through computer systems.

Geomatics has become indispensable for modern life. Topographic surveys are used in different applications such as (Alves and Silva, 2016):

- Mapping the Earth above and below sea level;
- Preparation of navigation charts for use in air, land and sea;
- Establishing public and private land property boundaries;
- Development of databases for land use and natural resources useful for environmental management;
- Determining facts about the Earth's size, shape, gravity, and magnetic fields;
- Preparation of charts of the Earth, other planets, stars and objects in the universe.

Detailed measurement applications can be performed, such as determining the geometry of individual plants for monitoring and precision management in agriculture using laser scanning (**LiDAR**)[1] associated with global navigation satellite system data information for spatial positioning (Figure 2.1).

2.4 Geomatics

The term "geomatics" is established as a quantitative and computation approach to the processing of spatial information, by combining traditional surveying and geography, influenced by theories and methods developed in the disciplines of computation and information science, called (geo)informatics (Sallis and Benwell, 1993). Geomatics can be defined as an interrelated approach to measuring, analyzing, managing, storing, and presenting spatial data descriptions and locations.

Geomatics was introduced by the Canadian Association of Aerial Surveyors to cover the disciplines of Topography, Remote Sensing and Geographic Information Systems (GIS). It is widely accepted in the United States, Canada, Australia, the United Kingdom and Brazil, including in sectors and disciplines of universities that have adopted Geomatics rather than Topography which motivated its practical form and the objective of the survey in recent years with an emphasis on geoinformatics. Recent technological advances have provided new tools for measurement and acquisition of information for computation and dissemination of information. Other factors are related to the demand for monitoring, management and regulation of land, water, air and other natural resources related to local, regional and global environments (Alves and Silva, 2016).

---

[1] https://youtu.be/QJGcX6harfY

FIGURE 2.1: Artist's conception of global positioning system Block II-F satellite in Earth's orbit. (Courtesy of NASA.)

2.5 Geomatics and Other Sciences

Geomatics is formed by a set of sciences such as (Silva and Segantine, 2015; Alves and Silva, 2016):

- Geodesy;
- Cartography;
- Topography;
- Measurements and units of measurement;
- Theory of errors and statistics;
- Photogrammetry;
- Remote sensing;
- Geographic information system;
- Global satellite positioning system.

With this, geomatics has been studied in many applied academic areas, such as (Alves and Silva, 2016):

- Agronomy;
- Agricultural engineering;
- Environmental engineering;
- Archaeology;
- Astronomy;
- Forest engineering;
- Geography;
- Geology;
- Geophysics;
- Architecture;
- Meteorology;
- Paleontology and seismology;
- Civil and military engineering;
- Zootechnics.

2.6   Geodetic and Plane Surveys

Geodetic and plane surveys are topographic survey classifications as the reference adopted on the basis of calculations of field measures. In the plane survey, except for leveling, it was assumed that the reference base for fieldwork was a horizontal surface. The direction of the plumb line and gravity are considered parallel along the raised region and all angles observed are presumably plane angles. For limited-sized areas, the ellipsoid surface was actually close to the plane. In an arc on the Earth's surface with 18.5 km, the horizontal distance between the ends of the arc was approximately 0.65 cm greater than the horizontal distance (McCormac et al., 2012; Alves and Silva, 2016). Therefore, it is evident that, except in surveys of large areas, the Earth's surface can be approximated by a plane, simplifying the calculations and techniques used. In general, algebra, analytical and plane geometry and plane trigonometry are used in the calculations of plane topographic surveys. Even for large areas, map projections allow plane topography calculations to be used (Ghilani and Wolf, 1989; Alves and Silva, 2016).

In the geodesic survey, the calculation of the Earth's curvature was considered in an ellipsoid with the approximate size and shape of the Earth. Geodesic methods are applied to determine relative positions of landmarks separated by large distances and to calculate the length and direction of large lines between these landmarks. The landmarks served as the basis for referencing other topographical surveys of smaller extents (Ghilani and Wolf, 1989; Alves and Silva, 2016) (Figure 2.2).

2.7   Types of Specialized Surveys

Many types of surveys have been specialized for specific purposes. The career in geomatics may vary according to the type of survey and mapping required, according to the following classifications in Table 2.1 (Ghilani and Wolf, 1989; Kavanagh and Slattery, 2015; Alves and Silva, 2016):

FIGURE 2.2: Curvature of the Earth considered in determining geodetic coordinates, distances and geodetic areas, giving greater accuracy in measurement of applied geodetic surveying.

TABLE 2.1: Surveys specialized in geomatics.

Type of Survey	Description
Control	Established horizontal and vertical network of landmarks to initiate other surveys. Many control surveys are performed with instruments
Topographic	Determination of locations of natural, artificial characteristics and altitude of locations used in mapping
Land registration	Established lines of borders and properties
Hydrographic	Determination of the margin and depth of rivers, streams, oceans, reservoirs and other water bodies
Alignment	Carried out to plan, design and build highways, pipelines and other linear projects
Construction	Determination of line, grid, elevation control, horizontal position, dimensions, and settings for construction operations
Survey of works built	Used to obtain the precise final locations of engineering work and the records of changes incorporated into the construction line

Type of Survey	Description
Mine	Carried out below and above the surface to guide the realization of tunnels and other operations associated with mining
Solar	Used to map solar servitudes, obstructions according to solar angles to meet requirements for zoning boards and insurance companies
Industrial	Extremely accurate measurement method for low tolerance manufactured processes

Surveys are classified according to the altitude level used to obtain the data, such as:

- Surface survey;
- Aerial survey;
- Satellite survey.

In the surface survey, measurements are performed with equipment for surface readings, such as automatic levels and electronic total station instruments. Aerial surveys are performed by photogrammetry or remote sensing. In the photogrammetry, cameras are used attached to airplanes or drones to obtain images of remote areas, while remote sensing used cameras and other types of sensors on board spacecraft or satellites. Aerial methods have been used in specialized types of surveys, except for optical alignment. Photographs taken from the surface are also used. Satellite surveys included surface measurements performed by GNSS receivers, or the use of satellite images in mapping and monitoring large extents on the Earth's surface (Alves and Silva, 2016).

2.8   Federal Surveying and Mapping Agencies

Some government agencies carry out work in support of topographic surveys in Brazil, such as (Alves and Silva, 2016):

- Instituto Brasileiro de Geografia e Estatística (IBGE);
- Instituto Nacional de Colonização e Reforma Agrária (INCRA);
- Instituto Nacional de Pesquisas Espaciais (INPE);
- Ministério do Meio Ambiente;
- Instituto Brasileiro do Meio Ambiente e dos Recursos Naturais Renováveis (IBAMA);
- United States Geological Survey (USGS);
- National Aeronautics and Space Administration (NASA);
- European Space Agency (ESA).

Documents from the Brazilian Association of Technical Standards, such as NBR13133, establishes the conditions required for the execution of topographic surveys in Brazil (ABNT, 1994).

2.9   Professional and Surveying Science Organizations

Professional organizations are used to facilitate communication among surveyors, to update ethical standards in the practice of surveying in favor of knowledge progress. Some international organizations are (Alves and Silva, 2016):

- American Congress on Surveying and Mapping(ACSM);
- American Society for Photogrammetry and Remote Sensing (ASPRS);
- Surveying and Geomatics Educators Society (SAGES);
- Urban and Regional Information Systems Association (URISA);
- Canadian Institute of Geomatics (CIG);
- International Federation of Surveyors (FIG).

Some scientific journals generate knowledge and updates in geomatics, such as (Alves and Silva, 2016):

- *Tectonophysics*;
- *Studia Geophysica et Geodaetica*;
- *Remote Sensing of Environment*;
- *Journal of Geodesy*;
- *International Journal of Health Geographics*;
- *ISPRS Journal of Photogrammetry and Remote Sensing*;
- *Computers & Geosciences*;
- *Earth-Science Reviews*;
- *Geomorphology*;
- *International Journal of Geophysics*;
- *Science of the Total Environment*;
- *Journal of Geophysics and Engineering*;
- *Advances in Space Research*;
- *Atmospheric Research*;
- *International Journal of Applied Earth Observation and Geoinformation*;
- *Survey Review*;
- *Environmental Monitoring and Assessment*;
- *Applied Geomatics*;
- *Measurement*.

2.10 Future Challenges in Geomatics

Technological advances enabled greater demand to obtain data with high standard of accuracy and the processing of information in computer systems. With this, in a matter of a few years, the demand for surveyors and geomatics professionals varied rapidly.

Some future challenges are (Bhattacharjee and Clery, 2013; Alves and Silva, 2016):

- Update the spatial reference systems of horizontal and vertical control point networks;
- Plan and design the expansion of urban and rural areas;
- Plan and design fast transit systems connecting large cities;
- Determine environmental impacts and perform risk analysis;
- Do accurate research on surface deformation to monitor existing structures such as dams, bridges and buildings;
- Search for new maps and update existing products on planets, moons, stars and other extraterrestrial objects (Figure 2.3).

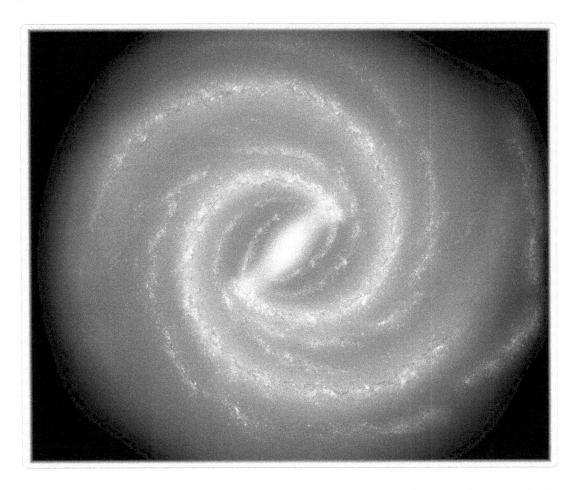

FIGURE 2.3: Structure of the Milky Way, including the location of the spiral arms and other components such as the bulge deduced from survey data from ESO's VISTA telescope at the Paranal Observatory. (Courtesy of NASA/JPL-Caltech/ESO/R. Hurt.)

2.11   R Software and R Packages for Geomatics

R is considered a powerful computation language for geomatics with thousands of geographic functions. In R you can obtain support for vector and raster data with visualization possibilities, statistical and geospatial analyses, packages and statistical methods. In open source GIS, such as 'QGIS', geographic analyses are accessible globally. In GIS there was a tendency to emphasize graphical user interfaces with minimal reproducibility, although many can be used in the command line. R was used with the command-line interface and maximum reproducibility (Lovelace et al., 2019b).

R has been used by researchers to create R packages across all disciplines, with thousands of R packages available on the Comprehensive R Archive Network **CRAN**[2] (Tippmann, 2015). This book applied R in questions related to surveying technology and science (Figure 2.4).

---

[2]http://cran.r-project.org/

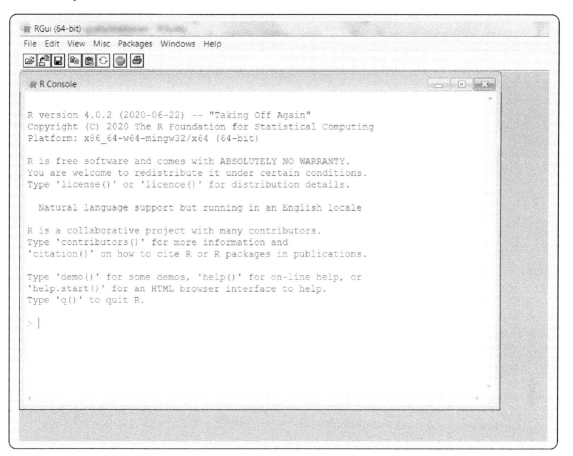

FIGURE 2.4: R console used for science and technology applications of geomatics.

2.12 Computation

Several packages for measurements, mapping and geospatial analysis are available in R. The `measurements` is the package used for measuring physical units in computation practice of this chapter. Other packages in advanced geomatics spatial analysis for data interpolation and mapping are `geoR` and `gstat`. The `maptools` package can be used to read shapefile vector files. The `sp` and `sf` packages are also used in vector geometries usage situations. Packages for matrix files, digital image processing, and map algebra are, for example, `raster` and `RStoolbox`. Cartographic projections and geometric operations can be used in the `rgdal` package. Spherical trigonometry calculations can be performed in the `geosphere` package. We can use `ggmap`, `rasterVis`, `tmap`, `leaflet` and `mapview` for viewing and mapping obtained results (Lovelace et al., 2019b). Therefore, R packages can be applied in scientific research on geomatics.

We presented some basic topics of R package installation settings and some units of measure used in evaluations of distance, angles, and conversion of units into products or rates, scales, and graphical error. Finally, we demonstrated how to evaluate the version of the operating system and R used throughout the book, as well as a method for searching R packages by keywords.

2.12.1   Setting up directory in R

The `getwd` function is used to evaluate the working directory in R. The `setwd` function is used to perform directory configuration in R and to address new working directory.

```
getwd()
setwd("E:/Aulas/Topografia/markdown/Cap1")
```

2.12.2   Installing R packages

The `install.packages` function is used to install the `measurements` package in the R console.

```
install.packages("measurements")
```

2.12.3   Enabling R packages

The `library` function is used to enable the `measurements` package in the R console.

```
library(measurements)
```

2.12.4   Measurement units

Measurement is used to compare one quantity with another, of the same nature, used as a standard, from a measurement method and procedure. Therefore, measurement is the set of operations required to determine the final value of a quantity. In possession of a measurement pattern, a unit of measure is defined. Other definitions are shown in Table 2.2 (Silva and Segantine, 2015).

TABLE 2.2: Definitions of terms used in measurement.

Term	Definition
Measurement method	Logical sequence of operations described to perform the measurements
Measurement procedure	Set of operations described to perform measurements according to an established method
Measuring equipment	Device used to perform measurements to provide information about the physical value of a measured variable
Calibration	Set of operations performed to compare measured values with a known accuracy pattern
Metrology	Science that groups knowledge and techniques to measure and interpret measurements performed

A "quantity" is defined as a property of a phenomenon, body, or substance, with magnitude that can be expressed as a number and a reference. This reference can be a measurement unit, a measurement procedure, a reference material, or a combination of such. A measurement system consists of base units for base quantities, and derived units for derived quantities (Pebesma and Bivand, 2021).

The metric measurement system used in Brazil is called the International System of Units (SI) (INMETRO, 2012); however, other systems are developed for similar purposes, such as the English metric system. Measurement magnitudes are provided in specific units. In the topographic survey, the most used units are length, area, volume and angle. In SI, the measurement units of distances determined by topography instruments are millimeter, centimeter, meter and kilometer, equivalent to 0.001, 0.01, 0.1, 1 and 1000 m, respectively (Alves and Silva, 2016).

SI consists of seven base units: length (meter, m), mass (kilogram, kg), time (second, s), electric current (ampere, A), thermodynamic temperature (Kelvin, K), amount of substance (mole, mol), and luminous intensity (candela, cd). Derived units are composed of products of integer powers of base units, as speed (m s^{-1}), density (kg m^{-3}) and area (m^2) (Göbel et al., 2006; Pebesma and Bivand, 2021).

Regarding the area unit, the most used unit is the square meter (m^2). For large areas of land, the area is generally accounted for hectares (ha), which in turn amounts to 10,000 m^2. In surveying by satellite images, the km^2 unit has generally been adopted. For volumes, the cubic meter unit (m^3) has been adopted. Angle measurements are subdivided into degree (°), minutes (′) and seconds (″), as proposed in the sexagesimal system. However, radian is also accepted as a unit of measurement for angles. A radian is the angle subtended by an arc of a circle with length equal to the radius of the circle. Therefore, 2 π rad is equal to 360° and 1 rad, equal to 57° 17′44.8" or 57.2958°. Other divisions of angles used to divide a circle by other systems are the grade or gon, dividing the circle into 400 parts and the mils, used by the U.S. military to divide the circle into 6,400 units (Ghilani and Wolf, 1989). Although the exact value of a measured quantity is never known, it can be known exactly what the sum of a group of measures is. For example, the sum of the three inner angles of a triangle should be 180°. For a rectangle, the sum of the inner angles must be 360°. If the angles of a triangle have been measured with the total value of approximately 180°, you must eliminate the measurement errors adjusting or revising each angle (Ghilani and Wolf, 1989).

The metric system was originally developed in the 1790s in France. In the same period, the French Academy of Sciences defined the metro as 1/10000000 of the length of the meridian that passed through Paris, departing from Ecuador to the Pole. In later measures, it was found that the measure adopted 0.2 mm lower than the one originally proposed. In the 1960s and 1970s, significant effort was made to adopt the SI in the United States as a system of weights and measures. This initiative would facilitate the implementation of treaties and trade between countries. However, there was a lot of resistance from agencies and people from some states, cities and localities, as well as from some enterprises. As a result, the SI has not yet been made official in the U.S. The confusion caused by the lack of standardization of the measurement system contributed to the collision of the Mars Orbiter satellite on Mars, in 1999. The $125 million dollar satellite collision occurred due to the use of U.S. system units by contracted companies which were mistaken for SI units by the Propulsion Laboratory of the National Aeronautics and Space Administration (NASA) (Ghilani and Wolf, 1989). Other measurement units and relationships used in geomatics are described in Table 2.3 (Alves and Silva, 2016).

Absolute location in position and time requires a fixed origin with possibilities to measure other absolute space-time points defined by a datum. A datum involves more than one geospatial dimensions. The combination of a datum and a measurement unit of scale defines a reference system (Pebesma and Bivand, 2021).

In several R packages, there are unit conversions. For example, with the `measurements` package (Birk, 2019) we can obtain a collection of tools to facilitate the work with physical measurements and unit conversions, as an example by the function `conv_unit`. The conversion values are defined based on international measurement authorities. While a lot of effort has been made to make conversions as accurate as possible, you should check to ensure that the conversions are accurate enough for geomatics applications (Birk, 2019).

TABLE 2.3: Measurement units, conversion factors and important numbers in survey.

Unit of Measurement	Description
Length	1 millimeter (mm) = 1000 micrometers (μm); 1 centimeter (cm) = 10 mm; 1 meter (m) = 100 cm; 1 kilometer (km) = 1000 m
Area	10000 m^2 = 1 hectare (ha)
Volume	1 liter (L) = 1000 milliliters (mL); 1 m^3 = 1000 L
Mass	1 kilogram (kg) = 1000 grams (g); 1 ton (ton) = 1000 kg
GPS signal	C/A code = 1.023 MHz; P code = 10.23 MHz; code L1 = 1575.42 MHz; L2 code = 1227.60 MHz; L5 code = 1176.45 MHz
Electronic distance	299792458 m s^{-1} = speed of light or electromagnetic energy in a vacuum; 1 Hertz (Hz) = 1 cycle per second; 1 kilohertz (kHz) = 1000 Hz; 1 megahertz (MHz) = 1000 kHz; 1 gigahertz (GHz) = 1000 MHz; 1.0003 = approximate index of atmospheric refraction; 760 mm of mercury = standard atmospheric pressure
Angles	1 circumference = 360° = 2π radians; 1 degree (1° = 60') (minutes); = 60" (seconds); 1° = 0.017453292 radians; 1 radian = 57.29577951° = 57°17'44.806" ; 1 radian = 206264.8062"; 1 circumference = 400 grads; tan 1" = sen 1" = 0.000004848; π = 3.141592654
Error analysis	68.3% = 1 standard deviation (σ) ; 0.6745 = (σ) 50% error; 1.6449 = (σ) 90% error; 1.9599 = (σ) 95% error
Taping	0.0000116 = expansion coefficient of steel tape for 1°C; 2000000 kg cm^{-2} = Steel elasticity module; 20°C = standard temperature for measuring tape
Error analysis	68.3% = 1 standard deviation (σ); 0.6745 = (σ) 50% error; 1.6449 = (σ) 90% error; 1.9599 = (σ) 95% error
Taping	0.0000116 = expansion coefficient of steel tape for 1°C; 2000000 kg cm^{-2} = Steel elasticity module; 20°C = standard temperature for measuring tape
Geodetic measurements	6371000 m = average radius of the Earth; 10000 km = distance from Equator to pole, used as original base for meter length

Absolute location in position and time requires a fixed origin with possibilities to measure other absolute space-time points defined by a datum. A datum involves more than one geospatial dimensions. The combination of a datum and a measurement unit of scale defines a reference system (Pebesma and Bivand, 2021).

In several R packages, there are unit conversions. For example, with the `measurements` package (Birk, 2019) we can obtain a collection of tools to facilitate the work with physical measurements and unit conversions, as an example by the function `conv_unit`. The conversion values are defined based on international measurement authorities. While a lot of effort has been made to make conversions as accurate as possible, you should check to ensure that the conversions are accurate enough for geomatics applications (Birk, 2019).

Another R package, the `units` (Pebesma et al., 2019), provides a class for maintaining unit metadata, for numeric data with associated measurement units. Operations on objects of this class retain the unit metadata and provide automated dimensional analysis based on dimensions taken into consideration in computations and comparisons, eliminating a whole class of potential scientific programming mistakes (Pebesma et al., 2016). Applications using the units package are presented in chapters 12 and 13.

2.12.5 Evaluate unit of measure

The units of measure of length, angle, area, and volume are obtained with the conv_unit_options function of the measurements package.

```
names(conv_unit_options)
```

```
##  [1] "acceleration" "angle"       "area"        "coordinate"  "count"
##  [6] "duration"     "energy"      "file_size"   "flow"        "length"
## [11] "mass"         "power"       "pressure"    "speed"       "temperature"
## [16] "volume"
```

```
conv_unit_options
```

```
## $acceleration
##  [1] "mm_per_sec2"   "cm_per_sec2"   "m_per_sec2"    "km_per_sec2"
##  [5] "grav"          "inch_per_sec2" "ft_per_sec2"   "mi_per_sec2"
##  [9] "kph_per_sec"   "mph_per_sec"
##
## $angle
## [1] "degree" "radian" "grad"   "arcmin" "arcsec" "turn"
##
## $area
##  [1] "nm2"      "um2"      "mm2"      "cm2"      "m2"       "hectare"
##  [7] "km2"      "inch2"    "in2"      "ft2"      "yd2"      "acre"
## [13] "mi2"      "naut_mi2"
##
## $coordinate
## [1] "dec_deg"     "deg_dec_min" "deg_min_sec" "dms"         "DMS"
##
## $count
## [1] "fmol" "pmol" "nmol" "umol" "mmol" "mol"
##
## $duration
##  [1] "nsec"      "usec"      "msec"      "sec"       "min"       "hr"
##  [7] "hour"      "day"       "wk"        "week"      "mon"       "yr"
## [13] "year"      "dec"       "decade"    "cen"       "century"   "mil"
## [19] "millenium" "Ma"
##
## $energy
## [1] "J"    "kJ"   "erg"  "cal"  "Cal"  "Wsec" "kWh"  "MWh"  "BTU"
##
## $file_size
##  [1] "byte" "KB"   "kB"   "MB"   "GB"   "TB"   "PB"   "bit"  "Kbit" "kbit"
## [11] "Mbit" "Gbit" "Tbit" "Pbit"
```

```
## 
## $flow
##  [1] "ml_per_sec"  "ml_per_min"  "ml_per_hr"   "l_per_sec"   "l_per_min" 
##  [6] "LPM"         "l_per_hr"    "m3_per_sec"  "m3_per_min"  "m3_per_hr" 
## [11] "Sv"          "gal_per_sec" "gal_per_min" "GPM"         "gal_per_hr"
## [16] "ft3_per_sec" "ft3_per_min" "ft3_per_hr"
## 
## $length
##  [1] "angstrom"    "nm"          "um"          "mm"          "cm"    
##  [6] "dm"          "m"           "km"          "inch"        "ft"    
## [11] "foot"        "feet"        "yd"          "yard"        "fathom"
## [16] "mi"          "mile"        "naut_mi"     "au"          "light_yr"
## [21] "light_year"  "parsec"      "point"
## 
## $mass
##  [1] "Da"          "fg"          "pg"          "ng"          "ug"  
##  [6] "mg"          "g"           "kg"          "Mg"          "Gg"  
## [11] "Tg"          "Pg"          "carat"       "metric_ton"  "oz"  
## [16] "lbs"         "short_ton"   "long_ton"    "stone"
## 
## $power
##  [1] "uW"          "mW"          "W"           "kW"          "MW"        
##  [6] "GW"          "erg_per_sec" "cal_per_sec" "cal_per_hr"  "Cal_per_sec"
## [11] "Cal_per_hr"  "BTU_per_sec" "BTU_per_hr"  "hp"
## 
## $pressure
##  [1] "uatm"  "atm"   "Pa"    "hPa"   "kPa"   "torr"  "mmHg"  "inHg"  "cmH2O"
## [10] "inH2O" "mbar"  "bar"   "dbar"  "psi"   "PSI"
## 
## $speed
##  [1] "mm_per_sec"  "cm_per_sec"  "m_per_sec"   "km_per_sec"  "inch_per_sec"
##  [6] "ft_per_sec"  "kph"         "km_per_hr"   "mph"         "mi_per_hr" 
## [11] "km_per_day"  "mi_per_day"  "knot"        "mach"        "light"
## 
## $temperature
## [1] "C" "F" "K" "R"
## 
## $volume
##  [1] "ul"        "ml"        "dl"        "l"         "cm3"       "dm3"    
##  [7] "m3"        "km3"       "us_tsp"    "us_tbsp"   "us_oz"     "us_cup" 
## [13] "us_pint"   "us_quart"  "us_gal"    "inch3"     "ft3"       "mi3"    
## [19] "imp_tsp"   "imp_tbsp"  "imp_oz"    "imp_cup"   "imp_pint"  "imp_quart"
## [25] "imp_gal"
```

```
conv_unit_options$length
```

```
##  [1] "angstrom"    "nm"          "um"          "mm"          "cm"    
##  [6] "dm"          "m"           "km"          "inch"        "ft"    
## [11] "foot"        "feet"        "yd"          "yard"        "fathom"
## [16] "mi"          "mile"        "naut_mi"     "au"          "light_yr"
## [21] "light_year"  "parsec"      "point"
```

```
conv_unit_options$angle
```

```
## [1] "degree" "radian" "grad"    "arcmin" "arcsec" "turn"
```

```
conv_unit_options$area
```

```
##  [1] "nm2"      "um2"      "mm2"      "cm2"      "m2"       "hectare"
##  [7] "km2"      "inch2"    "in2"      "ft2"      "yd2"      "acre"
## [13] "mi2"      "naut_mi2"
```

```
conv_unit_options$volume
```

```
##  [1] "ul"       "ml"       "dl"       "l"        "cm3"      "dm3"
##  [7] "m3"       "km3"      "us_tsp"   "us_tbsp"  "us_oz"    "us_cup"
## [13] "us_pint"  "us_quart" "us_gal"   "inch3"    "ft3"      "mi3"
## [19] "imp_tsp"  "imp_tbsp" "imp_oz"   "imp_cup"  "imp_pint" "imp_quart"
## [25] "imp_gal"
```

2.12.6   Convert distance measurements

In geomatics, the SI base length unit is called "meter". The conv_unit function are used to convert measurement units between the obtained distances in mm, cm, dm, m, km and inch.

```
conv_unit(2.54, "cm", "inch")
```

```
## [1] 1
```

```
conv_unit(1, "m", "mm")
```

```
## [1] 1000
```

```
conv_unit(1, "m", "cm")
```

```
## [1] 100
```

```
conv_unit(1, "m", "dm")
```

```
## [1] 10
```

```
conv_unit(1, "m", "m")
```

```
## [1] 1
```

```
conv_unit(1, "m", "km")
```

```
## [1] 0.001
```

2.12.7   Convert angle measurements

Considering plane angular measurements, the measurement units defined by the SI are: radian, degree, and grade (gon). For the sexagesimal degree unit, the degrees, minutes and seconds are described with symbols placed in the upper right part of the corresponding number, °, ′, ″, respectively. To perform calculations with angles, it may be necessary to transform degrees, minutes and seconds in decimal degrees and vice versa using the conv_unit function.

```
conv_unit(c("10 20 30"), "deg_min_sec", "dec_deg")
```

```
## [1] "10.3416666666667"
```

```
conv_unit(c("10.53 20.01 30.06"), "deg_min_sec", "dec_deg")
```

```
## [1] "10.87185"
```

```
conv_unit(c("10 35.40","20 38.763"), "deg_dec_min", "deg_min_sec")
```

```
## [1] "10 35 24"                "20 38 45.7799999999988"
```

2.12.8   Convert units to products or rates

The conversion of units into products or rates can be performed with the conv_dim function. For example, it is possible to determine how many minutes it took to travel 100 m at 2 m s^{-1}.

```
conv_dim(x = 100, x_unit = "m", trans = 2, trans_unit = "m_per_sec",
         y_unit = "min")
```

In this case, 0.833 minutes are required.

2.12.9 Surface measurements

For surface measurement, some additional units of centiare, are, and hectare are shown in Table 2.4 (Silva and Segantine, 2015).

TABLE 2.4: Units of surface measurement.

Unit	Acronym	Description
1 centiare	ca	1 m^2 (1 x 1 m)
1 are	a	100 m^2 (10 x 10 m)
1 hectare	ha	10000 m^2 (100 x 100 m)

In the conversion of 6000 m^2 to ha, it is enough to divide the value in square meters per 10000, obtaining 0.6 ha.

```
# Conversion from 6000 square meters (m2) to ha

ha<-10000
m2<-6000

#1ha---10000m^2
#   x---6000m^2

x<-m2/ha
x
```

```
## [1] 0.6
```

In Brazil, the hectare unit is considered official, but other old surface units, as the alqueire, can be used regionally in the country (Table 2.5) (Borges, 2013; Silva and Segantine, 2015).

TABLE 2.5: Examples of regional surface measures used in Brazil.

Regional Measure	Area (ha)
Alqueire Paulista	2.4200
Alqueirão	19.3600
Alqueire Baiano	9.6800
Alqueire Mineiro	4.8400
Alqueire Goiano	4.8400
Alqueire do Norte	2.7225

The alqueire also designates units of measure of area of land that could be sown with an alqueire of seed.

2.12.10   Scale

After taking measurements on the ground and calculating the coordinates of the vertices, it may be necessary to physically represent the measurements on paper. The denominator of the scale (E) is the constant relationship between the distances measured on the terrain (D) and its distance on the representing paper (d), as a proportion:

$$E = \frac{D}{d} \tag{2.1}$$

The scale can be numeric 1:500, graphical (bar scale on the map) and nominal (written in full).

The most common scales used in topography for surveying environments are shown in Table 2.6 (Borges, 2013).

TABLE 2.6: Common scales used in planimetric surveys.

Environment	Scale
Urban lots	1:100 to 1:200
Urban subdivisions and streets	1:1000
Rural properties	1:1000 to 1:5000
Large regions	>1:5000

In altimetry, scales are generally different to represent horizontal and vertical values in order to highlight level differences, so the vertical scale can be larger than the horizontal, for example, using 1:1000 horizontal scale and 1:100 vertical scale (Borges, 2013).

2.12.11   Graphic resolution

Graphic resolution is the smallest magnitude represented in a drawing by scale. The graphical error is the smallest possible graphic magnitude of viewing with the naked eye in magnitude of 0.2 mm or 0.0002 m. With the denominator of the drawing scale (E), you can calculate the smallest possible dimension to be represented (d) (Silva and Segantine, 2015):

$$d = 0.0002E \tag{2.2}$$

2.12.12   Evaluate information about R packages used

The R package `sf` (Pebesma et al., 2021) is installed and enabled to evaluate information about R packages used in the preparation of the book.

The `install.packages` function is used to install the `sf` package in the R console.

```
install.packages("sf")
```

The `library` function is used to enable the `sf` package in the R console.

```
library(sf)
```

The print function is used to evaluate information about the operating system and version of R used.

```
print(version)
```

```
##                         _
## platform       x86_64-w64-mingw32
## arch           x86_64
## os             mingw32
## system         x86_64, mingw32
## status
## major          3
## minor          5.2
## year           2018
## month          12
## day            20
## svn rev        75870
## language       R
## version.string R version 3.5.2 (2018-12-20)
## nickname       Eggshell Igloo
```

The print function allows to obtain the windows version used, the base R packages, and other R packages installed on the computer.

```
print(sessionInfo())
```

```
## R version 3.5.2 (2018-12-20)
## Platform: x86_64-w64-mingw32/x64 (64-bit)
## Running under: Windows 10 x64 (build 19041)
##
## Matrix products: default
##
## locale:
## [1] LC_COLLATE=Portuguese_Brazil.1252  LC_CTYPE=Portuguese_Brazil.1252
## [3] LC_MONETARY=Portuguese_Brazil.1252 LC_NUMERIC=C
## [5] LC_TIME=Portuguese_Brazil.1252
##
## attached base packages:
## [1] stats     graphics  grDevices utils     datasets  methods   base
##
## other attached packages:
##  [1] measurements_1.4.0 gridExtra_2.3      raster_3.1-5
##  [4] readr_2.0.1        tidyr_1.1.3        ggplot2_3.3.5
##  [7] rgdal_1.4-8        sp_1.4-1           rnaturalearth_0.1.0
## [10] COVID19_2.3.2      dplyr_1.0.7        tmap_3.3-2
## [13] sf_1.0-2
```

```
##
## loaded via a namespace (and not attached):
##  [1] bit64_0.9-7        vroom_1.5.4       viridisLite_0.3.0  shiny_1.7.0
##  [5] assertthat_0.2.1   yaml_2.2.0        pillar_1.6.2       lattice_0.20-41
##  [9] glue_1.4.2         digest_0.6.25     RColorBrewer_1.1-2 promises_1.1.0
## [13] colorspace_1.4-1   htmltools_0.5.2   httpuv_1.6.1       XML_3.98-1.20
## [17] pkgconfig_2.0.3    stars_0.5-3       bookdown_0.23.4    purrr_0.3.4
## [21] xtable_1.8-4       scales_1.0.0      later_1.0.0        tzdb_0.1.2
## [25] tibble_3.1.4       generics_0.1.0    ellipsis_0.3.2     withr_2.4.2
## [29] leafsync_0.1.0     cli_3.0.1         magrittr_2.0.1     crayon_1.4.1
## [33] mime_0.7           evaluate_0.14     fansi_0.4.0        lwgeom_0.1-7
## [37] class_7.3-14       tools_3.5.2       hms_1.1.0          lifecycle_1.0.0
## [41] stringr_1.4.0      munsell_0.5.0     compiler_3.5.2     e1071_1.7-3
## [45] rlang_0.4.11       classInt_0.4-3    units_0.6-4        grid_3.5.2
## [49] tmaptools_3.1-1    dichromat_2.0-0   rstudioapi_0.13    htmlwidgets_1.5.3
## [53] crosstalk_1.0.0    leafem_0.1.6      base64enc_0.1-3    rmarkdown_2.9
## [57] gtable_0.3.0       codetools_0.2-15  abind_1.4-5        DBI_1.1.1
## [61] R6_2.5.1           knitr_1.33        bit_1.1-15.2       fastmap_1.1.0
## [65] utf8_1.1.4         KernSmooth_2.23-15 stringi_1.4.3     parallel_3.5.2
## [69] Rcpp_1.0.4.6       vctrs_0.3.8       png_0.1-7          leaflet_2.0.4.1
## [73] tidyselect_1.1.1   xfun_0.24
```

The print function is used to evaluate R package installation directory on the computer.

```
print(.libPaths())
```

```
## [1] "C:/Users/UFLA/OneDrive/Documentos/R/win-library/3.5"
## [2] "C:/Program Files/R/R-3.5.2/library"
```

The packageVersion function is used to evaluate specific R package version installed on the computer. In this case, the sf version is obtained.

```
packageVersion("sf")
```

```
## [1] '1.0.2'
```

2.12.13   Evaluate the existence of an R package with keywords

The search for information about an R package can be carried out based on keywords on the subject of interest through the sos package (Graves et al., 2020). As an example, the search for keywords is performed to evaluate packages available for application in different themes.

The sos package is installed and enabled as previously described, on the R.

```
install.packages("sos")
```

```
library(sos)
```

The `findFn` function is used to search for existing packages on a thematic subject of interest, in this case, with the keywords: latitude, longitude and cartesian.

```
findFn("latitude longitude cartesian")
```

In this research, 48 functions related to the keywords are found, in different packages, in addition to a description and a link to access more detailed information on the Internet.

2.13 Solved Exercises

2.13.1 Can geomatics and topography be considered synonymous?

A: Both terms can be used, but the term geomatics is more comprehensive than topography because it encompasses the use of geocomputation techniques and remote sensing imagery in topography.

2.13.2 Recent technological developments in measurements and topographic surveys have determined the disuse of the term "topography", which has been replaced by geomatics?

A: The terms "topography" and "geomatics" are synonymous; however, geomatics arose due to the change of technology and methodology of surveys with the support of other sciences, such as electronics, informatics, geophysics, engineering, meteorology and astromomia, but has not yet replaced but complemented translating a differentiated emphasis from traditional topography.

2.13.3 What determined the emergence of the first surveyors who were called string stretchers in Egypt around 1400 BC?

A: Herodotus divided Egypt into tax collection plots, but the floods of the Nile River determined the need for a new demarcation of the land, emerging the need for surveyors.

2.13.4 How did Eratosthenes determine the dimensions of the Earth using solar measurements from different locations, around 200 BC?

A: Eratosthenes measured the angle formed between the shadow of the sun in a bulkhead in Alexandria and Siene, in order to determine the length of the arc between the two cities, which was later extended to the entire Earth's circumference.

2.13.5 Why do geodetic surveys have a higher order of accuracy than plane surveys?

A: Geodetic surveys consider the Earth's curvature in calculations. In plane surveys, the horizontal land surface was considered plane, resulting in errors for surveys of large areas (>8 km). When considering the Earth's curvature, geodetic surveys can be applied to both large and small areas.

2.13.6 How can a geodetic control network be used to establish vertical and horizontal reference control for other surveys?

A: Control surveys are generally used to correct data collected with GNSS instruments from topographic landmarks materialized in the terrain, with coordinates of longitude and latitude, and altitude known from measurements made with high standard accuracy and precision.

2.13.7 Have the safety standards in topographic surveys in Brazil been specifically applied to topography, and do they have legal protection for the injured professional?

A: In Brazil, security in topographic surveys is regulated by Regulatory Norms, such as NR 18 which establishes guidelines for the implementation of control measures and preventive safety systems in the processes (MTE, 2015), largely applicable to work environment in the construction industry. In these rules, there is no approach related to topography, which by its nature also does not fit into rural work, being unprotected from legal protection if considered in terms of specific legislation.

2.13.8 Is the search for new maps and product updates of already existing planets, moons, stars and other extraterrestrial objects a future challenge in topography?

A: Topography is in the middle of a revolution where the processes of measuring, recording, processing, storing, retrieving and sharing data take place. Therefore, technological advances in society contribute to wide demand for obtaining data with a high standard of accuracy and the processing of information, in order to update and generate mapping products in different areas of science.

2.13.9 The length of a fence on the ground has a value of 5 cm measured on a map on the 1:1000 scale. Determine the fence length.

A: The estimated fence length on the ground is 50 m.

```
#Data
E<-1000
d<-0.05
#Fence distance on the ground
D<-E*d
D
```

```
## [1] 50
```

2.13.10 A terrain measurement length of 500 m is mapped on the 1:2000 scale. Determine the length of the terrain measurement on the map.

A: The length of the terrain measurement on the map is 0.25 m or 25 cm.

```
#Data
E<-2000
D<-500
#Distance drawn on paper
d<-D/E
d
```

```
## [1] 0.25
```

2.13.11 Represent the length of 324 m on a 1:500 scale in the drawing.

A: The length of the terrain measurement in the drawing is 0.648 m or 64.8 cm.

```
#Data
E<-500
D<-324
#Distance drawn on paper
d<-D/E
d
```

```
## [1] 0.648
```

2.13.12 In a mapping, vertices are found spaced 820 m apart on the ground and 37 cm in the drawing. Determine the scale.

A: The scale is 1:2216.2.

```
#Data
d<-0.37
D<-820
#Distance drawn on paper
E<-D/d
E
```

```
## [1] 2216.216
```

2.13.13 Determine the smallest possible dimensions of the terrain representation on the scales 1:100, 1:1000 and 1:10000 considering the graphical error 0.0002.

A: Considering the graphical error 0.0002, the smallest possible dimensions of the terrain representation on the scales 1:100, 1:1000 and 1:10000 are 2, 20 and 200 cm, respectively.

```
#Data
E1<-100
E2<-1000
E3<-10000
e<-0.0002
#Distances
d1<-e*E1
d1
```

```
## [1] 0.02
```

```
d2<-e*E2
d2
```

```
## [1] 0.2
```

```
d3<-e*E3
d3
```

```
## [1] 2
```

2.14   Homework

Choose four solved exercises presented by the teacher and improve the answer with more details on the subject. Present the references used in the consultation.

2.15   Resources on the Internet

As a study guide, slides and illustrative videos are presented about the subject covered in the chapter as shown in Table 2.7.

TABLE 2.7: Slide shows and video presentations on geomatics and measurement units.

Guide	Address for Access
1	Slides on Introduction to Geomatics and Measurement units[3]
2	Units of Measure: Unit Conversion/Dimensional Analysis[4]
3	Units of Measure: Scientific Measurements & SI System[5]
4	What is geomatics?[6]
5	Future Challenges in Geomatics: Farewell to Asteroid Bennu[7]

2.16 Research Suggestion

The development of scientific research on geomatics is stimulated by the activity proposals that can be used or adapted by the student to assess the applicability of the subject matter covered in the chapter (Table 2.8).

TABLE 2.8: Practical and research activities used or adapted by students using geomatics and units of measure.

Activity	Description
1	Identify and select a problem to be studied with geomatics and R
2	List the variables of interest for analysis in geomatics and what units of measurement will be used
3	Identify an area that can be measured in the field to perform measurement evaluations with geomatics instruments

2.17 Learning Outcome Assessment Strategy

Perform a summary of the chapter, "Introduction to Measurement Units", on a single A4 page in order to show the student's abilities to summarize a subject presenting key points considered of greater importance today.

---

[3] http://www.sergeo.deg.ufla.br/geomatica/book/c2/presentation.html#/
[4] https://youtu.be/jTR9ly5x6q4
[5] https://youtu.be/oAtDAoqdExw
[6] https://youtu.be/pHr1uDXlQtM
[7] https://youtu.be/RkJo6BXfbmA

3

Theory of Measurement Errors

DOI: 10.1201/9781003184263-3

3.1 Learning Questions

The emergent learning questions answered through reading the chapter are as follows:

- How to distinguish direct and indirect measurements.
- What possible types of errors can be observed in topographic measurements?
- How to determine the source and magnitudes of error propagation in geomatics.
- How to select instruments and procedures to minimize errors within acceptable limits in geomatics.
- How to evaluate the difference of mistakes, systematic and random errors.
- How is probability theory used in geomatics?
- What are the advantages and disadvantages of measurements with electronic equipment?
- How to apply basic statistics to a series of data obtained by survey.

3.2 Learning Outcomes

The learning outcomes expected from reading the chapter are as follows:

- Know the types of direct and indirect measurements.
- Analyze the possible types of errors in direct measurements, including source, magnitude and form of error propagation.
- Evaluate systematic mistakes and errors knowing accuracy, precision and resolution according to an adopted standard.
- Briefly know the theory of probability and its laws, the definition of the most probable value and determination of residual.
- Know the advantages and disadvantages of measurements with electronic equipment.
- Describe a series of data obtained by surveying with electronic measurement using basic statistics.

3.3 Introduction

Performing measurements has been the main interest of a surveyor. In geomatics, the measurement of quantities with exact or true value cannot be determined, such as distances, heights, volumes, directions and weight. The more accurate the equipment used, the closer the result of a measurement to the estimated exact value, but it will never be possible to determine the actual value of the measure (McCormac et al., 2012).

After the measurements, subsequent calculations and analyses are performed, generating satisfactory results. To perform satisfactory observations, it is necessary to combine human skill and good equipment. In addition, it became necessary to analyze the different types of existing errors, the expected magnitudes and the form of propagation, in order to select instruments and procedures implemented for error reduction according to acceptable limits, considering a reference standard. Sophisticated computers and programs are typically used to plan projects and investigate the distribution of errors after obtaining the results of the survey carried out (Figure 3.1) (Alves and Silva, 2016).

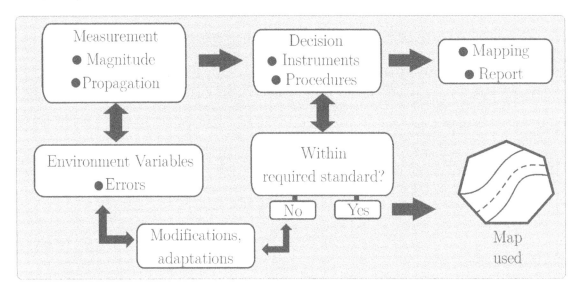

FIGURE 3.1: Decision-making process based on the evaluation of measurement errors.

3.4   Direct and Indirect Observations

Observations can be made directly or indirectly in geomatics. Considering direct measurements, a measurement instrument is applied and adjusted directly over the measured location. As an example, we can mention the use of a measuring tape over the ground when carrying out the measurement. In indirect measurement, it is not possible to apply a measurement instrument directly over the location. The answer is obtained by inference, based on the relationship of other observed variables with the variable of interest. With this, it is possible to verify the distance of a line between two points separated by obstacles or at great distances by trigonometric functions based on reading a grade rod performed through the telescope of a theodolite, as well as by an active electromagnetic sensor, which emitted a signal from the origin and received another signal reflected from a prism at the desired target, performing the indirect distance measurement (Alves and Silva, 2016). Direct and indirect measurements showed errors. The way in which measurement errors are combined to generate error responses is termed "error propagation" (Ghilani and Wolf, 1989).

3.5   Measurement Error Sources

By definition, an error is the difference between an observed value of a quantitative measure and its true value, or (Ghilani and Wolf, 1989)

$$E = X - \overline{X} \tag{3.1}$$

where E is the error of an observation, X, the observed value, and $\overline{X}$, the true value.

When performing topographic measurements, it can be unconditionally stated that (Ghilani and Wolf, 1989):

- No observation is accurate;
- All observations contain errors;
- The exact error present is always unknown.

The accuracy of observations depends on the scale division size, the configuration of the equipment used and human limitations in estimating values less than one-tenth of the scale division. Even for the best equipment developed, observations closer to their true value will never be accurate (Ghilani and Wolf, 1989).

3.6   Types of Errors

Measurement errors can be subdivided into (Schofield et al., 2007):

- Mistakes;
- Systematic errors;
- Random errors.

Mistakes are defined as errors caused by problems of lack of attention, carelessness, fatigue, lack of communication or lack of knowledge. Other examples are: mistaken registration of numbers; reading from a counterclockwise angle, while the reading should be clockwise; aiming at the wrong target and recording the wrong distance. Coarse mistakes must be detected carefully by checking survey data. Small mistakes are difficult to be detected, as they mix with measurement errors (Alves and Silva, 2016).

Systematic errors are known as biases or trends, resulting from factors that are part of the measurement system and included environment, instrument and observer. If the conditions of the measurement system remained constant, systematic errors also remained constant. If the conditions are modified, the magnitude of the systematic error is also modified. As systematic errors tend to accumulate, they are also named cumulative errors (Ghilani and Wolf, 1989). If the conditions that determined the occurrence of systematic errors are known, a correction can be calculated and applied to the observed values (Alves and Silva, 2016).

Random errors remained at the measured values after eliminating mistakes and systematic errors. These errors are caused by factors beyond the observer's control and obey the laws of probability. Random errors are also named accidental, as they are present in the survey measurements. The magnitudes and algebraic signs of random errors occurred at random. Thus, it is not possible to calculate or eliminate random errors; however, it is possible to estimate them using least squares adjustment procedures. Random errors are also named compensated errors, since they were partially eliminated from a series of observations (Alves and Silva, 2016).

3.7   Eliminating Mistakes and Systematic Errors

One of the best ways to identify mistakes can be done by comparing some measurements of the same magnitude. The elimination of error of the surveying can be observed by the outlier of the measurement dataset. Systematic errors can be calculated and corrected from knowledge about the source of systematic error (Alves and Silva, 2016).

3.8   Accuracy, Precision and Resolution

Discrepancy is defined as the difference between two observed values of the same measurement. A small discrepancy indicated that there is probably no mistake and the random errors are small. However, small discrepancies do not prevent systematic errors from occurring (Alves and Silva, 2016).

The difference between the observations of the same measurement can be classified according to the accuracy and precision of the measurement. Precision refers to the degree of refinement or consistency of a group of observations determined based on the size of discrepancy. If multiple observations of the same measurement are made, resulting in a small discrepancy, there is high precision. The degree of precision is dependent on the sensitivity of the equipment and the skill of the operator. Precision is obtained based on the reliability of the measurement process, not the measurement. The precision of a series of observations is represented by the standard deviation from the mean (Alves and Silva, 2016).

Accuracy refers to the proximity to the true value of a measurement performed. Accuracy is related to the reliability of the measurement obtained, and not to the measurement process used. The accuracy of a series of observations is defined by the standard deviation from the real value. The difference between accuracy and precision is illustrated based on a sniper's target. If all the shots are close and grouped, there is satisfactory precision in the operation. If the shots are made in the center of the target, there is satisfactory accuracy in the operation (Ghilani and Wolf, 1989) (Figure 3.2).

Another important concept to define the spatial quality of the survey is the spatial resolution. Spatial resolution is the smallest part identified when a measurement is made. The resolution can be better understood with the example of a target divided into different unit intervals. In this case, there will be greater resolution on the target with a greater number of subdivisions (Figure 3.3).

FIGURE 3.2: Accuracy and precision of an archer when the objective is to hit the center of the target in which good precision and good accuracy (left), bad precision and bad accuracy (center), and good precision and bad accuracy (right) are obtained.

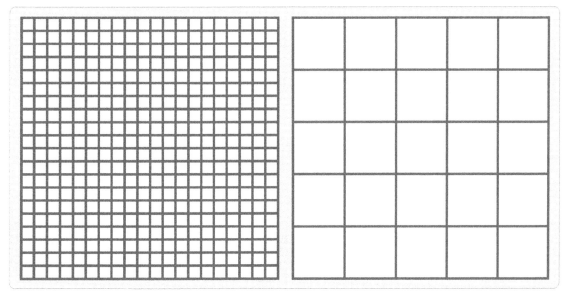

FIGURE 3.3: Resolution of images used in surveying with different pixel sizes. In the image on the left there is a greater number of subdivisions and higher resolution.

3.9 Probability Theory

Events that occur at random or randomly can be characterized by probability theory. A probability can be defined by the rate relative to the number of times a result can occur in relation to the total number of possibilities. In general, if a m result does not occur n times, the probability of occurrence, p, of this result is (Ghilani and Wolf, 1989; Alves and Silva, 2016):

$$p = \frac{m}{(m+n)} \tag{3.2}$$

The probability of occurrence of any result should vary between 0 and 1, with 0 being the indication of impossibility and 1 being the absolute certainty. The sum of the probabilities of a result occurring or failing must be 1.

Considering that perfect measurements cannot be made, random errors will occur in the observations made. The magnitude and frequency of occurrence of these errors can be represented within the laws of probability. Disregarding the mistakes and systematic errors that can occur in the database, only random errors are obtained from the measurement results by means of statistical indices that determine the measurement uncertainties.

3.10   Most Probable Value

Although the true value of a measurement is never known, the most probable value can be quantified after making repeated observations. Repetitions are measures taken in excess to determine an average measurement, based on the most probable value, or the simple arithmetic mean of the measurements (Ghilani and Wolf, 1989; Alves and Silva, 2016):

$$\overline{M} = \sum_{i=1}^{n} \frac{M}{n} \tag{3.3}$$

where $\overline{M}$ is the most probable value, that is, the sum of the individual measurements, M in relation to the total observations, n.

3.11   Residuals

After determining the most probable value of a measurement, it is possible to calculate the residuals. A residual is simply the difference between the most probable value of a measurement and any other measured value in the measurement (**Figure** 3.4) (Ghilani and Wolf, 1989; Alves and Silva, 2016):

$$v = M - \overline{M} \tag{3.4}$$

where v is the residual of any M observation and $\overline{M}$ the most likely value of the measurement. Residuals are theoretically identical to the errors, except that residuals can be calculated and errors cannot, since the true values are never known. Thus, the term residual is used instead of error, when determining the adjustment of survey data.

3.12   Random Errors

Mistakes and systematic errors must be detected and removed before the analysis of random errors in a database. The histogram is defined as a barplot representing the size of observations in relation to the frequency of occurrence. This procedure enabled to obtain a visual impression of the distribution pattern of the observations or residuals (**Figure** 3.5) (Ghilani and Wolf, 1989; Alves and Silva, 2016).

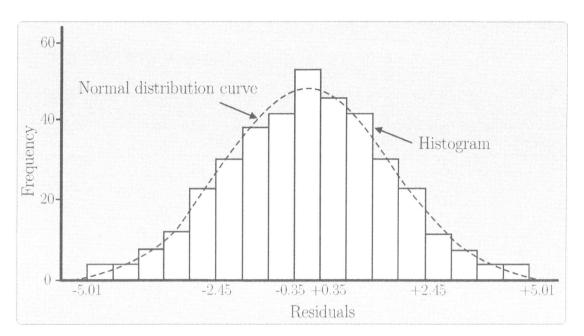

FIGURE 3.4: Frequency histogram with normal distribution curve of residuals from horizontal distances measured with total electronic station.

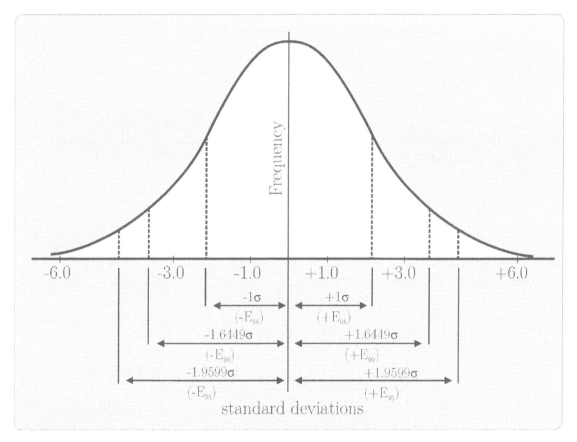

FIGURE 3.5: The standardized normal distribution curve used to compare the accuracy of surveys.

3.12.1   General probability laws

Some general probability laws are established in the theory of errors (Ghilani and Wolf, 1989):

- Small-value residuals occur more frequently than large-value residuals, i.e., small residuals are more probable;
- Large residuals are less frequent and therefore less probable. In errors with normal distribution, typically larger values are more mistakes than random errors;
- Positive and negative residuals occur with equal frequency and with the same probability, that is, the most probable value of a group of repetitions of measurements performed with the same equipment and procedure is the mean.

3.12.2   Precision measurements

The dispersion magnitude is an indication of the relative dispersion of observations, indicated by the standard deviation and variance (Schofield et al., 2007):

$$\sigma = \pm\sqrt{\frac{\sum_{i=1}^{n} v^2}{n-1}} \tag{3.5}$$

where σ is the standard deviation of a group of observations of the same measurement, v, the residual of an individual observation, $\sum_{i=1}^{n} v^2$, the sum of squares of the individual residuals, and, n, the number of observations. The variance is equal to σ^2.

A graph of the normal distribution curve enabled to observe multiple values of standard deviation correspondence with the percentage of the total area below the normal distribution curve. The value of one standard deviation corresponds to $\sim 68\%$ of the area below the normal distribution curve (**Figure** 3.6) (Ghilani and Wolf, 1989).

Based on the value of the standard deviation, it is possible to establish limits within which the occurrence of the observations is expected. For measurement residuals with normal distribution, the probability of an error of any percentage of probability is determined by (Ghilani and Wolf, 1989):

$$E_p = C_p \sigma \tag{3.6}$$

where E_p is a certain percentage of error and C_p is the corresponding numerical factor obtained by the error ratio and the percentage of the area below the normal distribution curve.

Thus, the equations for errors of 50, 68, 90 and 95% probability of occurrence are (Ghilani and Wolf, 1989):

$$E_{50} = \pm 0.6745\sigma \tag{3.7}$$

$$E_{68} = \pm 1\sigma \tag{3.8}$$

$$E_{90} = \pm 1.6449\sigma \tag{3.9}$$

$$E_{95} = \pm 1.9599\sigma \tag{3.10}$$

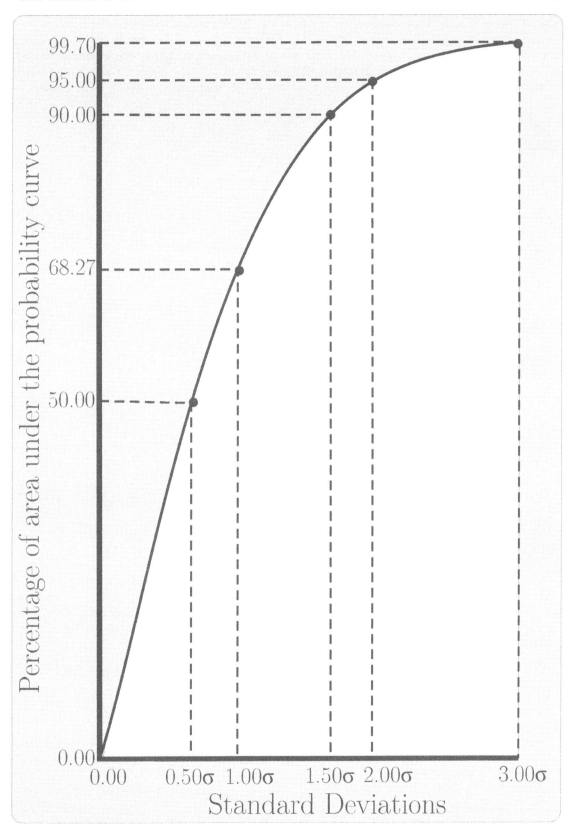

FIGURE 3.6: Relationship between error and percentage of the area below the normal distribution curve.

The 50% error is named "probable error". In this case, limits are established in which the observations occur in 50% of the repetitions, that is, they presented the same probability of falling within or outside that limit. The 90 and 95% errors are normally used to specify the accuracy required in survey designs. The 95% error is named "2 sigma (2σ) error". The 3 sigma (3σ) error is widely used as a criterion for rejecting individual observations of a dataset, considered as a mistake.

3.13   Error Propagation

Since all observations contain errors, any measurements made can probably contain errors. The process of evaluating errors in quantities calculated based on the observed values is named "error propagation". The general law of propagation of variances can be used to characterize the propagation of errors by mathematical equations. In topographic surveys, this equation can be simplified, as the observations are mathematically independent (Alves and Silva, 2016).

3.13.1   Sum error

Whereas the sum of independent observations a, b, c, ..., n is equal to Z, the equation for determining the Z measure is (Ghilani and Wolf, 1989):

$$Z = a, b, c, ..., n \tag{3.11}$$

The error propagated in relation to the sum of the measurements containing different random errors is the error of a sum (E_{sum}):

$$E_{sum} = \pm\sqrt{E_a^2 + E_b^2 + E_c^2 + ... + E_n^2} \tag{3.12}$$

3.13.2   Series error

Similar series of measures, such as angles within a closed polygon or distances measured under the same condition, are read so that each observation has the same error regarding a measure, but with a series of measures. In this case, random errors tend to accumulate in proportion to the square root of the number of measurements. This type of error can only be considered if all measurements are made with equal precision and reliability (McCormac et al., 2012).

The total error of the sum of all these measures in series is named error of a series (E_{series}) (Ghilani and Wolf, 1989):

$$E_{series} = \pm\sqrt{E_1^2 + E_2^2 + E_3^2 + ... + E_4^2} \; = \; \pm\sqrt{nE^2} \; = \; \pm E\sqrt{n} \tag{3.13}$$

where E_1, E_2, ..., E_n represent the error of each observation and, n, the number of observations.

3.13.3   Error of a product

The equation for propagating the error of a product (E_{prod}) between the AB observations are (Ghilani and Wolf, 1989):

$$E_{prod} = \pm\sqrt{A^2 E_B^2 + B^2 E_A^2} \tag{3.14}$$

where E_A and E_B are the errors of A and B, respectively.

The propagation error of a product can be used in areas with sides A and B observed from a rectangular plot of land with the errors E_A and E_B, respectively. The product of the AB sides defines the parcel area.

3.13.4 Error of a mean

The error of a mean (E_m) calculation enabled to determine the error of a series of observations with the same error for each observation (Ghilani and Wolf, 1989):

$$E_m = \pm\frac{E_{series}}{n} = \pm\frac{E\sqrt{n}}{n} = \pm\frac{E}{\sqrt{n}} \tag{3.15}$$

The error of a mean can be applied to specific probability percentages so that errors of 68% (E_{68m}), 90% (E_{90m}) and 95% (E_{95m}) of an average were (Ghilani and Wolf, 1989):

$$E_{68m} = \sigma_m = \pm\frac{\sigma}{\sqrt{n}} = \pm\sqrt{\frac{\sum_{i=1}^{n} v^2}{n(n-1)}} \tag{3.16}$$

$$E_{90m} = \pm\frac{E_{90}}{\sqrt{n}} = \pm 1.6449\sqrt{\frac{\sum_{i=1}^{n} v^2}{n(n-1)}} \tag{3.17}$$

$$E_{95m} = \pm\frac{E_{95}}{\sqrt{n}} = \pm 1.9599\sqrt{\frac{\sum_{i=1}^{n} v^2}{n(n-1)}} \tag{3.18}$$

3.14 Electronic Data Acquisition

Data collectors can be interoperable with modern survey instruments, automatically collecting and storing data in files compatible with computers (Figure 3.7). Data transfer can be carried out using memory cards, cables and Bluetooth. The biggest advantages of automatic data storage are (Alves and Silva, 2016):

- Avoid mistakes when reading and/or recording field observations manually;
- Reduce the time for processing, viewing and filing field notes in the office;
- Increase the amount of information stored in digital data collectors.

The disadvantages of automatic data storage are (Alves and Silva, 2016):

- The need for a qualified operator to operate the instrument;
- Data may be accidentally deleted;
- The need for batteries for the operation of digital data collectors.

FIGURE 3.7: LT30 TM data collector, with 806 MHz processor, with the SurvCE program, used to record and calculate surveying measurements with satellite positioning systems.

3.15 Computation

As a computation practice, a graph is performed in R in order to characterize the errors of 1σ, 2σ and 3σ used in geomatics. Then, the occurrence of random errors is analyzed by means of a topographic survey database with 100 repetitions of measurement of slope separating two vertices with an electronic total station. Dataset normality is analyzed as well as the calculation of summary statistics, residuals and probability errors of horizontal distance measurements.

3.15.1 Errors of 1σ, 2σ and 3σ

The errors of 1σ, 2σ and 3σ represent 68, 95 and 99.7% of the area under the normal frequency distribution curve for measurement residuals. Based on this information, a graph is constructed to illustrate this concept (Figure 3.8).

```
plot(seq(-3.2,3.2,length=50),dnorm(seq(-3,3,length=50),0,1),type="l",xlab="",
    ylab="",ylim=c(0,0.5))
segments(x0 = c(-3,3),y0 = c(-1,-1),x1 = c(-3,3),y1=c(1,1))
text(x=0,y=0.45,labels = expression("99.7% of data considering 3" ~ sigma))
arrows(x0=c(-2,2),y0=c(0.45,0.45),x1=c(-3,3),y1=c(0.45,0.45))
segments(x0 = c(-2,2),y0 = c(-1,-1),x1 = c(-2,2),y1=c(0.4,0.4))
text(x=0,y=0.3,labels = expression("95% of data considering 2" ~ sigma))
arrows(x0=c(-1.5,1.5),y0=c(0.3,0.3),x1=c(-2,2),y1=c(0.3,0.3))
segments(x0 = c(-1,1),y0 = c(-1,-1),x1 = c(-1,1),y1=c(0.25,0.25))
text(x=0,y=0.15,labels = expression("68% of data considering 1" * sigma),
    cex=0.9)
```

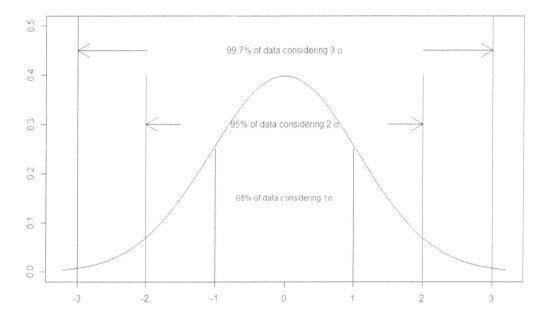

FIGURE 3.8: Using R codes to illustrate the concept of sigma errors.

A database with 100 repetitions of slant distance measurement with an electronic total station is used to analyze the occurrence of random errors assuming that there are no mistakes and systematic errors. The database is related to the x vector, including the distance values.

3.15.2 Import database with horizontal distance measurements through electronic total station

Data with 100 values of horizontal distance measurements obtained with the same methodology through Ruide total electronic station are stored in x.

```
x <- c(166.9200308039,166.9170649536,166.9190421871,166.9180535704,
       166.9190421871,166.9180535704,166.9150877201,166.9200308039,
       166.9171881078,166.9190421871,166.9170649536,166.9103942717,
       166.9191685993,166.9171913760,166.9160795788,166.9152141527,
       166.9111365241,166.9091593023,166.9111365241,166.9131137460,
       166.9220112443,166.9220112443,166.9141023569,166.9160795788,
       166.9229998552,166.9141023569,166.9180568006,166.9160795788,
       166.9131137460,166.9132369294,166.9112597061,166.9091593023,
       166.9082938711,166.9160795788,166.9142255410,166.9150909678,
       166.9200340225,166.9141023569,166.9150909678,166.9229998552,
       166.9190454115,166.9160795788,166.9141023569,166.9141023569,
       166.9141023569,166.9150909678,166.9190454115,166.9101479132,
       166.9102710944,166.9150909678,166.9142255410,166.9131137460,
       166.9152141527,166.9191685993,166.9162027644,166.9201572110,
       166.9121251351,166.9160795788,166.9131137460,166.9150909678,
       166.9141023569,166.9122483177,166.9111365241,166.9191685993,
       166.9231230459,166.9150909678,166.9142255410,166.9132369294,
       166.9142255410,166.9152141527,166.9131137460,166.9142255410,
       166.9150909678,166.9121251351,166.9141023569,166.9122483177,
       166.9122483177,166.9122483177,166.9141023569,166.9141023569,
       166.9121251351,166.9112597061,166.9121251351,166.9111365241,
       166.9180568006,166.9190454115,166.9171913760,166.9190454115,
       166.9249770771,166.9229998552,166.9190454115,166.9102710944,
       166.9112597061,166.9081706914,166.9162027644,166.9121251351,
       166.9162027644,166.9190454115,166.9191685993,166.9102710944)
```

3.15.3   Evaluation of the frequency distribution of the distance measurement data

To evaluate whether the distribution of the histogram presented a bell shape may be vague. A more sensitive graphical technique for assessing data normality may be through a normal probability plot. The normal probability plot is one of the observed values of the variable versus the normal scores of the expected observations for a variable with the standard normal distribution. If the variable is normally distributed, the normal probability plot should be approximately linear i.e., close to a straight line. In R, the functions qqnorm and qqline are used to perform the normal probability or Q-Q plot (Figure 3.9) (Hartmann et al., 2018).

```
qqnorm(x, main = '')
qqline(x, col = 3, lwd = 2)
```

3.15.4   Determination of the mean

The mean slant distance of 166.9154 m represents the most probable value by dividing the sum of observations by the total number of observations.

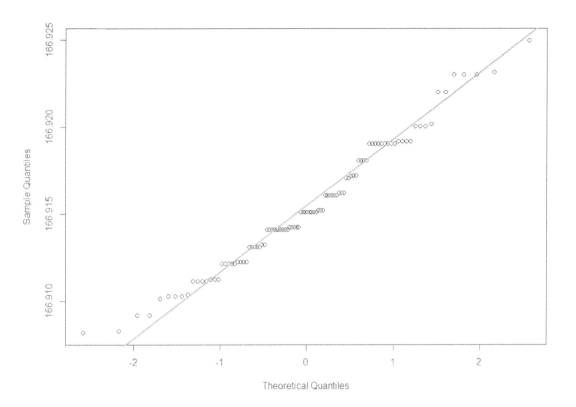

FIGURE 3.9: Using Q-Q plot to evaluate the normality of the slant distance measurement data.

```
mean <- mean(x)
mean
```

```
## [1] 166.9154
```

3.15.5 Median determination

The median of 166.9151 m represents the value of the data medium.

```
median <- median(x)
median
```

```
## [1] 166.9151
```

3.15.6 Mode determination

Mode is the most frequently occurring value in a dataset. Along with mean and median, mode is a statistical measure of central tendency in the dataset. The mode is determined by a function followed by the application in the distance measurement database. The mode determined on slant distance measurements is 166.9141 m, that is, the value with the most frequency occurrence in the database.

```
mode <- function(x) {
    ux <- unique(x)
    ux[which.max(tabulate(match(x, ux)))]} #Function
mode <- mode(x) #Mode calculation
mode
```

```
## [1] 166.9141
```

3.15.7   Variance determination

Through variance, the dispersion value of the measurement dataset is characterized in relation to the mean. The variance is 0.0000136435 m^2.

```
variance <- var(x)
variance
```

```
## [1] 1.326435e-05
```

3.15.8   Determination of standard deviation

The standard deviation is a measure used to quantify the magnitude of variation or dispersion of the data. The standard deviation of the analyzed dataset is 0.003642025 m.

```
sd <- sqrt(var(x))
#or
sd <- sd(x)
sd
```

```
## [1] 0.003642025
```

3.15.9   Histogram determination in the residuals

The histogram of residuals of the database is performed adopting the average value as the most probable value for residual calculation of each distance measurement (Figure 3.10).

```
v<-x-mean
hist(v, breaks=10, col="grey90", xlab="Residual",
     main="", xlim=c(-0.010, 0.010))
```

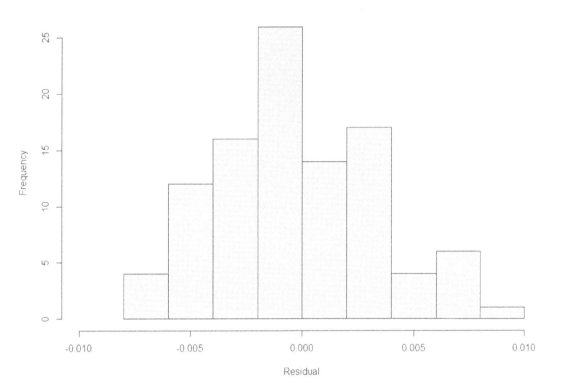

FIGURE 3.10: Determination of the frequency distribution histogram of measurement data.

3.15.10   Adding a normal curve to the histogram of residuals

A distribution curve from normal to residual values is adjusted. In this case, the data distribution pattern is close to the normal curve, with a bell-like shape (Figure 3.11).

```
hv<-hist(v, breaks=10, col="grey90", xlab="Residual (m)",
    main="",
    xlim=c(-0.010, 0.010))
xvfit<-seq(min(v),max(v),length=40)
yvfit<-dnorm(xvfit,mean=mean(v),sd=sd(v))
yvfit <- yvfit*diff(hv$mids[1:2])*length(v)
lines(xvfit, yvfit, col="black", lwd=2)
```

It should be noted that other R packages, such as ggplot2, can be used to make histograms of real and simulated data.

3.15.11   Determination of probability errors of horizontal distance measurements

The errors at 50, 68, 90, 95 and 99.7% probability of horizontal distance measurements with the Ruide electronic total station are determined by mathematical multiplication operations, obtain-

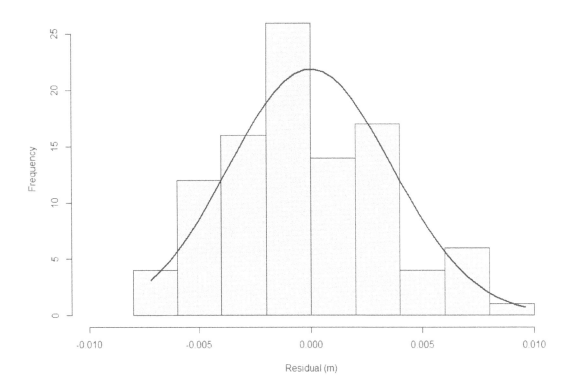

FIGURE 3.11: Normal distribution curve fitting to the measurement residual values.

ing the values of ±0.002456546, ±0.003642025, ±0.005990767, ±0.007138005 and, ±0.01092607 m, respectively.

```
E50<-0.6745*sd(x)
E50
E68<-1*sd(x)
E68
E90<-1.6449*sd(x)
E90
E95<-1.9599*sd(x)
E95
E99<-3*sd(x)
E99
```

3.15.12   Determination of probability confidence intervals for horizontal distance measurements

The confidence intervals at 50, 68, 90, 95 and 99.7% probability for horizontal distance measurements with the Ruide electronic total station are determined by sum and multiplication operations of the errors previously obtained, obtaining interval values of 166.9129 to 166.9178 m, 166.9117 to 166.9190 m, 166.9094 to 166.9214 m, 166.9082 to 166.9225 m, and 166.9044 to 166.9263 m, respectively.

```
c(mean(x)- E50, mean(x)+ E50)
c(mean(x)- E68, mean(x)+ E68)
c(mean(x)- E90, mean(x)+ E90)
c(mean(x)- E95, mean(x)+ E95)
c(mean(x)- E99, mean(x)+ E99)
```

3.15.13 Determination of probability errors of the mean of horizontal distance measurements

The errors of the mean of 50, 68, 90, 95 and 99.7% probability of horizontal distance measurements with the Ruide electronic total station are determined with equations for each specified probability error of the mean, obtaining the values of ± 0.0002456546, ± 0.0003642025, ± 0.0005990767, ± 0.0007138005 and, ± 0.001092607 m, respectively.

```
E50m<-0.6745*sd(x)/sqrt(length(x))
E50m
E68m<-1*sd(x)/sqrt(length(x))
E68m
E90m<-1.6449*sd(x)/sqrt(length(x))
E90m
E95m<-1.9599*sd(x)/sqrt(length(x))
E95m
E99m<-3*sd(x)/sqrt(length(x))
E99m
```

3.16 Solved Exercises

3.16.1 Suppose a line is observed in three sections, with the individual parts equal to 241.25 ±0.02, 236.45 ±0.03, and 261.32 ±0.01 m. Determine the total length of the line and the standard deviation.

A: Based on the error determination of a sum, the total length is 739.02 m and the standard deviation is ± 0.03741 m.

```
length<-241.25+236.45+261.32
length
Esum<-sqrt(0.02^2+0.03^2+0.01^2)
Esum
```

3.16.2 Consider that a distance of 100 m can be demarcated with an error of ±0.02 m using a specific survey methodology. Determine the demarcation error of 25000 m using the same methodology.

A: Based on the error of a series, the error for demarcating 25000 m is ±0.3162 m.

```
n<-25000/100
n
Eserie=0.02*sqrt(250)
Eserie
```

3.16.3 The distance of 15000 m is demarcated with an error of ±0.10 m. Determine the measurement accuracy of 100 m in length ensuring that the error will not exceed the permitted limit.

A: Whereas $n=15$, the average error for measuring 100 m is ±0.025 m.

```
n<-1500/100
n
Em<-0.10/sqrt(15)
Em
```

3.16.4 The sides A and B of a plot measured with 95% probability error are 552.46 ±0.053 and 605.08 ±0.072 m, respectively. Calculate the parcel area with the expected error of 95% probability.

A: The calculated area of the parcel is 334282.5 m². The expected area error is ±51.09 m². Therefore, the area should be between 334231.4 and 334333.6 m².

```
area<-552.46*605.08
area
Eprod<-sqrt(552.46^2 * 0.072^2 + 605.08^2 * 0.053^2)
Eprod
sup_int<-334282.5+51.09457
sup_int
inf_int<-334282.5-51.09457
inf_int
```

3.17 Homework

Choose two exercises presented by the teacher and solve the questions with different input values. Compare the results obtained. If possible, use a real database measured in the field with an available instrument.

3.18 Resources on the Internet

As a study guide, slides and illustrative videos are presented about the subject covered in the chapter as shown in Table 3.1.

TABLE 3.1: Slide shows and video presentations on error theory.

Guide	Address for Access
1	Slides on error theory in geomatics[1]
2	Error theory in surveying[2]
3	Errors and sigma[3]
4	Theory of errors[4]

3.19 Research Suggestion

The development of scientific research on geomatics is stimulated by the activity proposals that can be used or adapted by the student to assess the applicability of the subject matter covered in the chapter (Table 3.2).

TABLE 3.2: Practical and research activities used or adapted by students using theory of measurement errors for surveying applications.

Activity	Description
1	Define if your research will be based on the improvement of an existing method by analysis of survey errors or if it will refer to measurements made in the field, such as the demarcation of a contour polygonal in the area of interest
2	Research how error theory can be used to analyze uncertainties in topographic surveys according to previously established standards
3	Use a surveying instrument provided by the teacher and perform measurements with repetitions to analyze results with error theory

3.20 Learning Outcome Assessment Strategy

Perform a summary of the chapter, "Theory of Measurement Errors with Geomatics and R", on a single A4 page in order to show the student's abilities to summarize a subject presenting key points considered of greater importance today.

---

[1] http://www.sergeo.deg.ufla.br/geomatica/book/c3/presentation.html#/
[2] https://youtu.be/VULaVSumGJk
[3] https://youtu.be/LVqhn9kVst4
[4] https://www.youtube.com/watch?v=5H7VzrxkJc0

4

Angle and Direction Observations

4.1 Learning Questions

The emergent learning questions answered through reading the chapter are as follows:

- How to differentiate between horizontal and vertical angles in geomatics.
- What angles are needed to determine direction in geomatics?
- How are the systems for the angular reading of mechanical optical instruments and electronic angle measurement built?
- How is the magnetic field modeled on Earth?
- What are the variations of the magnetic field and how does it interfere with the use of the compass?
- How are the R software and R packages used for recording and converting angles, performing mathematical operations and trigonometric calculations with angles, determining angles in geometric figures and the angular variation of the Earth's magnetic field?

4.2 Learning Outcomes

The learning outcomes expected from reading the chapter are as follows:

- Understand the horizontal or vertical angles measured in geomatics as a function of the observation reference plane in different measurement units.
- Understand the horizontal angles necessary for determining bearing and azimuth from calculations involving internal angles, external angles, deflection angles, and the direction of a line, recognizing possible measurement errors.
- Know an angular reading system for mechanical optical instruments and electronic angle measurement.
- Understand how the magnetic field works, its variations and how it interferes with the use of compass and the difference between magnetic and geodetic meridian.
- Use the R Program and the `oce`, `LearnGeom` and `circular` packages for converting different angle systems, mathematical operations with angles, trigonometric calculations, determination of angles in geometric figures, determination of the Earth's magnetic field by model at different times and graphical evaluation of the direction bearing quadrant.

DOI: 10.1201/9781003184263-4

4.3 Introduction

Human beings oriented themselves by rudimentary methods using angle measurements. In 4000 BC, Egyptians and Arabs tried to work out a calendar by observing the sun and an angle corresponding to the movement of the Earth's orbit. In 1700 BC, the Babylonians used decimal systems and sexagesimal fractions to measure an arc of a circle (Silva, 2003). Determining the location of points and the orientation of lines can be accomplished by observing angles and directions. In topographic surveys, directions are obtained based on azimuths and bearings with respect to an alignment and a starting reference (Ghilani and Wolf, 1989; Alves and Silva, 2016).

The angles measured in the topographic survey are classified as horizontal or vertical, depending on the reference plane of observation. Horizontal angles are the basic observations necessary for determining bearing and azimuth in topographic survey. Vertical angles are used in trigonometric leveling, stadia and for slope correction of horizontal distances. With theodolite and electronic total station, both horizontal and vertical angles are measured. In digital equipment, we can choose the type of vertical angle, whether the positive direction of the horizontal angle is clockwise or counterclockwise and whether the unit of measurement is sexagesimal or grad (Silva, 2003; Ghilani and Wolf, 1989; Alves and Silva, 2016).

The basic requirements for determining angles are (Figure 4.1) (Alves and Silva, 2016):

- Determination of an initial or reference line;
- Sighting direction;
- Angular distance or angular value.

4.4 Angle Measurement Units

The sexagesimal system is used in Brazil, the United States and many other countries as a reference for measuring angles based on degrees, minutes and seconds, with the last unit divided into decimals. The grad or gon is commonly used in Europe. The angle unit in radians is most commonly used in computation and geodetic calculations (Ghilani and Wolf, 1989; Alves and Silva, 2016).

4.5 Horizontal Angle Types

A horizontal angle is defined by a dihedral angle between two vertical planes passing through the ends of two alignments (Figure 4.2) (Silva, 2003). The types of horizontal angles observed in surveying are (Ghilani and Wolf, 1989):

- Internal angles;
- External angles;
- Deflection angles.

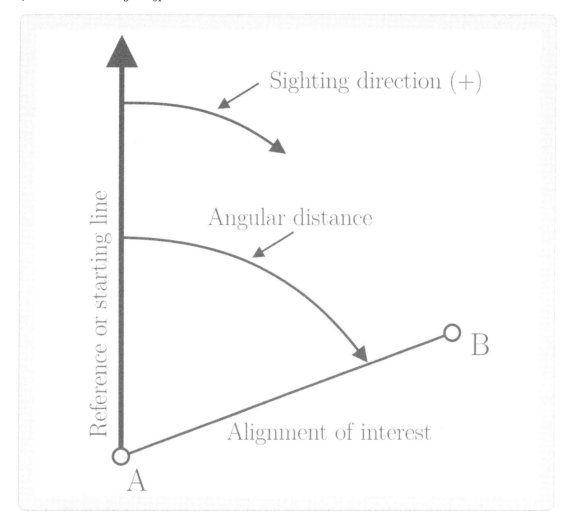

FIGURE 4.1: Basic requirements for determining an angle.

The internal angles are observed inside a closed polygon. The angle at each vertex is measured. A check can be performed on these values, because the sum of all interior angles of polygon must equal to 180° (Alves and Silva, 2016):

$$\sum_{n}^{i=1} Ai = (n-2)180°$$ (4.1)

where Ai is the internal angle and n is the number of vertices.

Polygons are commonly used in boundary survey and other work to describe objects such as buildings in closed geometric figures. Surveyors use the term "closed polygon" when referring to this type of survey. The direction of determining angles, to the right or left, in the direction of walking, can determine errors of mistake, if the procedure performed is not systematic (Alves and Silva, 2016).

External angles, located on the outside of a closed polygon, are complementary to internal angles. The advantage, by looking at the internal and external angles is the possibility to check the results, because the sum of these angles must be 360°. In this case, the sum of the external angles of a closed polygon is (Alves and Silva, 2016) (Figure 4.3):

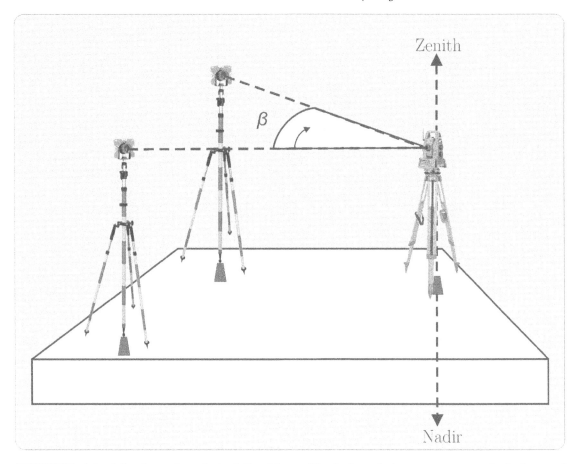

FIGURE 4.2: A horizontal angle is defined by a dihedral angle between two vertical planes of two alignments.

$$\sum_{n}^{i=1} Ai = (n+2)180°$$ (4.2)

Deflection angles can be observed from an extension of the previous line toward the subsequent or forward station. These angles have been used for surveying long linear alignments of routes. Deflection angles can be observed to the right (clockwise) or to the left (counterclockwise) depending on the direction of the route. Clockwise angles are considered positive and counterclockwise are negative. The deflection angles are always smaller than 180° and can be identified according to the direction, right or left (Figure 4.4).

4.6   Direction of a Line

The direction of a line can be defined by the horizontal angle between the alignment and an arbitrary reference line called a "meridian". Different meridians are used to specify directions (Ghilani and Wolf, 1989; Alves and Silva, 2016):

- Geodetic or true;

- Astronomical;
- Magnetic;
- Grid;
- Register;
- Relative or arbitrary.

The geodetic meridian defines the north-south reference line that passes through an average position of the Earth's geographic poles. The mean position of the poles was defined between 1900 and 1905 years. The oscillation of the Earth's axis of rotation determined the temporal variation in the position of the Earth's geographic poles. The astronomical meridian is the reference line that passes through an instantaneous position of the Earth's geographic poles. The term "astronomical meridian" is derived from the way meridians are determined when demarcating positions of celestial objects. Geodetic and astronomical meridians are very similar, so that minor corrections are needed to convert geodetic meridians to astronomical ones. The magnetic meridian is defined by a suspended magnetic needle influenced by the Earth's magnetic field. In surveys based on a plane coordinate system, usually a grid or graticule squared meridian is used as a reference. The north grid is the direction in geodetic north to a selected central meridian, remaining parallel to or over the entire filled area in a plane coordinate system. In boundary or boundary surveys, the term "meridian of a record" is the directional reference recorded on documents from previous survey of a particular area. Another similar term is "registered meridian", used in the description of a parcel of land recorded in the register of a land property. The relative or arbitrary meridian can be established by determining any arbitrary direction. Thereafter, the direction of all other lines are determined relative to the arbitrary meridian (Ghilani and Wolf, 1989; Alves and Silva, 2016).

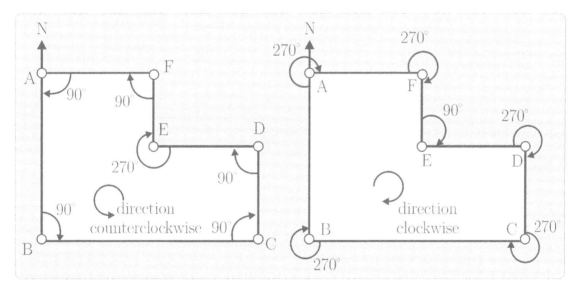

FIGURE 4.3: Closed polygon with interior angles measured counterclockwise (left) and exterior angles measured clockwise (right).

4.7 Azimuths

Azimuths are horizontal angles observed clockwise from any reference meridian, ranging from 0 to 360°. In plane surveys, azimuths are usually observed from the north, but astronomers and

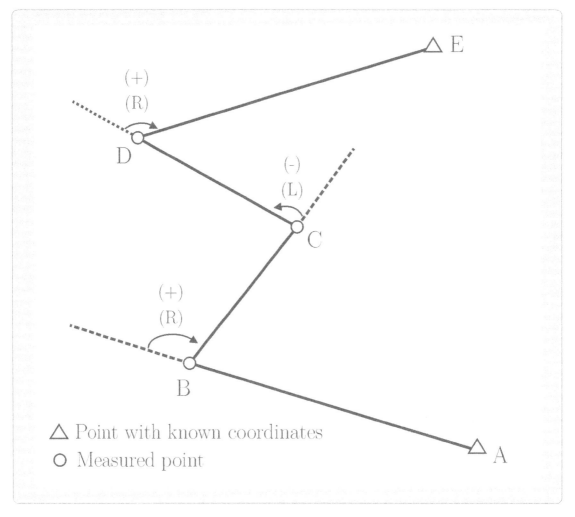

FIGURE 4.4: Angles of deflection used for surveying long linear alignments of routes.

military personnel adopted the south as their reference direction. Azimuths can be geodetic, astronomical, magnetic, grid, register, and relative, depending on the reference meridian used. From a line determined in the forward direction of the survey, the forward azimuth is determined, and for the reverse direction, the reverse azimuth is obtained. On planar surveys, azimuths from forward have been converted to azimuths from backward and vice versa by adding or subtracting 180°. For example, if the azimuth AB is 60°, the azimuth BA is 60° + 180° = 240°. If the azimuth BA is 240°, the azimuth AB is 240° - 180° = 60°. Azimuths can be read directly on the graduated circle of a total station after proper orientation of the instrument. This can be done by sighting across a line of a known azimuth with its indexed value on the circle and then turning to the desired course. Azimuths have been used in boundary, topographic, control, and other types of surveys, as well as in computation determinations (Ghilani and Wolf, 1989; Alves and Silva, 2016) (Figure 4.5).

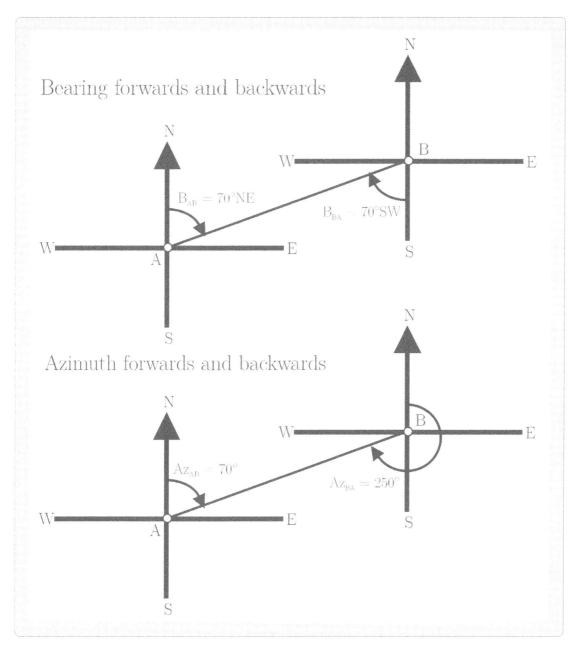

FIGURE 4.5: Graphical comparison of forward and backward bearing and azimuth.

4.8 Bearing

Bearing is another type of system for determining the direction of lines. The bearing of a line is defined as the horizontal angle between a reference meridian and an alignment, according to the corresponding cardinal directions north, east, south, and west, commonly denoted by their initials N, E, S, and W, respectively. The angle is observed from north to south, in the east or west direction, to obtain a reading from 0° to 90°. The letters N or S and E or W are combined after the angle to designate the location of the quadrant giving direction to the alignment to represent intercardinal directions northeast (NE), southeast (SE), southwest (SW), and northwest (NW) (Ghilani and Wolf, 1989; Alves and Silva, 2016). Similar to azimuths, bearings can also be measured forward or backward. Backward bearings must have the same numerical values as the forward ones, but the letters must be from opposite quadrants. Thus, if the bearing AB is 60° NE, the bearing BA is 60° SW. As defined for azimuths, the bearings can be: geodetic, astronomical, magnetic, grid, register and relative, depending on the reference meridian used (Ghilani and Wolf, 1989; Alves and Silva, 2016).

4.9 Azimuth and Bearing Comparison

Since bearings and azimuths are encountered in many survey operations, conversion between these two types of alignments can be useful, obtained by means of simple equations (Table 4.1) and graphical visualization (Ghilani and Wolf, 1989; Alves and Silva, 2016) (Figure 4.6).

TABLE 4.1: Comparison between bearings and azimuths.

Quadrant	Converting Azimuth to Bearing	Converting Bearing to Azimuth
I (NE)	Bearing = Azimuth	Azimuth = Bearing
II (SE)	Bearing = 180° − Azimuth	Azimuth = 180° − Bearing
III (SW)	Bearing = Azimuth − 180°	Azimuth = Bearing + 180°
IV (NW)	Bearing = 360° − Azimuth	Azimuth = 360° − Bearing

4.10 Azimuth Computation in Traversing

Surveys using polygons require the calculation of azimuths or directions. A polygon is a sequence of connected lines in which the lengths and angles are noted at the points where the lines joined or intersected vertices. Traversing lines can be used in many applications, such as surveying boundary lines between properties, surveying roads between towns, and others. Some professionals prefer to determine the direction of lines by azimuths, rather than directions, because azimuths are easier to work with by computation determinations, such as sine and cosine operations. In addition, directions require knowledge of quadrants and signs of trigonometric functions. To calculate directions in polygons, the direction of at least one side must be known or at least arbitrated. The angles between the sides of the polygon are measured, and from these values, the directions of the

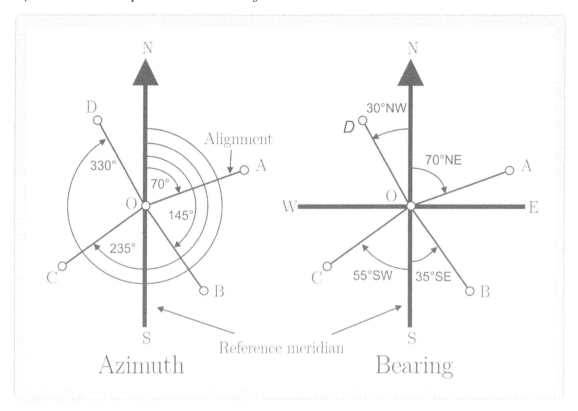

FIGURE 4.6: Graphical comparison between azimuth and bearing.

other sides are calculated. On a polygon where the interior angles are measured, the azimuth value of the previous alignment can be used to obtain the subsequent azimuth. Considering an *ABC* alignment, if the *BA* alignment has a bearing of 41°35′ SW, the azimuth *BA* can be obtained by adding 180° to the bearing, i.e., the azimuth *AB* is 180° + 41°35′ = 221°35 . If the interior angle of the alignment *ABC* is 129°11′, the azimuth *BC* can be obtained by adding the previous azimuth to the interior angle, i.e., the azimuth *BC* is 221°35 + 129°11′ = 350°46′. The process of adding or subtracting 180° to obtain the azimuth from backward and then adding the angle of each alignment can be repeated until the azimuth of the initial line is calculated, at the closure of the polygon. In this case, depending on the value of the angle of the alignment, an azimuth greater than 360° can be determined. Thus, the value of 360° must be subtracted from the total value, in order to obtain the azimuth value. The polygon angles can be determined to the total geometry of the figure before calculating the azimuths. In this case, the internal angles must be $(n - 2)\,180°$, where, n is the number of vertices. If any discrepancy occurs between the results of the sum of the probable polygon angle and the sum of the measured angles, some arithmetic error was made or the angles were not obtained properly. Bearings, rather than azimuths, are used predominantly for surveying divisions. This practice originated in the period when magnetic bearings of parcel boundaries were determined directly by a surveyor's compass. Later, other instruments were used to observe the angles, and the astronomical meridian came into use. The practice of using bearings for land surveying is still in use, because the boundaries of recent surveys must follow the original alignments, and it is necessary to understand magnetic directions and their variations (Ghilani and Wolf, 1989; McCormac et al., 2012; Alves and Silva, 2016).

4.11   Vertical Angles

The graduated circle of the measuring instrument can have three positions with origin in the vertical angle count. When the origin (zero) is in the position of zenith, the zero is adopted as zenith and the vertical angle is called zenith angle (Z). This type of angle can be the most usual in the equipment according to the most adopted form in each country. When the origin is in the horizontal position, zero is adopted in the horizontal plane and the vertical angle is named angle of height, elevation, or inclination (α). When the zero is in the nadir, the vertical angle is a denominated nadiral angle (η) (Silva, 2003) (Figure 4.7). Simply identifying the position of the origin when measuring vertical angles prevented the occurrence of gross errors in readings.

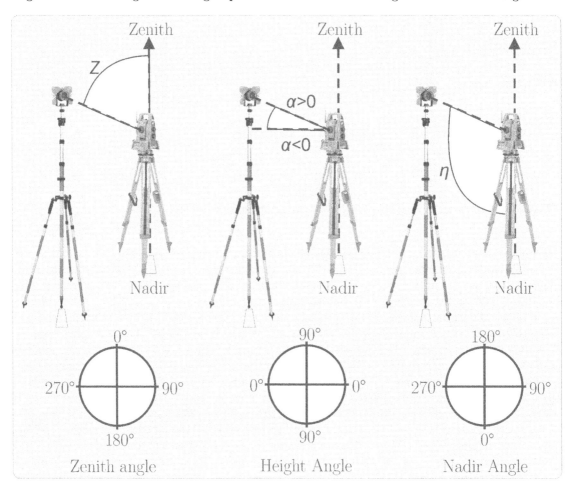

FIGURE 4.7: Types of vertical zenith angle (left), height angle (middle), and nadir angle (right).

4.12   Angular Reading System for Mechanical Optical Instruments

There are numerous angle reading systems developed for mechanical optical theodolites. The theodolites are built with strict quality control, so that the center of the horizontal circle must coincide with the major axis. The center of the vertical circle must coincide with the secondary axis. Thus, angular measurements can be referenced for angle readings (Barcellos, 2003).

4.13   Electronic Angle Measurement

The main physical components of an electronic angle measurement system are (Barcellos, 2003):

- A crystal circle with transparent light and opaque dark regions, encoded by a photolithography process;
- Light detecting photodiodes that pass through the graduated crystal circle.

The encoding and angle measurement can be based on the incremental and the absolute principle. In the incremental principle, the value is provided relative to an initial position. In the absolute principle, the angle value is provided for each position of the circle. In the incremental model, a glass disk with equally spaced opaque and transparent traces is used. A light is placed on one side of the circle and a photodetector on the other side, at positions LR and LS of the disk. This enables one to count the number of pulses when the theodolite rotates from one position to another to measure the angle. This number of pulses can be converted and displayed on a digital display (Kahmen and Faig, 1988).

In the absolute model, opaque trails arranged concentrically on a disk in the non-radial direction are used in the angle measurement. The number of trails is determined by the radius, not the perimeter. The value zero is obtained when the light does not pass through the disk and one when the light passes through. A series of radially shaped diodes can associate each position of the circle with a sequential binary code in decimal format (Barcellos, 2003).

In the TPS1100 electronic total station system, the glass disk carries only a grating line with codes on the positional information. The position is read by a CCD camera, and an 8-bit A/D converter provides the approximate position with 1 second accuracy. The fine measurement is performed by an algorithm that determines the average value between the center positions of each code line captured by the camera. At least 10 lines of code must be captured to determine the position; however, a simple measurement involves 60 lines of code, improving interpolation accuracy, redundancy, and replicability. This principle is applied for Leica total stations and theodolites (Zeisk, 1999).

The horizontal angle value is corrected before it is displayed or recorded by the instrument. The correction is calculated by parameters obtained from the vertical angle measurement, such as the last collimation error stored in the instrument, the verticality component of the transverse axis with respect to the line of sight. The vertical angle is corrected by the stored error index and the component of the vertical axis with respect to the line of sight. A vertical offset sensor monitors the variance components of the vertical axis. The reticle located on the prism is illuminated by an LED and imaged by a lens. The signal is deflected by a prism and subjected to double reflection on the liquid surface of a CCD camera. The triangular line pattern of the reticle enables to capture the vertical axis shift components by means of a unidirectional receiver (Zeisk, 1999).

The tilt sensor can be used to automatically compensate for instrument axis tilts (Barcellos, 2003).

4.14   Compass and Earth's Magnetic Field

Before the invention of transits, theodolites, and total station instruments, the direction of lines and angles was determined by means of compasses. The compass needle is free to rotate and align itself to the axis of the Earth's magnetic field, pointing in the direction of the magnetic meridian. The locations of the Earth's north and south geomagnetic poles are continually changing. It should be noted that the compass needle can be affected by local attractions such as anomalies caused by power lines, railroad tracks, and metal clasps (Ghilani and Wolf, 1989; Alves and Silva, 2016) (Figure 4.8).

FIGURE 4.8: Mobile compass (left) and analog handheld compass (right).

The Earth's magnetic forces not only align the compass needle, but also pull or sink one end below the horizontal position. The depth angle of the needle ranges from 0° near the Equator to 90° at the magnetic poles. In the Northern Hemisphere, the southern end of the needle was weighted with a small wire roller to balance the depth effect by keeping the compass horizontal. The position of the wire roller can be adjusted according to the latitude at which the compass is used (Ghilani and Wolf, 1989; Alves and Silva, 2016).

The Earth's magnetic field looks like a large magnetic dipole located at the center of the Earth, with a displacement of the Earth's axis of rotation. This field has been observed at about 200 magnetic observatories around the Earth, as well as at other temporary stations. The field strength and direction have been measured at each observation point based on many years of data. Models of the Earth's magnetic field were subsequently developed. These models are used to calculate magnetic declination and annual change, in view of the importance of these variables in surveys. The accuracy of the models is affected by several items, including observation locations, rock types and geological structures on the surface, and local attractions (Ghilani and Wolf, 1989; Alves and Silva, 2016).

4.15   Magnetic Declination and Meridian Convergence

Magnetic declination is defined as the observed horizontal angle from the geodetic meridian to the magnetic meridian. Navigators call this angle the "compass deviation". The military uses the term "deviation". East declination occurs when the magnetic meridian is east of geodetic north. A westerly declination occurs when the magnetic meridian lies west of geodetic north. Eastward and westward declinations are considered positive and negative, respectively. The relationship between geodetic north, magnetic north, and magnetic declination (md) can be obtained by the geodetic azimuth (ga) and magnetic azimuth (ma) (Ghilani and Wolf, 1989; Alves and Silva, 2016):

$$ga = ma + md \qquad (4.3)$$

Since the magnetic pole is constantly changing, magnetic declination is also changing. The magnetic declination of any location can be determined, eliminating the effect of any local attraction, by determining the meridian by astronomical or global navigation satellite system (GNSS) observations and then taking a compass reading with a sight on the determined meridian. Another way to determine the magnetic declination of a point can be done from isogonic and isophoric charts, which represent the temporal and spatial variation of magnetic declination, respectively (Figure 4.9) (ON, 2020). Computer programs can also determine magnetic declination, based on mathematical models. In this case, the variation of magnetic declination can be interpolated at any location between lines representing distinct declination values (Ghilani and Wolf, 1989). With spatial analysis of the geomagnetic field over time, crucial information about the Earth and the conditions of the outer mantle core can be obtained, and current measurements can be compared with historical records of geomagnetic distribution (Hernández-Quintero et al., 2020). Chernetsov et al. (2017) observed that experienced adult birds (*Acrocephalus scirpaceus*) use magnetic declination to migrate considering longitude variation, under absence of clouds.

Meridian convergence is the angular difference between the grid meridian of a plane projection and the geodetic meridian. Meridian convergence can be presented in topographic surveys to indicate the distance of the alignment from the grid's central meridian and the effect of distortion caused by converting the Earth's curvature to a plane surface.

4.16   Sources of Time Variation of Magnetic Declination

The temporal variation of magnetic declination can be secular, daily, annual, and irregular (Ghilani and Wolf, 1989; Alves and Silva, 2016). Secular variation is one of the most important variations of magnetic declination. Unfortunately, no physical law has been discovered to make long-period predictions of the secular variation, and its behavior in the past can be described only by tables and detailed data obtained by observations (Ghilani and Wolf, 1989; Alves and Silva, 2016). When re-drawing old lines based on compass or magnetic meridian, it was necessary to know the difference in magnetic declination at the survey date and at the present date (Ghilani

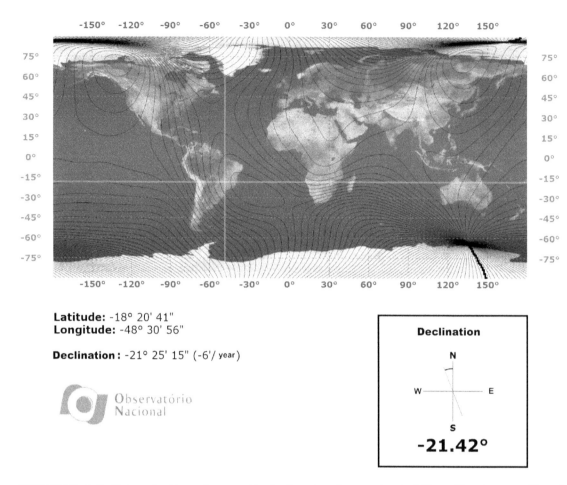

Latitude: -18° 20' 41"
Longitude: -48° 30' 56"

Declination: -21° 25' 15" (-6'/year)

Observatório
Nacional

FIGURE 4.9: Determination of magnetic declination for a point in Minas Gerais state, Brazil, by the mathematical model geomagnetic reference field, on September 15, 2020.

and Wolf, 1989; Alves and Silva, 2016). The daily variation of the compass needle determines a variation generally disregarded because it is within the expected error in the compass reading (Ghilani and Wolf, 1989; Alves and Silva, 2016). The annual variation determined by periodic oscillation is less than 1' of arc and can be neglected. This variation should not be confused with the annual variation determined by secular variation in a given year (Ghilani and Wolf, 1989; Alves and Silva, 2016). The irregular variation is determined by magnetic disturbances and storms that can cause small irregular variations of one degree or more (Ghilani and Wolf, 1989; Alves and Silva, 2016). In Brazil, the National Observatory (ON), created in 1827, is a research institute linked to the Ministry of Science, Technology and Innovation, working in the areas of Astronomy, Geophysics and Metrology in Time and Frequency. The activities of this institute include the training of researchers in graduate courses, the generation, conservation and dissemination of the Brazilian legal time and the dissemination of knowledge produced by specialized activities. On the National Observatory web page, it is possible to obtain isogonic and isophoric charts that represent the spatial and temporal variation of the magnetic declination of the Brazilian territory, respectively (Alves and Silva, 2016).

4.17   Computation of the Magnetic Declination

Computer programs can quickly provide magnetic declination values using models developed based on historical magnetic declination data and year-to-year variation obtained from observation stations around the world (Ghilani and Wolf, 1989; Alves and Silva, 2016).

Latitude and longitude values of the location and the date of the survey are provided as input data using the software. The results are presented for viewing and printing. Annual rates of change in magnetic declination can be determined using programs made available from the National Observatory, the National Geophysical Data Centers (NGDC) maintained by the National Oceanic and Atmospheric Administration (NOAA) (Alves and Silva, 2016) or using the R package oce (Kelley et al., 2020).

The magnetic declination in the Lavras city, Minas Gerais, Brazil is calculated by Magnetic Field Calculators available by the National Oceanic and Atmospheric Administration (NOAA) website (Figure 4.10) (NOAA, 2020a).

Another application of determining magnetic declination is presented in the computation practice using the `magneticField` function of the R package oce (Kelley et al., 2020).

4.18   Local Compass Attraction

Metallic objects and electric current can cause local attraction, affecting the main electromagnetic field. Local attraction can be detected using forward and backward readings from a compass alignment. If the angular measurements differ more than the errors of normal observations, it means that local attraction of the compass surveying occured (Ghilani and Wolf, 1989; Alves and Silva, 2016).

4.19   Mistakes in Measuring Angles

Some mistakes observed when using azimuths and bearings are (Ghilani and Wolf, 1989; Alves and Silva, 2016):

- Confusion between magnetic bearings and other reference bearings;
- Confusion between clockwise and counterclockwise marked angles;
- Swapping bearings for azimuths;
- Presenting bearings with angular value greater than 90°;
- Failing to include direction letters when setting bearings;
- Failing to change bearing letters when using a line's backward direction;
- Using the wrong angle when calculating bearings;
- Adopting a reference line that is difficult to reproduce;
- Taking readings in decimal degrees from a calculator, when it should be degrees, minutes and seconds;
- Failing to adjust polygon angles before calculating bearings or azimuths.

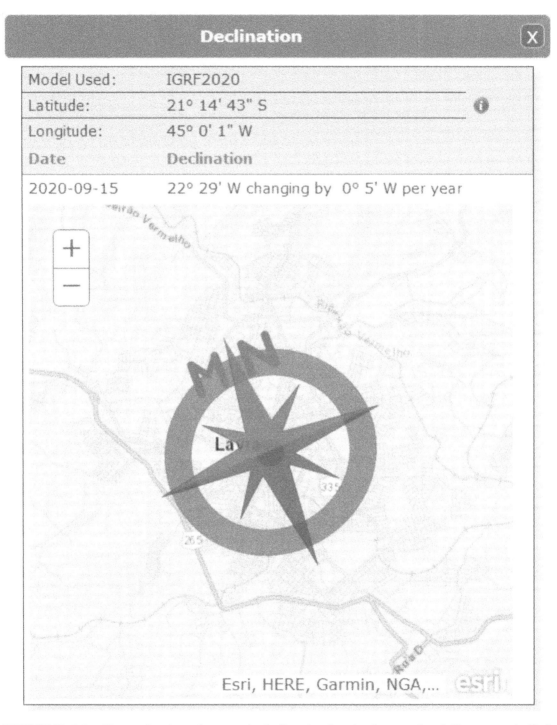

FIGURE 4.10: Determination of magnetic declination by the International Geomagnetic Reference Field model, in Lavras, Minas Gerais, Brazil, on September 15, 2020.

4.20   Computation

In this computation practice the R packages oce (Kelley et al., 2020), LearnGeom (Briz-Redon and Serrano-Aroca, 2020) and circular (Lund et al., 2017a) are used to perform conversions between different angle systems, mathematical operations with angles, trigonometric calculations, determination of angles in geometric figures, determination of the Earth's magnetic field by model at different times, and graphical evaluation of the bearing quadrant.

In the R package oce we can analyze oceanographic data and topographic data, using specific functions and graphical results (Kelley et al., 2020). The LearnGeom package includes a set of functions for learning and teaching basic plane geometry at the undergraduate level for students new to programming (Briz-Redon and Serrano-Aroca, 2020). The circular package includes functions for circular statistics and conversions between units of angles (Jammalamadaka and SenGupta, 2001).

4.20.1   Installing R packages

The install.packages function is used to install the oce, LearnGeom and circular packages in the R console.

```
install.packages("oce")
install.packages("LearnGeom")
install.packages("circular")
```

4.20.2   Enabling R packages

The library function is used to enable the oce, LearnGeom and circular packages in the R console.

```
library(oce)
```

```
## Error in get(genname, envir = envir) :
##    objeto 'testthat_print' não encontrado
```

```
library(LearnGeom)
library(circular)
```

First of all, angle conversions from degrees, minutes and seconds to decimal degrees and vice versa are performed. Then trigonometric calculations and measurements of angles on geometric figures are performed. Finally, magnetic declination is determined and mapped on the Earth and the bearing quadrant of alignments is evaluated in graphical format.

4.20.3   Conversions in the sexagesimal system

In converting angles in the sexagesimal system, division, sum, and multiplication operations are performed to convert angles. In converting angles to decimal degrees, the values for minutes and seconds are transformed into degrees by dividing the minute by 60 and the seconds by 3600. With the sum of all the values, the angle is determined in degrees. When determining decimal degrees for degrees, minutes and seconds, the decimal degrees are multiplied by 60 to obtain minute values, and the decimal minute is multiplied by 60 to obtain second values.

```
# Convert 65°20'30" to decimal degrees
65+20/60+30/3600
```

```
## [1] 65.34167
```

```
# Convert 65.34167 decimal degrees to degrees, minutes and seconds
65
```

```
## [1] 65
```

```
0.34167*60
```

```
## [1] 20.5002
```

```
0.5002*60
```

```
## [1] 30.012
```

4.20.4   Algebraic operations of angles

The algebraic operations of adding, subtracting, and dividing angles are performed. This required converting the angle values to decimal degrees before performing the operations.

```
# Add 65°20'30" and 180°
(65+20/60+30/3600) + 180
# Subtract 65°20'30" and 180°
(65+20/60+30/3600) - 180
# Divide 180° by 65°20'30"
180/(65+20/60+30/3600)
```

4.20.5 Conversions between degrees, radians and grads

The conversions between degrees, radians and grads are performed step-by-step, using the concepts involved in determining these units. The `rad` and `deg` functions are used to convert degrees into radians and radians into degrees, respectively.

```r
# Convert 120 degrees to radians
# pirad<-180
# xpirad<-120
x<-120*pi/180
x
```

```
## [1] 2.094395
```

```r
# or
rad(120)
```

```
## [1] 2.094395
```

```r
# Convert 2 pi radians to degrees
pirad<-180
2*pirad
```

```
## [1] 360
```

```r
# or
deg(2*pi)
```

```
## [1] 360
```

```r
# Convert 7/4 pi radians to degrees
pirad<-180
7/4*pirad
```

```
## [1] 315
```

```r
# or
deg(7/4*pi)
```

```
## [1] 315
```

```
# Convert 1 degree to degrees
# xdegree=60minutes
# ygrad=100minutes
# ygrad=(xdegree*100)/60; xdegree=(ygrad*60)/100
(1*100)/60
```

```
## [1] 1.666667
```

4.20.6   Trigonometric calculations with angles

Trigonometric calculations have also been used in surveying and geodesy. In R, trigonometric determinations must be performed with angles in radians, not degrees. Therefore, we must convert the angle to radian measure before calculating sine, cosine, tangent and arc tangent.

```
# Determine the cosine of 120 degrees
cos(120*pi/180)
```

```
## [1] -0.5
```

```
# Determine the cosine of 90 degrees
cos(90*pi/180)
```

```
## [1] 6.123032e-17
```

```
# or
cos(rad(90))
```

```
## [1] 6.123032e-17
```

```
# Determine the arc cosine of 6.123032e-17
acos(6.123032e-17)*180/pi
```

```
## [1] 90
```

```
# or
deg(acos(6.123032e-17))
```

```
## [1] 90
```

```
# Determine the sine of 120 degrees
sin(120*pi/180)
```

```
## [1] 0.8660254
```

```
# Determine the tangent of 45 degrees
tan(45*pi/180)
```

```
## [1] 1
```

```
# or
tan(rad(45))
```

```
## [1] 1
```

```
# Determine the arc tangent of 1
atan(1)*180/pi
```

```
## [1] 45
```

```
# or
deg(atan(1))
```

```
## [1] 45
```

4.20.7 Determining angles in geometric figures

If you have the geometry of the area of interest, angles can be determined using the three vertices that define the angular range. As an example, we can determine the angle between vertices of a triangle or other geometric figure of interest, in degrees (Figure 4.11).

```
# Define the grid
x_min <- -2
x_max <- 1
y_min <- -1
y_max <- 2
CoordinatePlane(x_min, x_max, y_min, y_max)
# Define triangle vertices
```

```
A <- c(-1,0)
B <- c(0,0)
C <- c(0,1)
# Draw angles
Draw(CreatePolygon(A, B, C), "lightgrey")
angle <- Angle(A, B, C, label = TRUE)
angle <- Angle(A, C, B, label = TRUE)
angle <- Angle(B, A, C, label = TRUE)
```

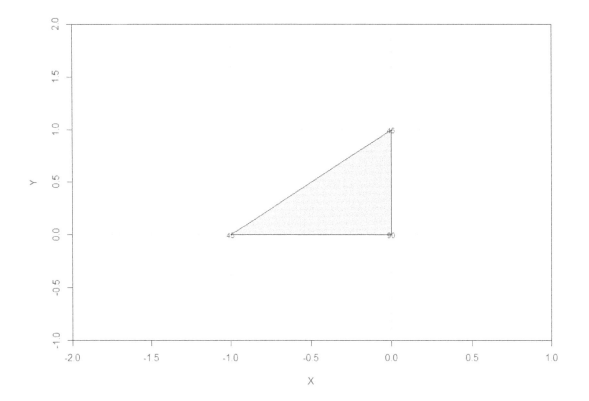

FIGURE 4.11: Determining and mapping angles in triangular geometry.

4.20.8   Determining the Earth's magnetic field

In the case of surveys where it is necessary to use a compass, it may be interesting to obtain the geodesic positioning of the area of interest by varying the magnetic field. Based on the geographic coordinates (latitude and longitude of the area of interest), the magnetic intensity, inclination and declination of Lavras, Minas Gerais, Brazil, are determined in the International Geomagnetic Reference Field 12th generation (IGRF-12) model on the date the function can be executed using the system date compared to the year 2000.

```
# IGRF-12 results for the current date
magneticField(-45.0014, -21.2485, Sys.Date())
```

```
## $declination
## [1] -22.79679
##
## $inclination
## [1] -38.14843
##
## $intensity
## [1] 23161.88
```

```
# IGRF-12 results for the Year 2000
magneticField(-45.0014, -21.2485, 2000)
```

```
## $declination
## [1] -20.52985
##
## $inclination
## [1] -31.05882
##
## $intensity
## [1] 23387.31
```

4.20.9   Mapping the Earth's magnetic field

Considering the interest to assess the variation of the Earth's magnetic field in different years, contour maps can be used to characterize the spatial and temporal variation of the magnetic declination. In this case, a comparative mapping of the Earth's magnetic declination was carried out between 2000 and 2020 in the Robinson pseudocylindrical 2D global cartographic projection (Figure 4.12).

```
par(mfrow=c(2,1))
# Mapping the Earth's magnetic declination in 2000
data(coastlineWorld)
par(mar=rep(0.5, 4))
mapPlot(coastlineWorld, projection="+proj=robin", col="lightgray")
# Create the declination matrix
lon <- seq(-180, 180)
lat <- seq(-90, 90)
dec2000 <- function(lon, lat)
    magneticField(lon, lat, 2000)$declination
dec <- outer(lon, lat, dec2000)
# Make contour maps for magnetic declination variations
mapContour(lon, lat, dec, col='black', levels=seq(-180, -5, 5),
           lty=3, drawlabels=FALSE)
```

```
mapContour(lon, lat, dec, col='black', levels=seq(-180, -20, 20))
mapContour(lon, lat, dec, col='black', levels=seq(5, 180, 5),
           lty=3, drawlabels=FALSE)
mapContour(lon, lat, dec, col='black', levels=seq(20, 180, 20))
mapContour(lon, lat, dec, levels=180, col='black', lwd=2,
           drawlabels=FALSE)
mapContour(lon, lat, dec, levels=0, col='black', lwd=2)
# Mapping the Earth's magnetic declination in 2020
data(coastlineWorld)
par(mar=rep(0.5, 4)) # no axes in the global projection
mapPlot(coastlineWorld, projection="+proj=robin", col="lightgray")
# Create the declination matrix
lon <- seq(-180, 180)
lat <- seq(-90, 90)
dec2000 <- function(lon, lat)
    magneticField(lon, lat, 2020)$declination
dec <- outer(lon, lat, dec2000)
# Make contour maps for magnetic declination variations
mapContour(lon, lat, dec, col='black', levels=seq(-180, -5, 5),
           lty=3, drawlabels=FALSE)
mapContour(lon, lat, dec, col='black', levels=seq(-180, -20, 20))
mapContour(lon, lat, dec, col='black', levels=seq(5, 180, 5),
           lty=3, drawlabels=FALSE)
mapContour(lon, lat, dec, col='black', levels=seq(20, 180, 20))
mapContour(lon, lat, dec, levels=180, col='black', lwd=2,
           drawlabels=FALSE)
mapContour(lon, lat, dec, levels=0, col='black', lwd=2)
```

4.20.10   Evaluating the quadrant of a bearing on the wind rose based on the vector projection

In topographic surveys it has been necessary to define the direction of alignments by azimuths and bearings. From the azimuth and distance angle values we usually calculate the partial and absolute projections of the mapped vertices. Based on the vector projection, it may be necessary to obtain the quadrant in which the vertex is located in order to perform inverse calculations according to the direction between the mapped vertices. The quadrant mapping of the direction relative to the vectors can be obtained with the as.windrose function in the R package oce (Figure 4.13).

```
par(mfrow=c(1,2))
# Quadrant of bearing on vector projection (-5, 10)
xcomp <- -5
ycomp <- 10
wr <- as.windrose(xcomp, ycomp)
plot(wr, col=c("black", "black", "black", "gray"))
# Quadrant of bearing on vector projection (5, 10)
xcomp <- 5
ycomp <- 10
```

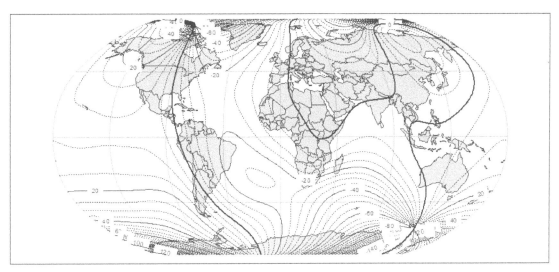

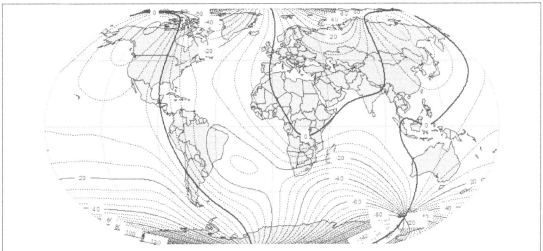

FIGURE 4.12: Comparative mapping of the Earth's magnetic declination in 2000 (top) and 2020 (bottom).

```
wr <- as.windrose(xcomp, ycomp)
plot(wr, col=c("black", "black", "black", "gray"))
```

4.21 Solved Exercises

4.21.1 The first direction of a boundary survey was defined as $32°13'$ NW. What is the equivalent azimuth?

A: Given that the bearing is in the northwest quadrant, the azimuth is 327°46.998" .

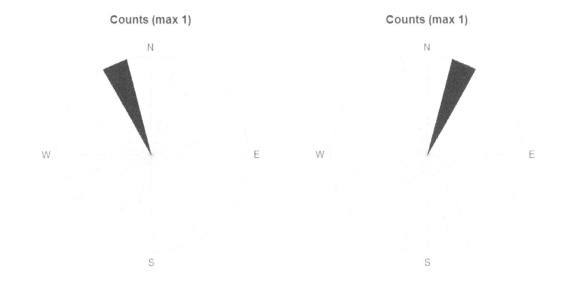

FIGURE 4.13: Bearing determination in the northwest (left) and northeast (right) quadrants.

```
360-(32+13/60)
```

```
## [1] 327.7833
```

```
0.7833*60
```

```
## [1] 46.998
```

4.21.2 **The magnetic bearing of a line on a rural property was recorded as $43°30'$ SE in 1862. At that time, the magnetic declination at the survey location was $3°15'$ W. What geodetic bearing would be needed for a plan to divide the property (Figure** 4.14**)?**

A: The geodetic bearing for dividing the property is $46°45'$ SE.

```
43+30/60+3+15/60
```

```
## [1] 46.75
```

```
0.75*60
```

```
## [1] 45
```

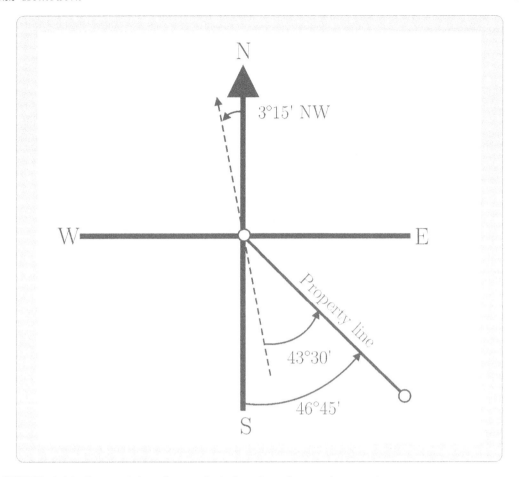

FIGURE 4.14: Determining the geodetic bearing of a rural property.

4.21.3 Calculate the magnetic declination of the city where you were born using Magnetic Field Calculators available at the NOAA website.

A: Access the website address at reference NOAA (2020a) and perform the survey.

4.21.4 What are the disadvantages of assuming a relative meridian at the beginning of a polygon survey?

A: The disadvantages are the inability to re-establish the points if the original points are lost and the non-conformity with other surveys of the same area.

4.22   Homework

Choose two exercises presented by the teacher and solve the questions with different input values. Compare the results obtained. If possible, use a real database measured in the field with an available instrument.

4.23   Resources on the Internet

As a study guide, slides and illustrative videos are presented about the subject covered in the chapter (Table 4.2).

TABLE 4.2: Slide shows and video presentations on measuring angles and magnetic declination.

Guide	Address for Access
1	Slides on angle measurement in geomatics[1]
2	Horizontal and vertical angle measurement with theodolite[2]
3	Horizontal angle measurement and recording[3]
4	Magnetic declination charts in the United States (1590-2020)[4]
5	Information on magnetic declination[5]
6	Measuring horizontal angles and distances using total station or theodolite[6]
7	Bearings and deflection angles[7]

4.24   Research Suggestion

The development of scientific research on geomatics is stimulated by the activity proposals that can be used or adapted by the student to assess the applicability of the subject matter covered in the chapter (Table 4.3).

TABLE 4.3: Practical and research activities used or adapted by students using angle measurement and magnetic declination.

Activity	Description
1	In the content about angles and magnetic declination, interest may arise in doing practical work based on the computation examples presented
2	Perform the analysis of the variation of magnetic declination at a location between different years. Compare and discuss the results obtained
3	Perform angle measurements using a sighting instrument. Also practice installing the equipment above a landmark and zeroing the instrument using a compass at magnetic north in order to obtain the magnetic azimuth of the measured angles

---

[1] http://www.sergeo.deg.ufla.br/geomatica/book/c4/presentation.html#/
[2] https://youtu.be/ckR-wBUTUjA
[3] https://youtu.be/7aYsAwXlZkg
[4] https://youtu.be/wL2a1jnE61E
[5] https://youtu.be/WwIKx96q8lE
[6] https://youtu.be/8HAz-DrC65k
[7] https://youtu.be/QY8K2logXA0

4.25   Learning Outcome Assessment Strategy

Perform a summary of the chapter, "Angle and Direction Observations with Geomatics and R", on a single A4 page in order to show the student's abilities to summarize a subject presenting key points considered of greater importance today.

5

Direct Distance and Angle Measurements

5.1 Learning Questions

The emergent learning questions answered through reading the chapter are as follows:

- How are diastimeters used in measuring alignments and angles in geomatics?
- How is distance measurement with taping performed on sloping terrain or over vegetation?
- How to transpose obstacles when measuring with taping.
- How can physical and environmental factors interfere with measuring results?
- How to calibrate taping measure for slope and horizontal alignment.
- How to draw a polygonal area, determining vertices angles and polygonal area from diastimeter measurements with R software, `circular` and `LearnGeom` packages.

5.2 Learning Outcomes

The learning outcomes expected from reading the chapter are as follows:

- Know the diastimeters used in measuring alignments and angles based on the law of cosines.
- Know how the distance measurement with taping on sloping terrain or over vegetation is performed, necessary adjustments in on-site measurements, such as the transposition of obstacles and the understanding of factors that interfere in the measurement such as temperature, pull and sag.
- Measure angles in a survey with taping, correction of slope and horizontal alignment.
- Draw a polygonal area, determining the angles of the vertices and the polygon area based on measurements taken with diastimeter, R software and R packages `circular` and `LearnGeom`.

5.3 Introduction

Distance is a numerical description of how far or close things are and is the most fundamental concept in geography. In the First Law of Geography, Waldo Tobler stated that "everything is related to everything, but things near are more related than things far away" (Lovelace et al., 2019b).

Distance measurement is a fundamental process among all survey measurements. In traditional surveys, even if many angles are measured, the length of at least one line must be determined as a supplement to the angles at the located points. In plane surveys, the distance between two points referred to the horizontal distance (H). Considering points at different altitudes, H is the horizontal length between the vertical lines at each point.

The instruments used for direct measurement of distances are called "diastimeters". In polygon surveying with diastimeter, alignments and angles can be measured using the cosine law. Another option is measuring perimeter lines and some internal lines based on geometry patterns looking to the map (Ghilani and Wolf, 1989; Alves and Silva, 2016).

The quantities measured in topographic survey can be linear and angular. The linear quantities are mainly horizontal distance (H), vertical distance or leveling difference (d). When we want to measure the slant distance AB or EF, it is necessary to know the angle of inclination α (Figure 5.1) (Garcia and Piedade, 1987).

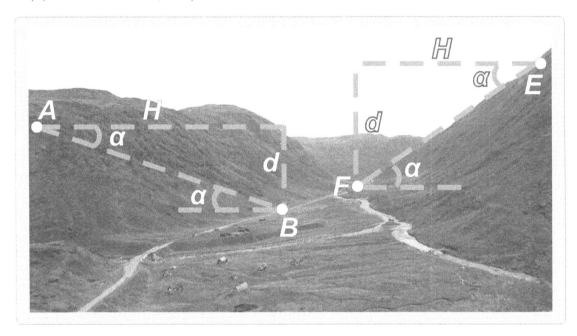

FIGURE 5.1: Measurement of horizontal distance (H) and level difference (d) in geomatics.

The most commonly used metric units for linear distance measurements can be obtained by different methods and equipment; however, the focus of this chapter is on distance and angle measurements in surveying with taping.

5.4   Measuring Distance with Taping

The most commonly used diastimeters in direct distance measurement are: tape, steel tape, and surveyor's chains. Tapes are made of polyvinyl chloride (PVC), fiberglass or steel, and graduated in meters, decimeters and centimeters, with lengths varying up to 50 m. Steel tapes are made of stainless steel blade, graduated in meters and decimeters with lengths ranging from 20 to 100 m, and are wound on a drum or cross. The surveyor's chain, when used, is made of steel or iron, joined together two-by-two by means of links. Generally, every one or two meters, there is a metal pendant that indicates the metreage. The most common chains are 20 m long. The distance between links is 20 cm (Garcia and Piedade, 1987).

Accessories used to make taping measurements are: pickets, stakes, range poles (ranging rod) and chaining pins or taping pins, to mark points and taping lengths. The purpose of the accessories is to enable locating and materialing topographic points on the terrain. Pickets measuring 15 to 30 cm should be 3 to 5 cm above the terrain. Stakes are used beside each picket to note the point number and should be 50 cm above the terrain (Figure 5.2) (Garcia and Piedade, 1987).

FIGURE 5.2: Steel, fiberglass and PVC tape, plumb bob, pickets, and tubular bubble stick level used for surveying.

In measuring distances with taping, a tape of known size is applied directly on a line for a number of repetitions, performing the following steps (Ghilani and Wolf, 1989; Alves and Silva, 2016):

- Stretch the tape measure over the surface;
- Apply tension to the ends;
- Leveling the tape measure, define the length to be measured;
- Read the marked value and record the distance.

In measuring horizontal distance between points A and B, we try to measure the projection A', B' on the topographic plane H' (Figure 5.3). Measuring the distance $A'B'$, one end of the diastimeter is placed at B'' and the other end is taken to point A'', keeping the tape measure horizontal. The alignment between points A and B must correspond to the line containing AB on the topographic plane (Garcia and Piedade, 1987; Alves and Silva, 2016).

When measuring distances greater than the length of diastimeters, range poles are used to avoid going out of alignment. In stakeout, three people, the backward, forward and intermediate marker, each with one marker, determine the forward alignment based on sighting at the intermediate and backward marker (Figure 5.4). Under high gradients, we can measure 5 to 10 m stretches at a time, to make it easier to keep the diastimeter horizontal (Alves and Silva, 2016; Garcia and Piedade, 1987).

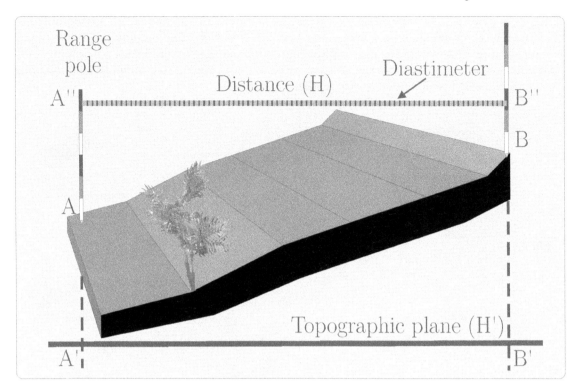

FIGURE 5.3: Horizontal distance measurement with a diastimeter.

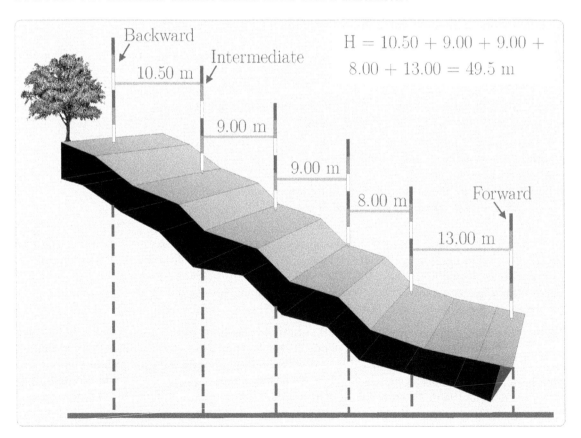

FIGURE 5.4: Measuring great distances with diastimeter.

5.5   Distance Measurement on Slant Terrain with Taping

On sloping terrain, three methods of measuring with a tape measure can be used (McCormac et al., 2012; Alves and Silva, 2016):

- The tape measure can be held horizontal by means of plumb lines;
- The tape measure can be held on the slope and then a slope correction is determined to obtain the horizontal distance;
- The slope distance and vertical angle can be measured for each slope followed by subsequent calculation of the horizontal distance.

5.6   Extension Alignment

Extending an existing alignment AB, an auxiliary surveyor stands at the starting point A and the direction is oriented according to the sight plane defined by the first two range poles. New range poles are placed in the same direction so that they are covered by the first two range poles. In case of large distances, the auxiliary surveyor responsible for orientation moves to the second-to-last placed range pole and continues to extend the alignment. The accuracy of the alignment decreases with the increasing number of necessary changes that the responsible auxiliary surveyor makes, due to human errors (Garcia and Piedade, 1987; Alves and Silva, 2016) (Figure 5.5).

5.7   Perpendicular Tracing with Taping

Plotting perpendiculars on the terrain has been necessary in different applications, such as demarcating an alignment perpendicular to an existing one or as an aid in tying up details of interest during a survey. Either right-angle or isosceles triangle demarcation methods can be used. To mark out a right angle using the right triangle, in practice a 12 m tape measure is used to form a triangle with 3, 4 and 5 m sides. One assistant surveyor at point C, on the line AB, 3 m from point A, holds the 0 m and the 12 m of the tape (initial handle), while another holds the 3 m over point A and a third holds the 7 m. When stretching the tape, the helper who holds the 7 m takes a new position which defines the angle of 90° with the AB alignment (Figure 5.6) (Garcia and Piedade, 1987; Alves and Silva, 2016).

In tracing the angle of 90° using the isosceles triangle, if at the alignment AB, it is desired to trace a perpendicular defined at point C. Equal distances are measured in the CA and CB directions, defining d and e points. Two assistants, one at d and one at e, hold the ends of the tape and a third holds the middle of the tape. By stretching the tape, the perpendicular direction is defined at point C and the middle of the tape, materializing point D at the perpendicular (Figure 5.7) (Garcia and Piedade, 1987; Alves and Silva, 2016).

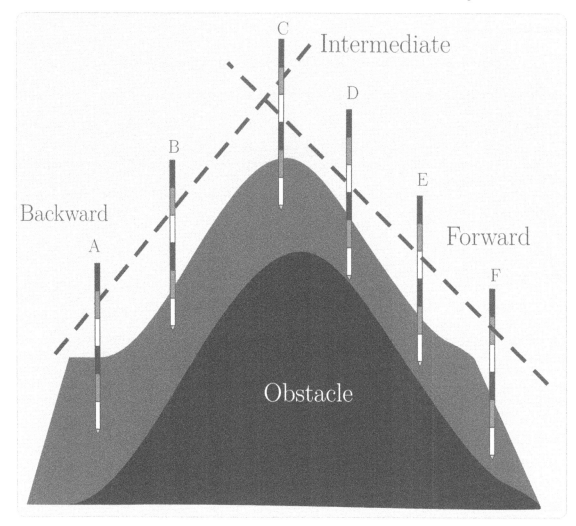

FIGURE 5.5: Extending the alignment of large distances using range poles.

5.8   Taping with Obstacles

5.8.1   Non-visible endpoints of alignment

In topographic work, measuring the distance between two non-visible points may be necessary, as under obstacles between the points, in front of buildings, vegetation or other obstacles. In this case, a procedure can be used to calculate the desired distance by knowing the sides of similar triangles. A point C is chosen from which the points A and B of the alignment to be measured are sighted. The distances CA and CB are measured. Obeying some ratio $1/2$ or $1/3$ of the measured alignments (CA and CB), points D and E are marked. The distance DE is measured (Figure 5.8) (Garcia and Piedade, 1987; Alves and Silva, 2016).

By the similarity of triangles formed, we have:

$$\frac{CD}{CA} = \frac{CE}{CB}; \frac{CD}{CA} = \frac{DE}{AB} \qquad (5.1)$$

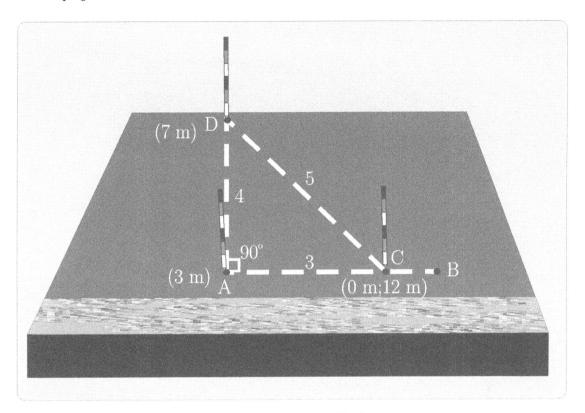

FIGURE 5.6: Using the right triangle to draw perpendiculars.

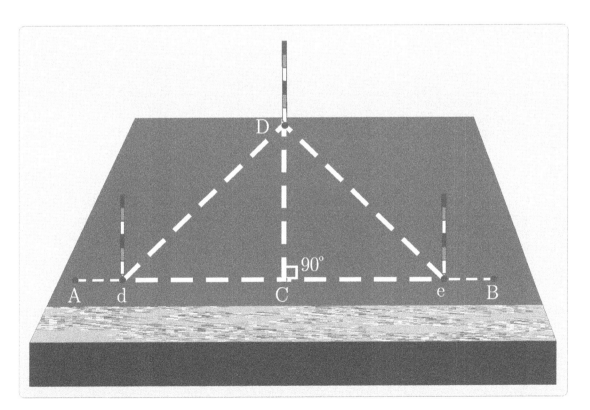

FIGURE 5.7: Using the isosceles triangle to define 90-degree angle alignment with diastimeters.

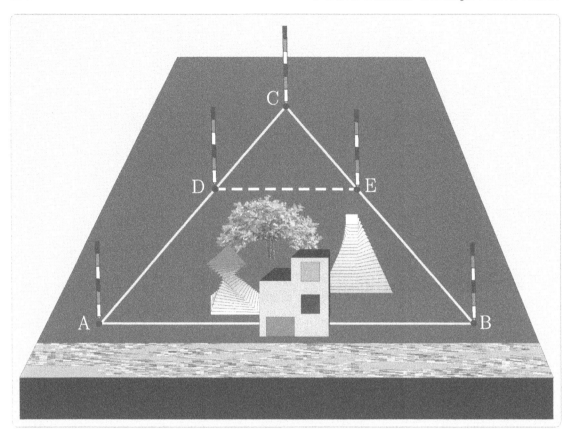

FIGURE 5.8: Using the similar triangle to measure the distance between points separated by obstacles.

where

$$AB = \frac{CA \; DE}{CD} \tag{5.2}$$

5.8.2 Visible endpoints of alignment

The measurement of a line-up that crossed a swamp, lake, pond, dam, building depression, or gully, required circumventing the obstacle using perpendiculars and parallels obtained by right angles. Right angles can be demarcated with chain and range poles, using a right-angled or isosceles triangle (Figure 5.9) (Garcia and Piedade, 1987; Alves and Silva, 2016).

5.9 Locating Details with Taping

In addition to measuring lines, it has almost always been necessary to locate a terrain accident and details, such as cultivated fields, buildings and roads. To do this, it becomes necessary to tie or reference this accident, detail or other geographic object using several points tied to a reference line (Garcia and Piedade, 1987; Alves and Silva, 2016).

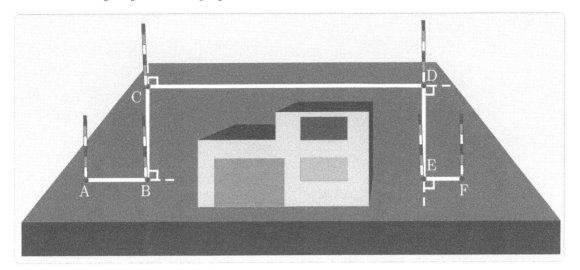

FIGURE 5.9: Using right angles for transposing obstacles from visible points.

5.10   Measuring Angles with Taping

In order to know the value of an angle using a diastimeter, we must determine the sides of any triangle containing the angle to be measured and apply the cosine law (Figure 5.10) (Garcia and Piedade, 1987; Alves and Silva, 2016):

$$b^2 = a^2 + c^2 - 2ac\cos\alpha \qquad (5.3)$$

$$\cos\alpha = \frac{a^2 + c^2 - b^2}{2ac} \qquad (5.4)$$

We can use $a = c$ so that the angle can be determined by chord table.

In the field survey, with taping and range poles we marked the horizontal distance in the BA direction and in the BC direction, defining points A and C. The tape line is stretched to connect points A and C, measuring the distance formed (chord). Then, using the cosine law, the angle is determined. It should be noted that, in this example, the resolution of the angle can be done in a simple way based on the 3, 4 and 5 catets of the right triangle. However, the law of cosines would be a solution for different types of triangles. Thus, to survey a polygon with diastimeter, we can either measure the alignments and angles using the law of cosines, or measure the perimeter lines and some internal lines forming triangles to make the plan. At the same time, a sketch of the field operations is made. During a survey, the use of a field notebook facilitates the recording of distance measurements with a diastimeter. In this case, an starlike symbol (*) can be used as an external angle symbol reference to the polygon (Table 5.1) (Alves and Silva, 2016).

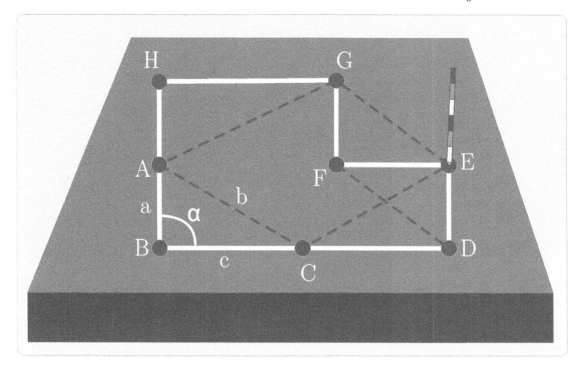

FIGURE 5.10: Surveying the polygon ABCDEFGHA with a diastimeter.

TABLE 5.1: Field notes of a polygon survey with diastimeter.

Alignment	H (m)	Chord (m)
AB	4	
BC	3	
AC		5
CD	4	
DE	3	
CE		5
EF	3	
DF	4	
FG	4	
GE		5*
GH	4	
HA	3	
GA		5

When surveying with taping we must observe if the area surveyed contains obstacles, such as walls of a building, or if it is an open area, such as a pasture field where you can establish a dynamic of pickets for cattle grazing from the taping survey. If there are building walls, it may be necessary to define external areas for area determination that should be excluded from the final polygon area. In the case of unobstructed areas, the inner areas of the polygon itself can be used to define angles and area.

5.11 Area Assessment with Taping Measurements

The knowledge of an area by means of graphic representation to define size, contour, relief, natural accidents, details of buildings, and its relative position on the Earth's surface have been frequent concerns of professionals responsible for urban, rural, and conservationist planning and projects, or simply for the need to know the elements that characterized a given area. When surveying a polygon with a diastimeter, the described process for measuring angles and distances of alignments can be carried out carefully and accurately with the use of range poles. The area can also be surveyed by measuring perimeter lines and some internal lines in order to form triangles that can be used to make the map. At the same time, a sketch of the field operations should be made. In general, the term "area" is defined as the region occupied inside the boundary of a plane object in space. The measurement is done in square units with the standard unit being square meters (m^2). For the area computation, there are pre-defined formulas for squares, rectangles, circle, triangles, and other geometric figures. The area of a triangle can be computed from different formulas and types of triangles. The area (A) of a triangle can be determined by looking at the polygon graph and calculating the area of a triangle BAC. Basically, the triangle area is equal to half of the base times height. Hence, to find the area of a tri-sided polygon, we have to know the triangle base (b) and height (h), based on the following equations (Garcia and Piedade, 1987; Alves and Silva, 2016):

$$A = \frac{bh}{2} \tag{5.5}$$

where b and h are the base and the height of the triangle, respectively.

The area of a triangle with three sides of different measures can be found using Heron's formula based on the perimeter of a triangle information. The perimeter of a triangle is the distance covered around the triangle, calculated by adding all three sides of a triangle. Heron's formula included two steps. In the first step, the semiperimeter of a triangle is found by adding all three sides of a triangle and dividing it by two. In the second step, the semi perimeter of triangle value is used to find the triangle area. These equations are used in situations where it is difficult to measure the height of the triangle. Thus, we can calculate triangle area using Heron's formula and the semiperimeter equation (p) (Garcia and Piedade, 1987; Alves and Silva, 2016):

$$A = \sqrt{p(p-a)(p-b)(p-c)} \tag{5.6}$$

where $p = \frac{a+b+c}{2}$ and a, b, c are the sides of the triangle.

In case the value of an angle is known, the area of the triangle can be calculated by (McCormac, 2007; Alves and Silva, 2016):

$$A = \frac{1}{2}acsen\alpha \tag{5.7}$$

After calculating the area of each triangle, the total area is obtained by adding up the areas of the triangles demarcated throughout the mapped region.

5.12   Taping Calibration and Measurement Correction

Studying measurements with tape measures has made it possible to understand the measurement process as a whole, regardless of the survey operation involved or the equipment used. The main situations where corrections may need to be applied in taping measurements are (McCormac, 2007; Alves and Silva, 2016):

- Taping length calibration;
- Temperature variations;
- Inclination and off-line alignment;
- Applying little tension on the ends causing sag in the tape measurement (Catenary);
- Excessive tension applied to the ends of the tape measure (pull).

After determining the error, taping actual distance (d) for a measured line can be obtained by incorporating corrections (c) of the measured distance (md):

$$d = md + \sum c \tag{5.8}$$

5.12.1   Length calibration of taping

When measuring a given distance with a longer taping, the sufficient length for the measurement will not be obtained and a positive correction must be made. That is, if the tape measure is longer, fewer lengths of taping are used to measure a distance than would be required for a shorter taping of the correct size. For a shorter taping, the reverse is true. Therefore, when a distance is measured with taping and the wrong tape size is subsequently found, we can adopt the rule that for a longer taping we should add a distance correction, and for a shorter taping we should subtract the correction value. However, care must be taken with this rule in the situation where you plan to measure a distance with a longer or shorter taping but still make the correct measurement. In this case, the rule is reversed so that when you plan to use a shorter taping you must add a correction factor and for a longer taping you must subtract a correction factor (McCormac, 2007; Alves and Silva, 2016).

An error caused by incorrect length of taping (C_L) (m) can be determined by (Ghilani and Wolf, 1989):

$$C_L = \left(\frac{l - l'}{l'}\right)L \tag{5.9}$$

where l is the actual length of the taping (m), l', the nominal length of the taping (m) and, L, the recorded measure of the total line length (m).

5.12.2   Temperature variations

Changes in taping length caused by temperature variations can be significant. For precision work, a temperature change of approximately 5°C will cause a length change of approximately 0.002 m on a 30 m taping. If a tape measure was used at 9°C to define a distance of 1000 m, and if the distance was checked the following summer with the same tape measure when the temperature reached 38°C, there will be a difference in length of 0.34 m caused by temperature change. The coefficient of linear expansion of the steel taping is 0.0000116°C. Thus, the correction of taping for temperature changes (C_T) is (McCormac, 2007; Ghilani and Wolf, 1989; Alves and Silva, 2016):

$$C_T = k(T_i - T)L \tag{5.10}$$

where k is the thermal coefficient of taping expansion and contraction, T_i, the estimated temperature (°C) at the time of measurement, T, the taping temperature (°C) under standard length conditions and, L, the observed line length (m).

Errors caused by temperature changes can be virtually eliminated by taking temperature measurements or using an Invar tape measure. Invar lengths have a very small coefficient of expansion and were useful for precise distance measurement work (McCormac, 2007).

5.12.3 Corrections for slope and horizontal alignment

Most of the taping measurements are performed keeping the tape in a horizontal position, avoiding the need to make corrections due to the slope of the terrain. However, taking several measurements in small sections can lead to an accumulation of random errors, reducing the accuracy of the measurement. Therefore, instead of segmenting the line into several segments, it may be more advantageous to measure the length of the slope and then calculate the difference in elevation (d) or the height angle (α). If the angle α is determined, the horizontal distance (H) between two vertices can be calculated using the relation (McCormac, 2007; Alves and Silva, 2016):

$$H = L cos\alpha \tag{5.11}$$

where L is the length of the slope (m) between the points and α the height angle (°) from the horizontal plane, usually obtained by clinometer or theodolite. If the elevation difference, d, between the two points is known, the horizontal distance can be calculated by the Pythagorean theorem (Ghilani and Wolf, 1989; Alves and Silva, 2016):

$$H = \sqrt{L^2 - d^2} \tag{5.12}$$

Another approximate equation obtained by expanding the Pythagorean theorem can be used in lower order surveys. Considering, c, the correction factor caused by the horizontal alignment or off-line alignment error and, $L - c$, the horizontal distance, H is (Garcia and Piedade, 1987; Ghilani and Wolf, 1989; Alves and Silva, 2016) (Figure 5.11):

$$(L - c)^2 + d^2 = L^2 \tag{5.13}$$

$$L^2 - 2Lc + c^2 + d^2 = L^2 \tag{5.14}$$

In practice, the c^2 term can be neglected,

$$2Lc = d^2 \tag{5.15}$$

$$c = \frac{d^2}{2L} \tag{5.16}$$

With this, the corrected horizontal distance can be obtained by (Ghilani and Wolf, 1989):

$$H = L - \frac{d^2}{2L} \tag{5.17}$$

where $\frac{d^2}{2L}$ is a vertical offset correction to be subtracted from the slope measurement to obtain the horizontal distance. This method is useful for obtaining quick estimates without complicated calculations or errors produced by large variations in slope.

Corrections for errors caused by taping inclination in the vertical plane can be calculated in the same way as horizontal alignment errors. This type of error has been found to be systematic and can be eliminated by careful alignment (Ghilani and Wolf, 1989).

5.12.4   Pull corrections

When a tape measure is stretched to a greater than standard tension, the length of the tape measure will be greater than the standard length. The modulus of elasticity of the tape regulates the possible amount of stretch. The pull correction (C_P) is the positive correction of the total elongation of the tape length as a result of the tension (m) and can be calculated by (Ghilani and Wolf, 1989; Alves and Silva, 2016):

$$C_P = (P_i - P)\frac{L}{AE} \tag{5.18}$$

where P_i is the pull applied to the tape at the moment of observation (kg), P, the standard tension (kg), A, the cross-sectional area (cm^2), L, the length of the observed line (m) and, E, the modulus of elasticity of the steel (kg cm^{-2}). Taping cross-sectional area can be obtained by the manufacturer information, by measuring the width and thickness or by the ratio between the total tape weight, length and unit weight of the steel.

5.12.5   Sag corrections

When a steel tape measure is held by the ends only, a curvature of the tape measure known as "sag" (catenary) is formed (Figure 5.12). The occurrence of sag determines a greater horizontal distance, because the distance recorded on the tape is greater than the distance between the two ends. The sag error (C_S) is the negative value of the catenary correction (m) and can be corrected by the equation (McCormac et al., 2012; Alves and Silva, 2016):

$$C_S = \frac{W^2 L_S^3}{24 P_i^2} \tag{5.19}$$

where L_S is the observed length (m) and, W is the taping weight (Kg m^{-1}). The sag can be reduced by applying greater tension at the ends and surveying at shorter intervals.

Laser interferometers and yaw sensors are currently being used to evaluate taping errors in calibration laboratories. A new yaw sensor has been designed to determine the error characteristics of measuring tapes. The results of a yaw error measurement sensor compared to a laser interferometer angle measurement system were satisfactory (Chinchusak and Tipsuwanporn, 2018).

The quality of an underwater archaeological survey with fiberglass taping was comparatively evaluated with 3D trilateration. Accuracy of 25 mm was observed in the measurements with taping. In the 304 measurements taken during testing, there was 20% error (Holt, 2003). This method of 3D trilateration has been used in marine archaeology (Rule, 1989) and the process of distance measurement was similar to measurements made by global positioning system receivers (UKOOA, 1994) and underwater acoustic positioning systems (Kelland, 1994).

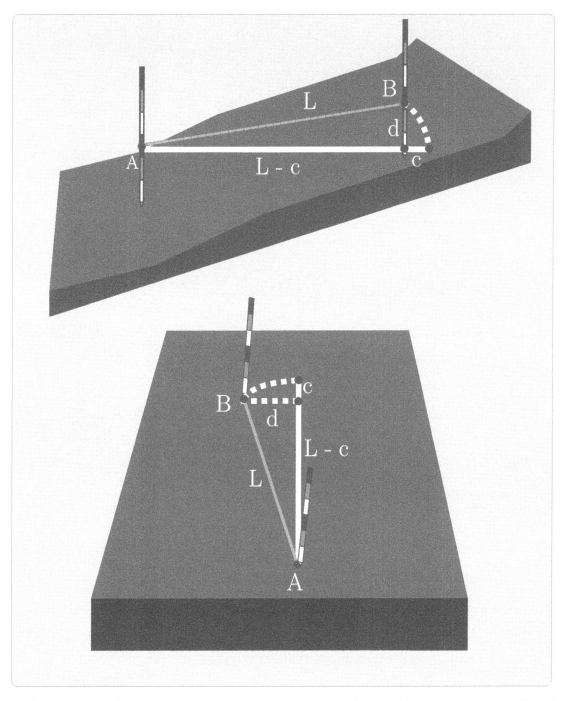

FIGURE 5.11: Correction of errors caused by taping inclination (not horizontal) (top) and off-line alignment (bottom).

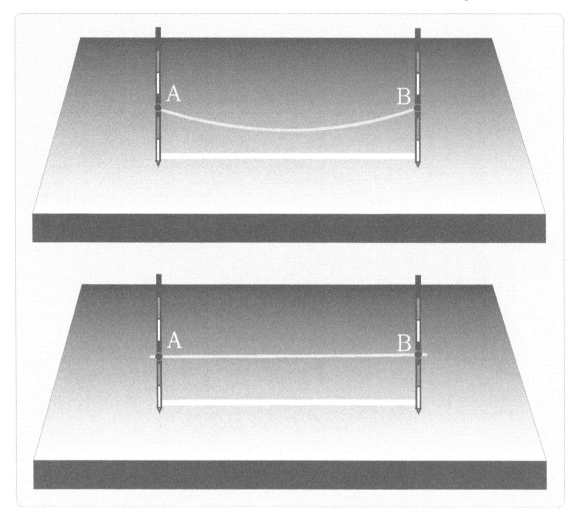

FIGURE 5.12: Catenary effect (sag) by low tension pull on the tape measure, making the observation larger than the distance between 2 points (top), and high tension pull applied to the diastimeter by the ends making the distance smaller than the distance between 2 points (bottom).

5.13   Summary of Errors in Taping Survey

The errors caused by surveying with a tape measure are classified as Natural (N), Instrumental (I), Human (H), Systematic (S) and Random (R). To reduce errors in precision work, we recommended that the same line be surveyed with different tapes at different times of the day and in alternate positions. An accuracy of 1/10000 can be obtained with attention to details of the survey (Table 5.2) (Ghilani and Wolf, 1989; McCormac et al., 2012; Alves and Silva, 2016).

TABLE 5.2: Summary of errors in taping surveying.

Error Type	Instrumental (I), Natural (N), Human (H)	Systematic (S), Random (R)	Error Magnitude
Tape length	I	S	0.01 m
Temperature	N	S, R	20 ° C
Pull	I	S, R	20 kg
Sag	N, H	S	20 cm at center
Alignment	H	S	1.1 m at one end
Unleveled tape	H	S	1.1 m elevation difference between ends of tape
Plumbing	H	R	0.01 m
Marking	H	R	0.01 m
Reading	H	R	0.01 m

5.14 Computation

As a computing practice, we proposed to draw a polygon in an area of 10 x 10 m based on taping measurements. Determine the angles of the vertices and the polygon area based on measurements taken with a diastimeter within this area. The angles are first determined by the R package LearnGeom (Briz-Redon and Serrano-Aroca, 2020) and then a check of the determined angles by the law of cosine is made using the package circular (Lund et al., 2017b). Finally, the area of the polygon is calculated and checked by the semiperimeter and the method of known angles.

5.14.1 Installing R packages

The install.packages function is used to install the circular and LearnGeom packages in the R console.

```
install.packages("LearnGeom")
install.packages("circular")
```

5.14.2 Enabling R packages

The library function is used to enable the circular and LearnGeom packages.

```
library(LearnGeom)
library(circular)
```

5.14.2.1   Define x, y-axis dimensions

A plane Cartesian coordinate axis x, y with minimum and maximum values is defined to draw the polygon measured with a diastimeter.

```
x_min <- 0
x_max <- 10
y_min <- 0
y_max <- 10
```

5.14.2.2   Draw the x, y-axis, the polygon and the angular variations of vertices of the polygon

The planar Cartesian coordinates x, y of vertices A, B, C, D, E, F, G, H are defined based on the field survey. The polygon with transparent interior is drawn with the `Draw` and `CreatePolygon` functions (Figure 5.13).

```
A <- c(1,4)
B <- c(1,1)
C <- c(4,1)
D <- c(8,1)
E <- c(8,4)
F <- c(5,4)
G <- c(5,8)
H <- c(1,8)
```

```
# Defining x,y-axis
CoordinatePlane(x_min, x_max, y_min, y_max)
# Drawing polygon inside the x,y-axis
Draw(CreatePolygon(A, B, C, D, E, F, G, H), "grey92", label=TRUE)
```

The internal angles of the vertices that defined the angular variations in the polygon are determined and mapped with the `Angle` function. In the case of the angle of vertex F, considering that the triangle EFG was external to the polygon, to obtain the internal angle in the polygon we subtracted 360° (Figure 5.14).

```
A <- c(1,4)
B <- c(1,1)
C <- c(4,1)
D <- c(8,1)
E <- c(8,4)
F <- c(5,4)
G <- c(5,8)
H <- c(1,8)
```

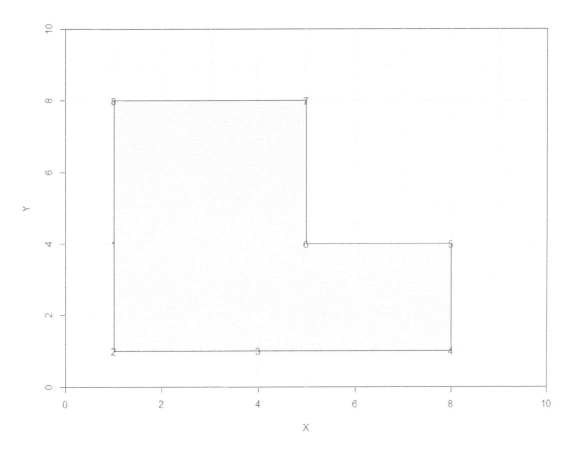

FIGURE 5.13: Defining polygonal vertices of field surveys with diastimeters.

```
# Defining x,y-axis
CoordinatePlane(x_min, x_max, y_min, y_max)
# Drawing polygon inside the x,y-axis
Draw(CreatePolygon(A, B, C, D, E, F, G, H), "grey92", label=FALSE)
# Mapping angles
anguloB <- Angle(A, B, C, label = TRUE)
anguloD <- Angle(C, D, E, label = TRUE)
anguloE <- Angle(D, E, F, label = TRUE)
anguloF <- 360 - Angle(E, F, G, label = TRUE)
anguloG <- Angle(F, G, H, label = TRUE)
anguloH <- Angle(G, H, A, label = TRUE)
# Determining angles
anguloB
anguloD
anguloE
anguloF
anguloG
anguloH
```

To check the angles using the cosine law, the line segments between each vertex of the polygon are created and determined.

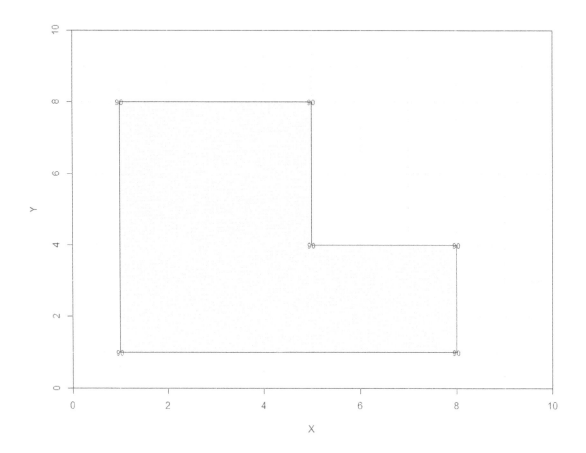

FIGURE 5.14: Definition of vertex angles of polygon surveyed in the field with diastimeter.

5.14.2.3 Define line segments

The `CreateSegmentPoints` function is used to create the line segments from vertices AB, BC, CD, DE, EF, FG, GH, HA.

```
sAB <- CreateSegmentPoints(A, B)
sBC <- CreateSegmentPoints(B, C)
sCD <- CreateSegmentPoints(C, D)
sDE <- CreateSegmentPoints(D, E)
sEF <- CreateSegmentPoints(E, F)
sFG <- CreateSegmentPoints(F, G)
sGH <- CreateSegmentPoints(G, H)
sHA <- CreateSegmentPoints(H, A)
```

5.14.2.4 Define chord line segments

The same procedure as above is used to create the chord segments *AC*, *CE*, *DF*, *EG*, *FH*, *GA*, in order to determine the internal angles of the polygon by the law of cosines.

```
sAC <- CreateSegmentPoints(A, C)
sCE <- CreateSegmentPoints(C, E)
sDF <- CreateSegmentPoints(D, F)
sEG <- CreateSegmentPoints(E, G)
sFH <- CreateSegmentPoints(F, H)
sGA <- CreateSegmentPoints(G, A)
```

5.14.2.5 Draw line segments

The line segments are drawn in the Cartesian coordinate plane with different colors (Figure 5.15).

```
# Create polygon named poly
poly<-CreatePolygon(A, B, C, D, E, F, G, H)
# Draw polygon with different colored lines
CoordinatePlane(x_min, x_max, y_min, y_max)
Draw(poly, c("grey92"))
Draw(sAB, "black", label = TRUE)
Draw(sBC, "red", label = TRUE)
Draw(sCD, "blue", label = TRUE)
Draw(sDE, "green", label = TRUE)
Draw(sEF, "orange", label = TRUE)
Draw(sFG, "yellow", label = TRUE)
Draw(sGH, "brown", label = TRUE)
Draw(sHA, "purple", label = TRUE)
```

5.14.2.6 Draw line segments and the chord-like segments

The chord-like line segments are drawn in the Cartesian coordinate plane with a gray colors. We noted that a chord is defined outside the polygon surveyed with the tape (Figure 5.16).

```
# Draw line segments
CoordinatePlane(x_min, x_max, y_min, y_max)
Draw(poly, c("grey92"))
Draw(sAB, "black", label = TRUE)
Draw(sBC, "red", label = TRUE)
Draw(sCD, "blue", label = TRUE)
Draw(sDE, "green", label = TRUE)
Draw(sEF, "orange", label = TRUE)
Draw(sFG, "yellow", label = TRUE)
Draw(sGH, "brown", label = TRUE)
Draw(sHA, "purple", label = TRUE)
# Draw chord segments
```

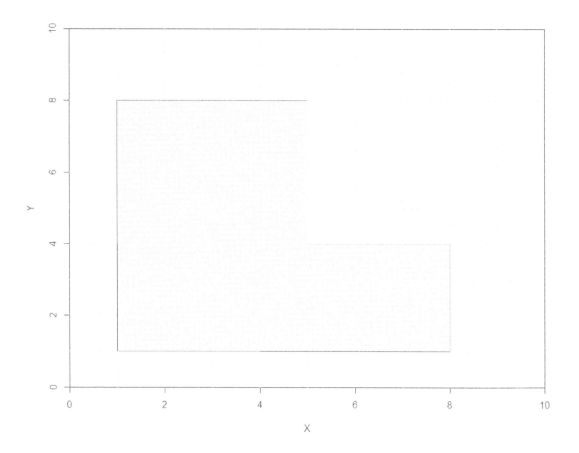

FIGURE 5.15: Determination of line segments with different colors between vertices of polygon surveyed in the field with diastimeter.

```
Draw(sAC, "gray", label = TRUE)
Draw(sCE, "gray", label = TRUE)
Draw(sDF, "gray", label = TRUE)
Draw(sEG, "gray", label = TRUE)
Draw(sFH, "gray", label = TRUE)
Draw(sGA, "gray", label = TRUE)
```

5.14.2.7   Determine the length of each line segment

Lengths of line segments AB, BC, CD, DE, EF, FG, GH, HA are determined with the Distance-Points function.

```
AB <- DistancePoints(A, B)
BC <- DistancePoints(B, C)
CD <- DistancePoints(C, D)
DE <- DistancePoints(D, E)
```

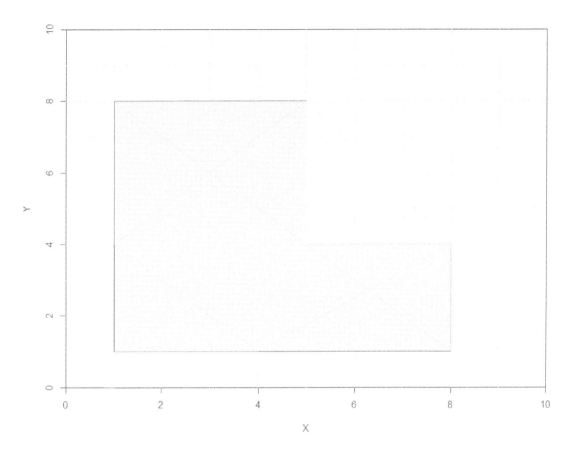

FIGURE 5.16: Determination of chord-like segments between vertices of polygon surveyed in the field with diastimeter.

```
EF <- DistancePoints(E, F)
FG <- DistancePoints(F, G)
GH <- DistancePoints(G, H)
HA <- DistancePoints(H, A)
AB
```

```
## distance
##         3
```

```
BC
```

```
## distance
##         3
```

CD

```
## distance
##        4
```

DE

```
## distance
##        3
```

EF

```
## distance
##        3
```

FG

```
## distance
##        4
```

GH

```
## distance
##        4
```

HA

```
## distance
##        4
```

5.14.2.8 Determine the length of the chord segments

Lengths of chord-type line segments AC, CE, DF, EG, FH, GA are determined with the DistancePoints function.

```
AC <- DistancePoints(A, C)
CE <- DistancePoints(C, E)
DF <- DistancePoints(D, F)
EG <- DistancePoints(E, G)
FH <- DistancePoints(F, H)
GA <- DistancePoints(G, A)
AC
```

```
## distance
## 4.242641
```

CE

```
## distance
##          5
```

DF

```
## distance
## 4.242641
```

EG

```
## distance
##          5
```

FH

```
## distance
## 5.656854
```

GA

```
## distance
## 5.656854
```

5.14.2.9   Determine the polygon angles by the law of cosine

The cosine and arc cosine of triangles B, C, D, E, F, G, H sides opposite the angles of interest
are determined by the law of cosines and `deg` and `acos` functions, respectively.

```
# Determine cosines of sides of triangles opposite to the angles
cosB<-(AB^2 + BC^2 - AC^2)/(2*AB*BC)
cosD<-(CD^2 + DE^2 - CE^2)/(2*CD*DE)
cosE<-(DE^2 + EF^2 - DF^2)/(2*DE*EF)
cosF<-(EF^2 + FG^2 - EG^2)/(2*EF*FG)
cosG<-(FG^2 + GH^2 - FH^2)/(2*FG*GH)
```

```
cosH<-(GH^2 + HA^2 - GA^2)/(2*GH*HA)
# Determine the arc cosine of angles in degrees
acosB<-deg(acos(cosB))
acosD<-deg(acos(cosD))
acosE<-deg(acos(cosE))
acosF<-deg(acos(cosF))
acosG<-deg(acos(cosG))
acosH<-deg(acos(cosH))
acosB
```

```
## distance
##       90
```

```
acosD
```

```
## distance
##       90
```

```
acosE
```

```
## distance
##       90
```

```
acosF
```

```
## distance
##       90
```

```
acosG
```

```
## distance
##       90
```

```
acosH
```

```
## distance
##       90
```

5.14.3 Determine the polygonal area by the semiperimeter method considering the vertices A, B, C, D, E, F, G, H

The polygon area is defined by the triangles BFD, DEF, FGH, HFB. Since the heights of these triangles are not known, the semiperimeter method method is used to determine each triangle area. The sum of each triangle area is used to obtain the total polygon area.

```
# The polygonal area is defined by triangles BFD, DEF, FGH, and HFB
# Determine distances
HF <- DistancePoints(H, F)
BF <- DistancePoints(B, F)
BD <- BC+CD
HB <- HA + AB
# Determine semiperimeter of each triangle within the polygon
BFD <- (BD + DF + BF) / 2
DEF <- (DE + EF + DF) / 2
FGH <- (FG + GH + HF) / 2
HFB <- (BF + HF + HB) / 2
areaBFD <- sqrt(BFD*(BFD - BD)*(BFD - DF)*(BFD - BF))
areaDEF <- sqrt(DEF*(DEF - DE)*(DEF - EF)*(DEF - DF))
areaFGH <- sqrt(FGH*(FGH - FG)*(FGH - GH)*(FGH - HF))
areaHFB <- sqrt(HFB*(HFB - BF)*(HFB - HF)*(HFB - HB))
areaBFD
```

```
## distance
##      10.5
```

```
areaDEF
```

```
## distance
##      4.5
```

```
areaFGH
```

```
## distance
##        8
```

```
areaHFB
```

```
## distance
##       14
```

```
areaABCDEFGH<-sum(areaBFD, areaDEF, areaFGH, areaHFB)
areaABCDEFGH
```

```
## [1] 37
```

The angles of triangles BFD, DEF, FGH, HFB are determined to calculate the polygon total area by the angular method. After the summation of each triangle area, the same area value obtained by both methods is observed.

```
# Check calculated area using angles
# Determine interior angles of vertices defining the polygon area
anguloBFD <- Angle(B, F, D)
anguloBFH <- Angle(B, F, H)
anguloFGH <- Angle(F, G, H)
anguloDEF <- Angle(D, E, F)
A1 <- 0.5*BF*DF*sin(rad(anguloBFD))
A2 <- 0.5*BF*HF*sin(rad(anguloBFH))
A3 <- 0.5*FG*GH*sin(rad(anguloFGH))
A4 <- 0.5*DE*EF*sin(rad(anguloDEF))
area<-sum(A1, A2, A3, A4)
area
```

```
## [1] 37
```

```
# check ok!
```

Based on this computation practice, we demonstrated that distance determinations from field measurements made with tape can be mapped and used to calculate angles and the area of closed polygons that are frequently used in surveys of small regions with low-cost geomatics equipment.

5.15 Solved Exercises

5.15.1 On a surveying with taping, the temperature and pull corrections:

    a. Can show the same sign. [X]
    b. Always present the same sign.
    c. Always have opposite signs.
    d. Always have positive signs.

5.15.2 The correction of slope on sloping terrain:

    a. Is positive.
    b. Is negative. [X]
    c. Can be positive or negative.
    d. Is zero.

5.15.3 If two points A and B are 55 m apart and have an elevation difference of 0.5 m, the slope correction of the measured length is:

    a. +0.0008 m.
    b. -0.0022 m. [X]
    c. -0.0125 m.
    d. +0.0010 m.

```
#H=L-d2/2L
L=55
d=0.5
d2=d^2
LL=2*L
c=-1*(d2/LL)
c
```

```
## [1] -0.002272727
```

```
#or
H=sqrt((L^2)- (d^2))
c<-H-55
c
```

```
## [1] -0.002272774
```

5.15.4 After performing a topographical survey with a 30 m steel tape, standardized at 20°C, pulled with 5.45 kg tension, we found that the real length of the tape was 30.012 m. The tape was stretched horizontally with a constant pull at the ends of 9.09 kg, while measuring a line from A to B, with 3 segments (Table 5.3). Apply the corrections for tape measure calibration, temperature, pull, sag, in order to determine the correct length of the distance measured from vertices A to D.

Required Information:

- The cross-sectional area of the tape measure: 0.050 cm^2;
- Weight of the tape measure: 0.03967 kg m^{-1};
- Thermal coefficient of expansion and contraction of the tape: 0.0000116;
- Modulus of elasticity of steel: 2,000,000 kg cm^{-2}.

TABLE 5.3: Field notes of a line measured with a steel tape.

Section	Recorded Distance (m)	Temperature (° C)
A - *B*	30.000	14.000
B - *C*	30.000	15.000
C - *D*	21.151	16.000

A: The calibration of the line length is obtained by:

```
CL<-((30.012-30)/30)*81.151
CL
```

```
## [1] 0.0324604
```

The temperature corrections for each segment are:

```
CTAB<-0.0000116*(14-20)*30
CTAB
```

```
## [1] -0.002088
```

```
CTBC<-0.0000116*(15-20)*30
CTBC
```

```
## [1] -0.00174
```

```
CTCD<-0.0000116*(16-20)*21.151
CTCD
```

```
## [1] -0.0009814064
```

```
SCT<-sum(CTAB, CTBC, CTCD)
SCT
```

```
## [1] -0.004809406
```

The pull correction is:

```
CP<-((9.09-5.45)/(0.05*2000000))*81.151
CP
```

```
## [1] 0.002953896
```

The sag correction is:

```
CSAC<- -2*(((0.03967^2)*(30^3))/(24*(9.09^2)))
CSAC
```

```
## [1] -0.04285279
```

```
CSCD<- -1*(((0.03967^2)*(21.151^3))/(24*(9.09^2)))
CSCD
```

```
## [1] -0.007508931
```

```
SCS<- sum(CSAC, CSCD)
SCS
```

```
## [1] -0.05036173
```

The calibrated distance AD is obtained by adding all the corrections:

```
AD<-81.151+CL+SCT+CP+SCS
AD
```

```
## [1] 81.13124
```

5.16   Homework

Choose one exercise presented by the teacher and solve the question with different input values. Compare the results obtained. Use a real database measured in the field with a tape type instrument. Determine the angles of the vertices and the polygon area.

5.17   Resources on the Internet

As a study guide, slides and illustrative videos are presented about the subject covered in the chapter in Table 5.4.

TABLE 5.4: Slide shows and video presentations on direct distance measurement.

Guide	Address for Access
1	Slides on taping measurement to define sample points[1]
2	Taping measurement with accessories[2]
3	Taping tips[3]
4	Measuring with taping in the field[4]
5	Corrections of field taping measure[5]
6	Tape and offset surveys[6]

5.18   Research Suggestion

The development of scientific research on geomatics is stimulated by the activity proposals that can be used or adapted by the student to assess the applicability of the subject matter covered in the chapter (Table 5.5).

TABLE 5.5: Practical and research activities used or adapted by students using direct distance measurement.

Activity	Description
1	In the content on direct distance measurement, interest may arise in doing practical work based on the computation examples presented
2	Perform a practical activity of measuring the distance between masonry walls of a building using taping
3	Propose a methodology of chord measurements and the cosine law to define the internal angles of the building from measurements of the external area made with taping

[1] http://www.sergeo.deg.ufla.br/geomatica/book/c5/presentation.html#/
[2] https://youtu.be/UMr9-SvoYVs
[3] https://youtu.be/jDfpll_I904
[4] https://youtu.be/BapoW7wY0yc
[5] https://youtu.be/BapoW7wY0yc
[6] https://youtu.be/ndMx_tcG798

5.19   Learning Outcome Assessment Strategy

Perform a summary of the chapter, "Direct Distance and Angle Measurements with Geomatics and R", on a single A4 page in order to show the student's abilities to summarize a subject presenting key points considered of greater importance today.

6

Stadia Indirect Measurements

DOI: 10.1201/9781003184263-6

6.1 Learning Questions

The emergent learning questions answered through reading the chapter are as follows:

- How to perform indirect measurement of horizontal distance and level difference between two points with mechanical optical instruments on plane or sloping terrain.
- How to calculate distances by stadimetric equation in R.
- How to create a planimetric database in R.
- How to export planimetric measurement results from R.

6.2 Learning Outcomes

The learning outcomes expected from reading the chapter are as follows:

- Perform indirect measurement of horizontal distance and level difference between two points with mechanical optical instruments on flat or sloping terrain.
- Calculate distances by stadimetric equation in R.
- Create a planimetric database in R.
- Export planimetric measurement results from R.

6.3 Introduction

Optical and mechanical indirect distance meters are called "tachymeters" or "tacheometers". Tachymetry is the procedure by which horizontal distances and elevation differences are determined using optical properties of a telescope and a measuring instrument, transit, theodolite, level, or tachymeter. The method has long been recognized as a simple and inexpensive tool for mapping areas of limited extent (Ali, 1995).

Indirect distance measurements are established using conventional optical instruments in stadimeter. Speed and accuracy are the great advantages of tacheometric surveys over direct distance measurement processes, as all measurements are performed by the instrument operator himself. The operator and assistant must be trained in the use and installation of the stadia, also known as "stadia rod". The assistant must install the stadia rod correctly on the point and keep it vertical (leveled) and without moving during the reading of the stadia lines (Alves and Silva, 2016).

6.4 Indirect Distance Measurement with Mechanical Optical Instruments

Indirect measurement with a theodolite by means of tacheometry, stadimeter or stadia is used as a quick method to determine horizontal distance and elevation of a point. The term "tacheometry" comes from the Greek and meant rapid measurements. The term "stadia" is the plural of the Greek word "stadium" (McCormac et al., 2012). Historical accounts indicated that the Scotsman James Watt was the developer of the stadimeter method in 1771 (Adrian Raymond Legault, 1956).

Stadia observations are obtained by sighting through a telescope with stadiametric lines of known spacing inside the telescope. The apparent length intercepted at the top and bottom of the stadia lines is read on a graduated rod or stadia rod, positioned vertically relative to the observed point. The distance from the telescope to the stadia rod is determined by similar triangle relationships (Ghilani and Wolf, 1989).

6.5 Indirect Distance Measurement on Flat Terrain

The difference between two readings of stadia lines intersecting a stadia rod is called "generator number". Stadia lines have been spaced such that at distances of 30, 60 and 80 m, the intersection at a vertical stadia rod generated the numbers 0.3, 0.6 and 0.8 m, respectively. Therefore, to determine a particular distance, the telescope is pointed over the stadia lines and the difference between the upper (F_S) and lower (F_I) stadia lines is multiplied generally by 100, according to the optical equipment configuration (Barcellos, 2003) (Figure 6.1).

The stadia principle can be demonstrated from the relation of similar triangles (Barcellos, 2003):

$$\frac{s}{f} = \frac{S}{H} \tag{6.1}$$

where S is the difference of readings on the stadia rod, f, the focal length, H, the distance to be determined and, s, the distance of stadimeter lines.

Isolating the distance (H) from the previous equation,

$$H = \frac{f}{s}S \tag{6.2}$$

Considering the relationship between the focal length (f) and the spacing of the reticle's lines (s), we have $\frac{f}{s} = 100$:

$$S = F_S - F_I \tag{6.3}$$

where F_S is the reading from the upper stadia line and, F_I, the reading from the lower stadia line.

Replacing S into the equation to determine distance,

$$H = 100(F_S - F_I) \tag{6.4}$$

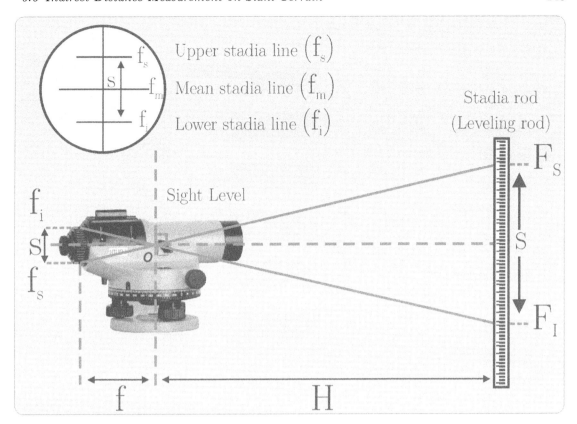

FIGURE 6.1: Indirect measurement of distances in the horizontal plane using conventional optical instruments (mechanical optical theodolites and levels).

The readings F_S and F_I are measured by a stadia rod, usually 3 or 4 m long, on which the meters, decimeters and centimeters are read directly and the millimeters estimated (Figure 6.2) (Barcellos, 2003).

6.6 Indirect Distance Measurement on Slant Terrain

When working on a sloping terrain, the vertical angle is used to calculate the horizontal component of the slant distance, as well as to determine the difference in height between two points (McCormac et al., 2012) (Figure 6.3).

In this case, the difference between the upper and lower stadia lines in the inclined measurement $(F_S - F_I)_i$ is defined by:

$$(F_S - F_I)_i = (F_S - F_I)\cos\alpha \tag{6.5}$$

The slant distance (L) can be obtained by:

$$L = 100(F_S - F_I)_i \tag{6.6}$$

By substituting the L value into the equation $H = L\cos\alpha$, we can define the slant distance measurement, for a height angle α by (Barcellos, 2003):

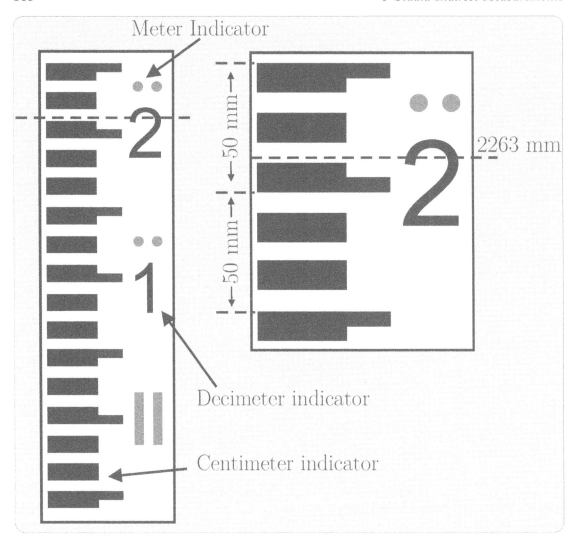

Meter Indicator

50 mm

50 mm

2263 mm

Decimeter indicator

Centimeter indicator

FIGURE 6.2: Reading system on stadia rod.

$$H = 100(F_S - F_I)cos^2\alpha \tag{6.7}$$

Similarly, the slant distance measurement for a zenith angle (z) can be obtained by (Barcellos, 2003):

$$H = 100(F_S - F_I)sen^2z \tag{6.8}$$

Ali (1995) evaluated the accuracy of five optical theodolites, Wild T16, T1, T2, Kern DKM-1, DKM-2 and an automatic level, Wild NA2, tested for horizontal distance and height accuracy measurement. In application areas with horizontal accuracy of approximately $\pm$ 30 mm in 100 m modern optical theodolites and levels suitable for the job were required and used as an alternative electronic distance measurement.

Similarly, Ali (2001) evaluated six simple electronic digital theodolites, three Sokkia instruments (DT6, DT5 and DT2), a Topon DT20, a Zeiss ETh4 and a Leica T1600, used comparatively in horizontal distance and level difference measurement using stadia lines etched into the telescope reticle of these theodolites. In both cases, semi electronic tacheometry provided much better

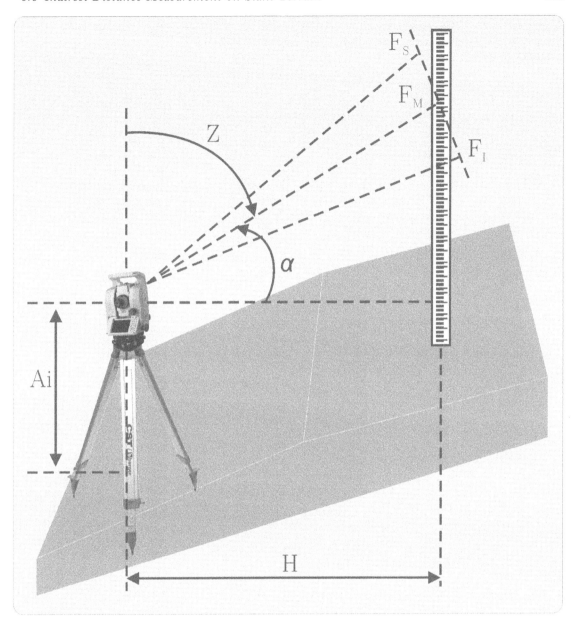

FIGURE 6.3: Reading stadia rod with slant theodolite sighting.

accuracy values (two to three times) than conventional optical surveying techniques. The high accuracy was probably attributed in part to the repeated measurements and the improved design of electronic theodolites used in the test. The range of accuracy values obtained was compatible with the requirements of surveying performed in civil, agricultural, and environmental engineering and other localized surveys that require positional accuracy values with moderate accuracy.

Methods for surveying and analyzing channels using topography were compared with a hand held stadiametric level, a laser distance meter, and a real-time kinematic global navigation satellite system (RTK-GNSS). The accuracy of this equipment was compared in determining slope and roughness of the small, dry and steep channel bed. The variability between four operators for each technique was also evaluated. The RTK-GNSS data was used as an accuracy reference. The inter-operator variability was found to be very low (coefficients of variation between 0.001 and

0.046) for most of the systems evaluated, except for the one-person laser measurement system. With two people performing the survey, the laser measurement method was as accurate as the level but provided advantages in difficult survey conditions, despite higher cost (Scott et al., 2016).

6.7 Computation

As a computation practice, we suggest to use field survey data of distances in a soccer field from the center to the edges in order to practice the creation of a `data.frame` database and distance determinations by the stadimeter equation. Merging new data into the `data.frame` and recording the results in a `.txt` file are also practiced.

R package `circular` is used in the computation practice for distance determinations through trigonometric functions (Jammalamadaka and SenGupta, 2001; Lund et al., 2017a).

6.7.1 Installing R packages

The `install.packages` function is used to install the `circular` package in the R console.

```
## install.packages("circular")
```

6.7.2 Enabling R packages

The `library` function is used to enable the `circular` package in the R console.

```
library(circular)
```

6.7.3 Import field notes from theodolite survey

Field note data of theodolite irradiation survey on a soccer field from the center of the field (O) to the borders (A, B, ..., M) are described with the following columns: ID = vertex ID; Az = Zenith angle in decimal degrees; AH = Horizontal angle in decimal degrees; F_I = Lower stadia line in mm; F_S = Upper stadia line in mm. The `data.frame` function is used to organize the readings taken in the topographic survey.

```
irr<-data.frame(ID=c('OA', 'OB', 'OC', 'OD', 'OE', 'OF', 'OG', 'OH',
                     'OI', 'OJ', 'OL', 'OM'),
AZ=c(90.7639, 90.9000, 91.3889, 91.1444, 90.8778, 90.5917, 90.6528,
     90.6833, 90.9361, 91.1111, 90.9444, 90.8750),
AH=c(2.4278, 21.1806, 47.6972, 164.0472, 175.1028, 188.5722,
     200.9833, 216.4611, 234.3111, 326.4361, 340.1611, 348.2583),
FI=c(100.0000, 100.0000, 100.0000, 100.0000, 100.0000, 200.0000,
     100.0000, 100.0000, 100.0000, 100.0000, 100.0000, 100.0000),
```

```
FS=c(845.0000, 800.0000, 590.0000, 662.0000, 810.0000, 1013.0000,
     948.0000, 912.0000, 760.0000, 685.0000, 785.0000, 822.0000))
```

6.7.4 Determine the horizontal distance by stadia

In possession of the table organized with the readings, the distance of each alignment in meters is determined. Therefore, the result of applying the stadimeter equation is divided by 1000.

```
H<-(100*(irr$FS-irr$FI)*(sin(rad(irr$AZ))^2)/1000)
```

6.7.5 Merging the results of calculated distances in the irradiation table

Horizontal distance determination results are joined to the original data as a new column through the cbind function.

```
irr<-cbind(irr, H)
```

6.7.6 Export the table in .txt

Finally, the write.table function is used to export the results as a table in .txt extension for use in further studies.

```
## write.table(irr, file = "E:/Aulas/Topografia/Aula5/irr.txt",
##             sep = " ", row.names = TRUE, col.names = TRUE)
```

6.8 Solved Exercises

6.8.1 Stadia is a form of tachyometric measurement based on:

    a. fixed generator number.
    b. fixed generator number angulation. [X]
    c. angular variation of the generating number.
    d. none of the previous alternatives.

6.8.2 The tacheometry survey method is generally preferred to:

    a. provide primary control.
    b. conduct large-scale surveys.
    c. establish points with greater accuracy.
    d. demarcate terrain with obstacles. [X]

6.8.3   A sight was targeted at a picket B from another picket A with altitude equal to 584.025 m. Determine the horizontal distance between vertices A and B.

Required information: Readings at B: $F_I = 0.417$ m (lower stadia line), $F_M = 1.518$ m (middle stadia line), telescope inclination angle, $\alpha = 5°30'$, descending.

A: The horizontal distance between vertices A and B is 218.1772 m.

```
# Horizontal distance between 2 points: H=100Scos^2alpha;S=2(Fm-Fi)
S<-2*(1.518-0.417)
S
```

```
## [1] 2.202
```

```
H<-100*S*(cos(rad(5.5)))^2)
H
```

```
## [1] 218.1772
```

6.9   Homework

Choose one exercise presented by the teacher and solve the question with different input values. Compare the results obtained. Use a real database measured in the field with mechanical optical instrument and stadia rod. Import the field notes in R to calculate distance with the stadia equation. Export the obtained results in a .txt file.

6.10   Resources on the Internet

As a study guide, slides and illustrative videos are presented about the subject covered in the chapter (Table 6.1).

TABLE 6.1: Slide shows and video presentations on stadia indirect measurements.

Guide	Address for Access
1	Slides on indirect distance measurement with mechanical optical instruments in geomatics[1]
2	Analog theodolite installation and leveling[2]
3	Installation and leveling of a digital theodolite in the field[3]
4	Digital theodolite measurement[4]
5	Theodolite angle and distance measurement applications[5]
6	Distance measurement with the stadia principle[6].
7	Reading stadia lines[7]
8	Leveling with stadia[8]
9	Reading stadia rod[9]

6.11 Research Suggestion

The development of scientific research on geomatics is stimulated by the activity proposals that can be used or adapted by the student to assess the applicability of the subject matter covered in the chapter (Table 6.2).

TABLE 6.2: Practical and research activities used or adapted by students using stadia indirect measurements.

Activity	Description
1	In the content on indirect distance measurement, you may be interested in doing practical work based on the computational examples presented
2	Perform horizontal distance measurements in the field using theodolite and stadia rod
3	Perform leveling difference measurements in the field using level and stadia rod

---

[1] http://www.sergeo.deg.ufla.br/geomatica/book/c6/presentation.html#/
[2] https://youtu.be/C78SuqmLPDE
[3] https://youtu.be/QUX9_1fRnlo
[4] https://youtu.be/BZi0owCSsso
[5] https://youtu.be/7h5CCl1JY7U
[6] https://youtu.be/RTdAD4hLhPo
[7] https://youtu.be/wYeAEpP1NYo
[8] https://youtu.be/tNRZPHLwC7k
[9] https://www.youtube.com/watch?v=fbwEORw1c9Y

6.12   Learning Outcome Assessment Strategy

Perform a summary of the chapter, "Stadia Indirect Measurements with Geomatics and R", on a single A4 page in order to show the student's abilities to summarize a subject presenting key points considered of greater importance today.

7

Electronic Distance and Level Measurements

7.1 Learning Questions

The emergent learning questions answered through reading the chapter are as follows:

- How is the development of the theory of propagation of electromagnetic energy?
- What are the principles of electronic distance measurement?
- How are electro-optical and total station instruments used in electronic horizontal and slant distance measurements, topographic leveling, and analysis of distance measurement errors?
- What are the principles about distance measurements and topographic leveling with satellite positioning systems (GNSS)?
- How to evaluate the proximity between surveyed points in the field.
- How to register field survey data from sampling grid points in R.
- How to determine distance matrix in R.
- How to perform hierarchical dissimilarity cluster analysis and distance matrix dendrogram in R.

7.2 Learning Outcomes

The learning outcomes expected from reading the chapter are as follows:

- Understand the theory of electromagnetic energy propagation and the principles of electronic distance measurement.
- Know the use of electro-optical instruments and total station in electronic measurements of horizontal and slant distance, topographic leveling and distance errors.
- Know distance and leveling measurements with GNSS.
- Register field survey data from sampling grid points in coffee plantation in R.
- Determine distance matrix.
- Perform hierarchical cluster analysis of dissimilarity and the dendrogram of the distance matrix.
- Evaluate the proximity between the points surveyed in the field using the R package `stats`.

7.3 Introduction

The new generation of electronic distance measuring instruments, combining digital theodolites and microprocessors, are called total station instruments. The measurement method is based on electromagnetic theory, leading to the emergence of electronic distance meters. With these instruments, the distance is measured using the wavelength as the basic unit of measurement. This enables one to measure the slant distance and the vertical angle, calculate the vertical and horizontal distance components, and present the results obtained in real time (Alves and Silva, 2016).

Some equipment is configured to collect and store the field data and transmit the information to computers, printers and other technologies, for further data processing in the office. These systems have also been called "field-to-finish systems", showing global acceptance, and have determined changes in traditional surveying.

After taking measurements, the distances can be described by a distance matrix. In a distance matrix we can obtain a number for the distance between all objects of interest and group them by a criterion of proximity of values.

7.4   Electromagnetic Energy Propagation

From early work on the atmospheric effect on light propagation (Barrell and Sears, 1939), empirical equations are proposed and later adapted for the calibration of electronic distance measuring instruments.

Electronic distance measurement is based on the rate and shape of the propagation of electromagnetic energy in the atmosphere. The propagation rate can be expressed by (Ghilani and Wolf, 1989; Alves and Silva, 2016):

$$V = f\lambda \tag{7.1}$$

where V is the electromagnetic energy velocity (m s^{-1}), f, the modulated frequency of the energy (Hz) and, λ, the wavelength (m). The hertz (Hz) is the unit of frequency equal to 1 cycle per second. The kilohertz (KHz), megahertz (MHz), gigahertz (GHz) and terahertz (THz) are equivalent to 10^3, 10^6, 10^9 and 10^{12} Hz, respectively.

The speed of electromagnetic energy in vacuum is 299,792,458 m s^{-1}. The speed of electromagnetic energy can be modified by atmospheric interference (Ghilani and Wolf, 1989; Alves and Silva, 2016):

$$V = \frac{c}{n_a} \tag{7.2}$$

where c is the speed of electromagnetic energy in vacuum and, n_a, the refractive index of the atmosphere. The value of n_a depended on the pressure and temperature, ranging from 1.0001 to 1.0005, with average of ~1.0003. Therefore, distances obtained by accurate electronic measurements must consider the temperature and pressure measurements in order to know n_a.

Temperature, atmospheric pressure and relative humidity influence the refractive index. Since the light source is composed of several wavelengths with different refractive indices, a set of waves will have a joint refractive index (group). The refractive value (N_g) of a group at standard atmospheric conditions in an electronic distance measurement is (Ghilani and Wolf, 1989; Alves and Silva, 2016):

$$N_g = 287.6155 + \frac{4.8866}{\lambda^2} + \frac{0.0680}{\lambda^4} \tag{7.3}$$

where λ is the wavelength of the light (μ).

Wavelengths light sources commonly used in electronic distance measurement were 0.6328 μm for red laser and 0.900 to 0.930 μm for infrared laser. It should be noted that the standard air conditions are 0.0375% carbon dioxide, temperature 0°C, pressure 760 mmHg, and 0% relative humidity.

The refractive indice of a group in the atmosphere (n_a), at the moment of observation, considering variations in temperature, pressure and humidity, can be calculated by (Ghilani and Wolf, 1989; Alves and Silva, 2016):

$$n_a = 1 + (\frac{273.15}{1013.25} \frac{N_g P}{t + 273.15} - \frac{11.27e}{t + 273.15})10^{-6} \qquad (7.4)$$

where e is the partial water vapor pressure (hPa) defined by temperature and relative humidity at the time of measurement, P, the atmospheric pressure (hPa) and t, the dry bulb temperature (°C). For conversion, 1 atmosphere is equivalent to 101.325 kPa or 1013.25 hPa or 760 mmHg.

The partial water vapor pressure, e (mmHg), can be calculated with satisfactory accuracy under normal operating conditions by (Ghilani and Wolf, 1989; Alves and Silva, 2016):

$$e = E \frac{h}{100} \qquad (7.5)$$

where $E = 10^{[7.5t/(237.3+t)]+0.7858}$ and h is the relative humidity (%).

It should be noted that the effects of relative humidity on wave transmission can be ignored in some precision work when using electronic distance meters with light in the near infrared region (Ghilani and Wolf, 1989; Alves and Silva, 2016).

A similar empirical equation is used to calculate the atmospheric correction in ppm (Δppm) of electronic distanciometer as a function of the atmospheric pressure and the humidity of the ambient air of propagation of electromagnetic energy (Silva and Segantine, 2015):

$$\Delta ppm = 281.5 - (\frac{0.29035P}{1 + 0.00366t}) + (\frac{11.27h}{100(273.16 + t)}10^x) \qquad (7.6)$$

where $x = [7.5t/(237.3 + t)] + 0.7857$ and, P, the pressure (mbar).

7.5 Principles of Electronic Distance Measurement

Despite the wide variety of electronic measuring instruments available on the market, there are basically two methods of wavelength measurement: the pulse method and the phase difference method.

7.5.1 Pulse method

In the pulse method, a short, intensive pulse of radiation has been emitted from the transmitter to the reflector. The reflected signal returns in a parallel path to the receiver. The distance is calculated by the speed of the signal multiplied by the time the signal completed the round-trip path (Figure 7.1) (Alves and Silva, 2016).

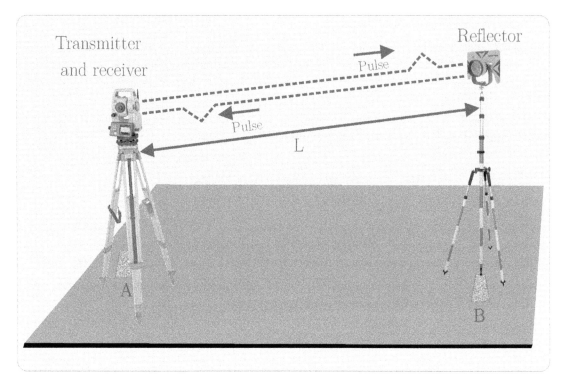

FIGURE 7.1: Principle of the pulse method for measuring distances.

Thus, we can deduce that (Barcellos, 2003):

$$2L = c\delta t \tag{7.7}$$

$$L = c\frac{\delta t}{2} \tag{7.8}$$

where t is the propagation time of the wave between the transmitter and the reflector, i.e., the outward and return of the signal, c, the speed of light in the propagating medium, and, L, the slant distance between the instrument and the target.

Therefore, considering the speed of propagation of light in the medium of 300000 km s^{-1}, and an error of $\pm$ 1 nanosecond (10^{-9} s) on the propagation time, will lead to an error of 15 cm in the distance measurement. Thus, the timer used in the instrument must have high measurement accuracy to avoid instrumental errors.

The pulse method was originated in hydrographic instruments using microwaves, but has been adapted for systems using laser propagation of electromagnetic waves. However, the phase difference method has been applied to most instruments using electromagnetic waves in the visible, infrared or microwave region.

7.5.2 Phase difference method

The way electromagnetic energy propagates through the atmosphere can be represented by a sinusoidal curve (Ghilani and Wolf, 1989; Alves and Silva, 2016) (**Figure 7.2**).

Wavelength regions or point positions along the wavelength have been determined by phase angles. A phase angle of 360° represented a full cycle, or a point at the end of a wavelength, while

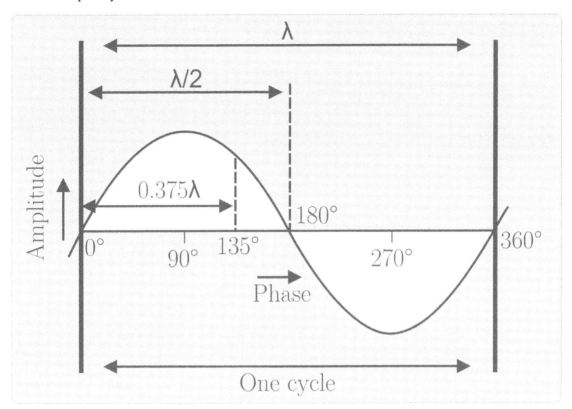

FIGURE 7.2: A wavelength of electromagnetic energy with the phase angles.

a wavelength half of 180° represented an average position. An intermediate position on the wavelength with a phase angle of 135° is 135°/360°, or 0.375 of a wavelength. After accurately knowing the wavelength of electromagnetic energy transmitted between two ends of a line, it is possible to perform calibrations in order to determine the distance between points by means of a carrier.

An analog wave-shaped signal can be modulated to represent the transmitted information. Considering two instruments leveled with a plumb line, optical plumb line or laser plumb line, the instrument at station A transmitted an electromagnetic carrier signal to station B. A precisely tuned wavelength reference frequency is superimposed or modulated on the carrier. A reflector at B returned the signal to the receiver, so that the path is the double distance between AB. The modulated electromagnetic energy can be represented by a series of sine waves, each with a wavelength λ (Figure 7.3) (Ghilani and Wolf, 1989; Alves and Silva, 2016).

The distance from vertices A to B (AB) can be determined by the number of wavelengths of the double path, multiplied by the wavelength (m), divided by 2. Since the measured distance is not always related to an integer number of wavelengths, we expected to get wavelength fractions when performing the measurements. Therefore, when performing a distance measurement by the phase measurement principle, the distance L is obtained by (Ghilani and Wolf, 1989; Alves and Silva, 2016):

$$L = \frac{n\lambda + p}{2} \tag{7.9}$$

where λ is the wavelength, n, the number of integer wavelengths and, p, the length of the wavelength fraction determined by the distance measuring instrument, based on the phase shift angle of the return signal (Figure 7.4).

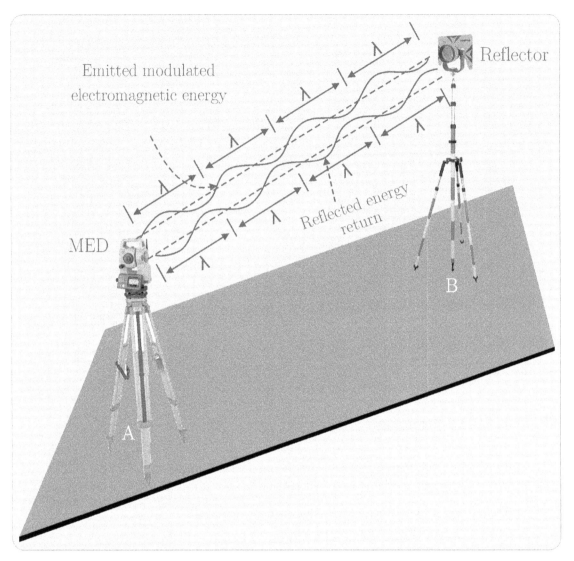

FIGURE 7.3: General electronic distance measurement procedure.

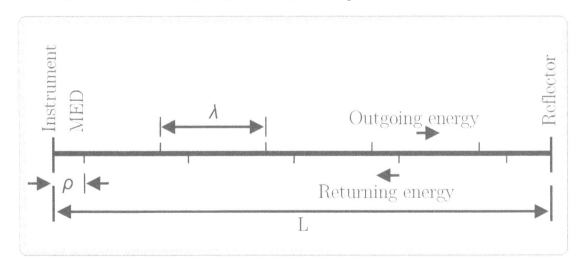

FIGURE 7.4: Principle of electronic distance measurement by phase difference.

Thus, it can be stated that for a wavelength of 20 m, the double distance traveled on the path determined an effective wavelength of 10 m. This is one of the fundamental wavelengths used by electronic distance meters that can be generated at a frequency of about 15 MHz. Equipment must transmit additional signals at different wavelengths in order to provide measurements on the order of a few millimeters to a few kilometers according to the instruments used (Ghilani and Wolf, 1989).

Distance determination by the phase difference method has been applied in most instruments using infrared, visible light or microwaves. The electronic instruments have, besides the devices for emission and reception of the electromagnetic waves, a device to measure the phase difference between waves. The phase difference, p, can be measured by analog and digital methods (Barcellos, 2003; Schofield et al., 2007) (Figure 7.5).

The type of modulation used in infrared instruments with electro-optical systems is the amplitude modulation of the measurement wave by varying the carrier wave. In this case the infrared beam can be controlled by small components, such as lenses, so that the beam transmitted by the instrument is highly collimated (Price and Uren, 1988; Barcellos, 2003) (Figure 7.6).

7.6 Electro-optical Instruments

Electronic distance measuring instruments used tungsten or mercury lamps; however, the equipment is bulky, used higher power source and short operating ranges, especially during the day, in view of excessive atmospheric scattering. Today, most electronic distance measuring instruments manufactured are electro-optical and transmit carrier signals with infrared or laser energy. The intensity of these signals can be modulated directly, simplifying the equipment. The instruments can use light derived from laser gas, and are smaller and portable, with the ability to take measurements over long distances during the day as well as at night. The transmitter uses a Gallium Arsenide (GaAs) diode that emits amplitude-modulated infrared light, being the wave source used in most electro-optical instruments with the main advantage that the output can be directly modulated in intensity. The radiation output is linearly related and stimulated to an applied current and the response time is very small (Barcellos, 2003) (Figure 7.7).

A crystal oscillator is used to precisely control the frequency to be modulated. The frequency modulation process can be associated with the passage of light through a funnel in which a damping plate rotates at a precisely controlled rate or frequency. When the damper plate is closed, no light passes through. Under opening of the plate, the light intensity increases up to a maximum phase angle value of 90°, under full opening. The intensity is reduced to zero again when the plate is closed under angle of 180°. The varying intensity and modulation amplitude is characterized by sine waves (Ghilani and Wolf, 1989; Alves and Silva, 2016).

The infrared beam can be variably attenuated by means of a filter disk to prevent receiver saturation caused by a strong return signal (near distance). The filter disk is positioned by a motor and controlled by a processor via an interface. Two filters cover the disk and attenuate the transmitted and received signals, each with 50% of all attenuation. At the end of the disk track, a mechanical stop limits the movement of the disk motor. A position detector is used for a quick setting of the amplitude of the received signals, and also for the anti-boomerang system to set the desired attenuation. The anti-boomerang signal occurs when a signal travels more than once the path between the optical transmitter, reflector and optical receiver, resulting in distance error. The position detector is a light barrier generated by an IR-Diode (LED) and a photodiode. The light illuminates the photodiode through a wedge-shaped diaphragm located at the edge of the filter disk circumference (Alves and Silva, 2016).

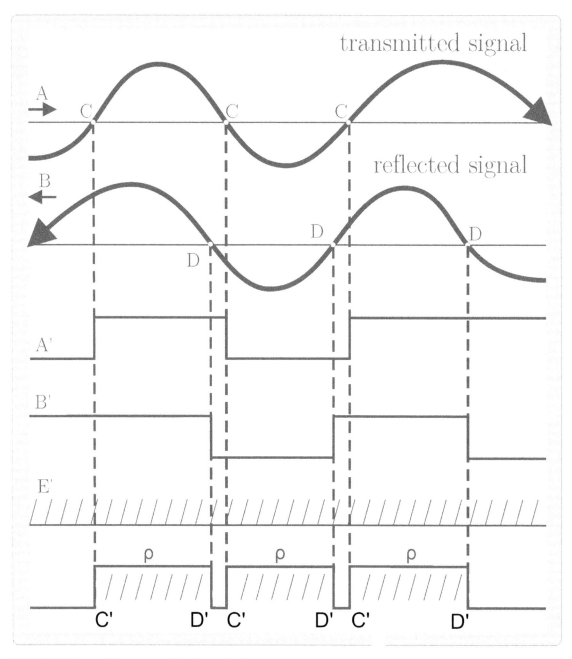

FIGURE 7.5: Schematic of a digital phase meter.

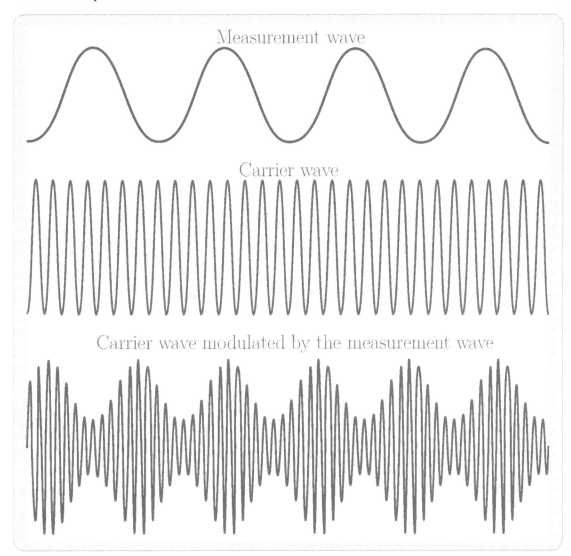

FIGURE 7.6: Modulation of the carrier wave by the measurement wave.

The current caused by the light is converted to direct current at a magnitude proportional to the actual position on the filter disk, and the voltage is read into the CPU via an AD converter. Upon initialization of the filter system, the CPU reads the threshold values and allocates the attenuation range to a position area from 0 to 255. This standardization is done to compensate for the tolerances of the light barrier if one of the limit values deviates more than a set standard. Then the accuracy of the filter's position location is checked against some pre-established position. If the difference is greater than a standard, there is new initialization of the motor and light barrier (Alves and Silva, 2016).

A beam splitter divides the light emitted from the diode into two distinct signals, called the external measurement beam and the internal reference beam. The external beam is directed to a reflector centered at a point on the opposite side of the measured line, by means of a bezel attached to the distance measuring instrument. A three-sided retroreflector type cube can be used to re-direct the outer, coaxial beam to the receiver. The internal beam passes through a variable density filter and is reduced to a level of intensity equal to that of the returned external signal, making a more accurate measurement possible. Both the internal and external signals pass

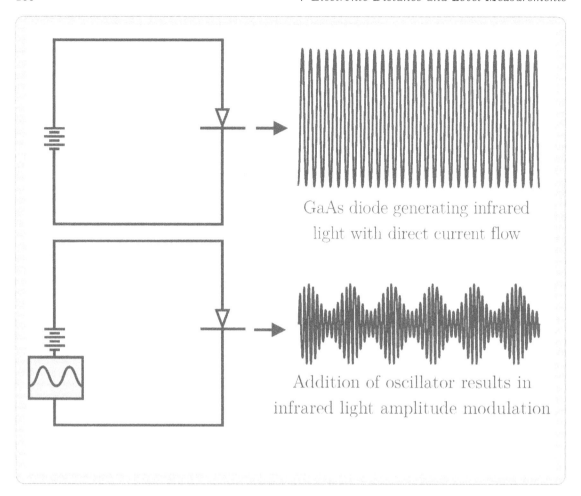

FIGURE 7.7: Amplitude modulation of GaAs diode.

through an interference filter to eliminate unnecessary energy, such as sunlight. The internal and external beams pass through analog to electrical energy conversion components, preserving the phase shift relationships obtained during the line path between the two measured ends. A phase metric converter is used to convert the phase difference into direct current, keeping the magnitude proportional to the difference. This current is connected to a metric annulator, set to annul the current. The wavelength fraction is measured during the current nullification process, converting the distance and giving the result (Ghilani and Wolf, 1989; Alves and Silva, 2016) (Figure 7.8).

To resolve the ambiguity of the number of full cycles the wave has undergone, electronic distance measuring instruments transmit frequencies in different modulations. For example, in an equipment with four frequencies, F_1, F_2, F_3 and F_4, modulated at 14.984 and 1.4984 and 149.84 and 14.984 KHz, with a refractive index of 1.0003, the corresponding effective wavelengths are 10, 100, 1000 and 10000 m, respectively. Considering the distance of 3867.142 m presented as the measured line result, the last four digits, 7.142 are obtained by phase shift measurement, while the wavelength of 10 m is transmitted at frequency F_1. Frequency F_2, with a wavelength of 100 m, is subsequently transmitted, generating the fraction value 67.14, followed by frequency F_3, with the value 867.1, and, F_4, with 3867, to complete the total value presented. With this, the high resolution measurement, close to 0.001 m, is guaranteed by the use of the 10 m wavelength, as well as the other wavelengths that made it possible to resolve the ambiguity when measuring the total distance (Ghilani and Wolf, 1989; Alves and Silva, 2016).

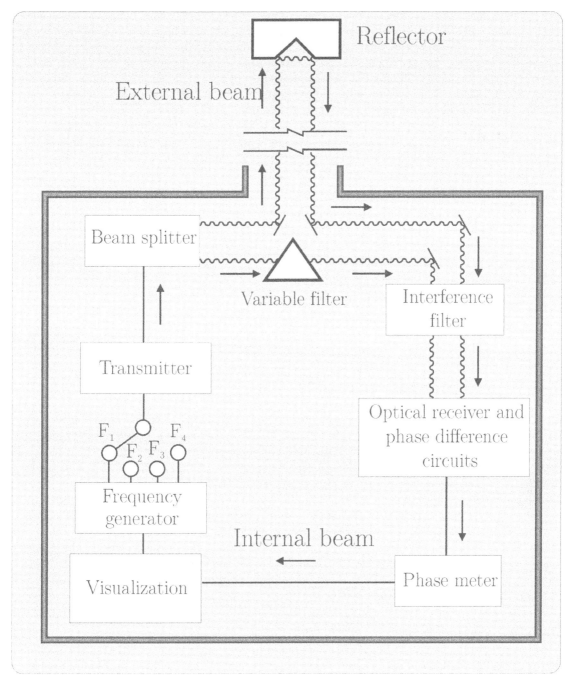

FIGURE 7.8: Diagram of the basic operation of an electro-optical electronic distance measuring instrument.

In older instruments, the frequency shift and null are obtained manually by screwing the device. Modern equipment has incorporated microprocessors to control the measurement process completely. After aiming the instrument at the reflector and starting the measurement, there is presentation of the final distance almost instantly on the display. Other changes in modern instruments have included electronic improvements to modulation amplitude control and replacement of the metric null with an electronic phase detector. These changes have contributed significantly to improving the accuracy of phase shift determination and reducing the number of different frequencies required for transmission. With this, less than two frequencies have been used in some instruments, one producing a short wavelength to obtain high-resolution digits under close measurement and another, with a long wavelength, for measurement over large distances. Thus, if there is no coincidence of digits measured by the two wavelengths used, modern instruments compare the occurrence of overlap and an error message can be displayed if the numbers are different. If the numbers are equal, the first four digits of the short wavelength and the first three digits of the long wavelength should be considered in the reading (Ghilani and Wolf, 1989; Alves and Silva, 2016).

Companies market sell instruments with accuracies ranging from $\pm$ (1 mm + 1 ppm) to $\pm$ (10 mm + 1 ppm). Accuracy in electronic distance measurement is divided into two parts, the first being a constant and the second proportional to the measured distance. The unit parts per million (ppm) is equivalent to 1 mm km^{-1}. With this, for a distance of 5000 m, an error of 5 ppm is equivalent to 5000 (5 $e10^{-6}$) or 0.025 m. Furthermore, modern instruments have combined the digital measurement of electronic theodolites with electronic distance measurement (Ghilani and Wolf, 1989; Alves and Silva, 2016).

7.7   Total Station Instruments

Total station instruments combine an electronic distance measuring instrument with an electronic digital theodolite and a computer in a single unit. These devices made it possible to automatically observe horizontal and vertical angles, horizontal and slant distance, and transmit the result to an internal computer in real time. Height and zenith angles can also be displayed. If the instrument is oriented in one direction and the coordinates of an occupied station are entered into the system, the coordinates of any point targeted can be immediately obtained. This data can be stored in the instrument or in data loggers, eliminating the manual recording process. Total station instruments can operate in tracking mode, also called "location mode". In this case, the required distance can be entered from the control panel and the bezel sighted in the appropriate direction. This function has been extremely useful in construction. Some total station instruments already have a robotic function for motion control, imaging, scanning, and measuring without a long range prism (2000 m), and have an internal processor with an operating system (Figure 7.9) (Ghilani and Wolf, 1989; Alves and Silva, 2016).

Automatic displacement monitoring of subway tunnels can be accomplished by means of robotic total stations and prisms as reflectors. With the use of robotic total stations it is possible to perform highly accurate displacement monitoring by measuring angles and distances without contact. By conducting an experiment in a tunnel, it is possible to verify the range and accuracy under monitoring with a robotic total station. By applying the developed monitoring system to Line 2 of the Guangzhou Metro enabled to confirm the robustness of the system (Zhou et al., 2020).

FIGURE 7.9: Leica Flexline TS09 plus electronic total station 1000 m range without prism.

7.8   Electronic Distance Measurement without Reflectors

In some electronic distance measuring instruments, the use of reflectors may be dispensable. These devices use pulse time varying infrared laser signals and can take measurements in prismless mode. However, with the use of the prism, these instruments can obtain length measurements over 3 km. The prismless mode is used to target inaccessible objects, such as construction details, faces of dams, retaining walls, structural elements of bridges, among others. This allows for greater survey speed and efficiency when measuring inaccessible features needed in projects (Figure 7.10) (Ghilani and Wolf, 1989; Alves and Silva, 2016).

Other electronic instruments are available for measuring without a prism by laser interferometry, such as electronic measuring tape. The digital laser distance measuring tape can measure from 500 mm to 50 m, with millimeter resolution (Alves and Silva, 2016).

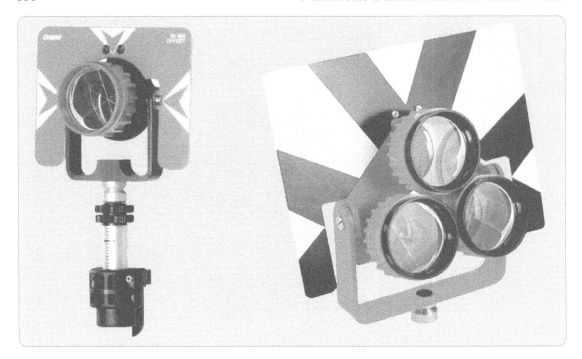

FIGURE 7.10: Triple (right) and single (left) prisms used to extend the capability of the electronic distance meter.

7.9   Electronic Measurement of Horizontal Distance

All electronic distance measuring devices measure the slant distance between two stations. The conversion from slant distance to horizontal distance can be based on elevation differences or zenith or vertical angles. Long distances must be treated by different ways, in view of the Earth's curvature. The horizontal distance calculation is performed internally by a microprocessor, but can be performed manually. Before performing the conversion to horizontal distance, the instrumental and atmospheric errors of the slant distance must be corrected (Ghilani and Wolf, 1989; Alves and Silva, 2016).

7.9.1   Horizontal distance correction by elevation difference

If elevation difference is used to calculate horizontal distance in field operations, the altitude of the electronic distance measuring instrument (h_e) and reflector (h_r) are measured and recorded. If the elevations between points A and B are known, the slope distance length will be converted into horizontal distance based on the value of the height difference between the instrument and the reflector (d) (Ghilani and Wolf, 1989; Alves and Silva, 2016) (**Figure** 7.11):

$$d = (elev_A + h_e) - (elev_B + h_r) \tag{7.10}$$

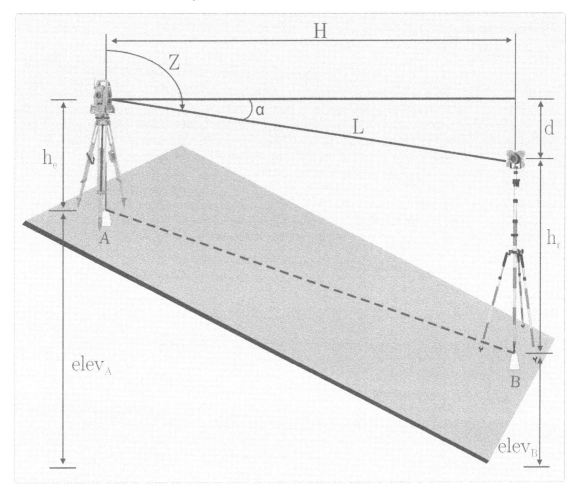

FIGURE 7.11: Conversion from slant distance to horizontal distance by means of an electronic distance meter.

7.9.2 Horizontal distance correction by vertical angles

If the zenith angle (z) is observed relative to the inclined path of the transmitted energy when measuring the slant distance L, then the horizontal distance can be obtained by (Ghilani and Wolf, 1989; Alves and Silva, 2016):

$$H = Lsen(z) \tag{7.11}$$

If the vertical angle α is observed, the equation $H = Lcos\alpha$ can be used in the conversion. In the case of no angle observation, the equation $H = \sqrt{L^2 - d^2}$ can be an alternative to determine the horizontal distance. In precision studies, vertical or height angles should be observed in the forward and reverse directions, followed by averaging the observations. The mean obtained from both ends will be used to compensate for the effect of curvature and refraction.

7.10   Digital Leveling

The digital level is an instrument that uses digital image processing to read a barcoded staff. The instrument outputs a signal pattern on a vertical ruler, and a correlation procedure in the instrument translates the reading into horizontal and vertical distance from the instrument to the levelling staff. The detector is a type of charge-coupled device (CCD) camera that transforms the black and white barcode into a binary code. The aperture angle of the instrument is very small, on the order of 1 to 2°, resulting in the imaging of a small section at close range and a larger section at maximum distance. The data are stored in the instrument in order to eliminate mistakes. The instrument is also affected by focus, vibrations, lighting, location of barcode coverage, collimation, and physical damage if pointed at the sun. The resolution of most of the instruments is 0.1 mm for height and 10 mm for distance, in measurements up to 100 m (Schofield et al., 2007) (Figure 7.12).

7.11   Distance and Level Measurements with GNSS

Global Navigation Satellite System (GNSS) enable to obtain coordinates of points based on electromagnetic waves measurements. The distance is obtained from a relative vector between the receiving antennas, transported to the topographic points on the terrain. The vector of distance calculated by the coordinates is equal to the slope distance between the points. Accuracies greater than $\pm (5 \text{ mm} + 2 \text{ ppm})$ can be obtained according to the instrument used, method of point positioning, satellite geometry at the time the data are obtained, and the methods of data adjustment and processing (Silva and Segantine, 2015).

For leveling with GNSS technology, point height values can be obtained by surveying in relative mode, post-processed or real-time kinematic (RTK) and under favorable conditions achieving centimeter and even millimeter accuracy. If leveling is performed in differential mode RTK, the survey is fast compared to other methods, but there may be accuracy degradation according to the satellites' geometry, the effects of the troposphere and ionosphere and the geoidal height variation. Additional information must be used to correct the values of the ellipsoidal heights measured by GNSS receiver antennas. If the altitude value of the point where the installed GNSS base receiver is known, this value indicates the geometric height at the other points measured with the remote receiver (rover). The heights will be related to the altitude value measured at the base. This procedure can be useful in regions with little variation in geoidal undulation. Other options may use a geoidal model incorporated in the GNSS processing software. We should be careful when choosing the global geoidal model because of data inconsistency in specific regions or estimation of geoid waves for other points by linear interpolation from points with known altitudes (Silva and Segantine, 2015).

7.12   Laser Leveling

Laser leveling can be used to generate a leveling plane by means of a visible laser beam that continuously rotates around a vertical axis that coincides with the vertical of the site. This enables an operator to determine height differences with a graduated staff between points in the

FIGURE 7.12: ZDL700 level with 3-second digital readout on double-sided levelling rod, accuracy of 0.7 mm per km, aluminum tripod.

survey region. A vertical, inclined or horizontal plane can be determined with laser leveling in the following applications (Silva and Segantine, 2015):

- Control of earthwork and excavation;
- Slope control;
- Ramp construction;
- Alignment and concreting control;
- Alignment tracing;
- Location and leveling gauge;
- Alignment of the facade;
- Leveling pre-cast slab.

7.13   Errors in Electronic Distance Measurement

Electronic measurement errors can be constant and scalar proportional to the observed distance. The constant error is the most significant at short distances. At large distances, the constant error becomes negligible and the proportional part is more important. The largest error components at an observed distance are the lack of centering of the instrument and reflector, the constant, and scalar errors of the distance measuring instrument. The error in a measured distance (E_d) can be obtained by (Ghilani and Wolf, 1989; Alves and Silva, 2016):

$$E_d = \pm\sqrt{E_i^2 + E_r^2 + E_c^2 + (ppmL)^2} \tag{7.12}$$

where E_i is the estimated centering error of the instrument, E_r, the estimated centering error of the reflector, E_c, the specified constant error of the equipment, ppm, the specified scalar error of the equipment, and, L, the measured slant distance.

When working with electronic distance measuring instruments, the sources of error can be human, instrumental, or natural (Alves and Silva, 2016).

7.13.1   Human errors in electronic distance measurement

Human errors in electronic distance measurement are related to (Alves and Silva, 2016):

- Improper configuration of instruments and reflectors;
- Incorrect measurement of instrument height and reflectors;
- Incorrect determination of atmospheric pressure and temperature.

These errors are random and can be minimized by careful work and use of quality barometers and thermometers. Mistakes in manual reading and recording can be common and costly. These errors can be eliminated by using data collection instruments and taking readings in different units for later comparison (Ghilani and Wolf, 1989; Alves and Silva, 2016).

An example of a common mistake can occur when setting the temperature and pressure of the electronic distance measuring instrument before taking the measurement. Consider the example of the refractive index calculated to be 1.0002672. If the effective wavelength for a standard atmosphere is 10 m, then the actual wavelength produced by the electronic meter is 10/1.0002672 = 9.9973 m. For an observed distance of 827.329 m, the error is -0.221 m:

$$e = (\frac{9.9973 - 10}{10})827.329 = -0.221 \text{ m} \tag{7.13}$$

```
# Wavelength
lambda<-10
lambda
```

```
## [1] 10
```

```
# Current Wavelength
la<-lambda/1.0002672
la
```

```
## [1] 9.997329
```

```
# Error in distance
e<-((la-lambda)/lambda)*827.329
e
```

```
## [1] -0.2210033
```

In this case, the effect of the failure to consider the current atmospheric condition, determined an accuracy (pr) of 1 m on 3743 m surveyed (1:3743):

```
pr<-(1/(0.221/827.329))
pr
```

```
## [1] 3743.57
```

7.13.2 Instrumental errors in electronic distance measurement

If an electronic distance measuring instrument is carefully adjusted and precisely calibrated, the instrumental errors will be small. To ensure accuracy and reliability, instruments should be checked against a first-order baseline at regular time intervals. Although the instruments are stable, occasionally the generation of wrong frequencies and improper tuning can occur. This will result in wavelengths that degrade the distance measurement in a similar way to using an incorrectly sized taping. Periodically checking the equipment against a calibrated baseline will make it possible to check for observation errors. The cubic corner reflectors are another source of instrumental error. Considering that light passes through glass at a slower speed than air, the effective center of the reflector will be behind the prism, not coinciding with the plumb line and causing the systematic error known as constant reflection. Considering that the reflector is composed of perpendicular faces, the light always crosses the total distance in the prism according to (Figure 7.13)(Ghilani and Wolf, 1989; Alves and Silva, 2016):

$$a + b + c = 2D \tag{7.14}$$

For a refractive index value of glass (n), which is higher than air, the speed of light in the prism is reduced to generate an effective distance (DE) of:

$$DE = nD \tag{7.15}$$

where n is the refractive index of the glass, of approximately 1.517 and, D, the prism depth. Under these conditions, for $D = 40$ mm, nD is 60.68 mm:

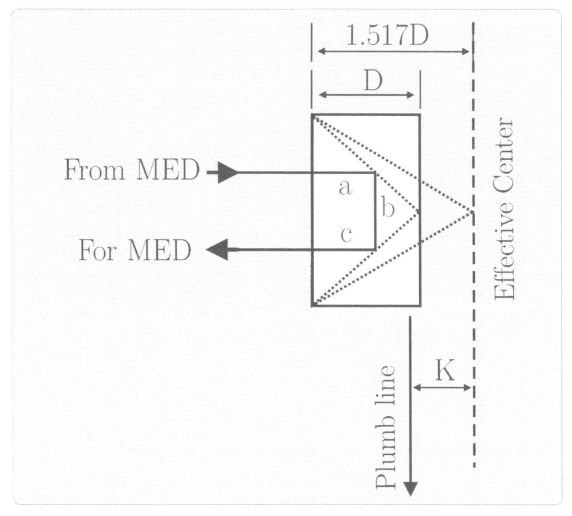

FIGURE 7.13: Schematic diagram of a reflector, where D is the prism depth.

```
D<-40
DE<-1.517*D
DE
```

```
## [1] 60.68
```

This enabled to determine the effective center and the reflector constant (K) of 20.68 mm:

$$K = DE - D \tag{7.16}$$

```
K<-DE-D
K
```

```
## [1] 20.68
```

The electric center of the electronic distance measuring instrument can be changed to a compensating value based on the reflector constant. However, if the measuring instrument is used with multiple reflectors, the offset value of each reflector must be subtracted from the observed distance to obtain the corrected values. Performing accurate baseline comparisons with observed distances can be used to determine a constant that can then be applied to subsequent observations for appropriate corrections. Although calibration using a baseline has been preferred, other methods can be used (Ghilani and Wolf, 1989; Alves and Silva, 2016).

7.13.3 Natural errors in electronic distance measurement

Natural errors in electronic distance measuring instruments have been related primarily to atmospheric variations in temperature, pressure and humidity, which affect the refractive index by modifying the wavelength of electromagnetic energy. The values of these measured variables must be used to correct the observed distances. Although relative humidity is dispensable for correcting distances measured with electro-optical instruments, this variable is important when microwave instruments are used. Electronic distance measuring instruments, such as total stations, have onboard microprocessors that make it possible to use atmospheric variables as input via keyboard, in order to correct the distances after making the observations, but before publishing the results. Companies have developed tables and graphs to assist in this process. The magnitude of errorin electronic distance measurement due to atmospheric and temperature variation can be up to 50 ppm, with an error of 1 m for a distance of 200 m (Figure 7.14), depending on the equipment used (Ghilani and Wolf, 1989; Alves and Silva, 2016).

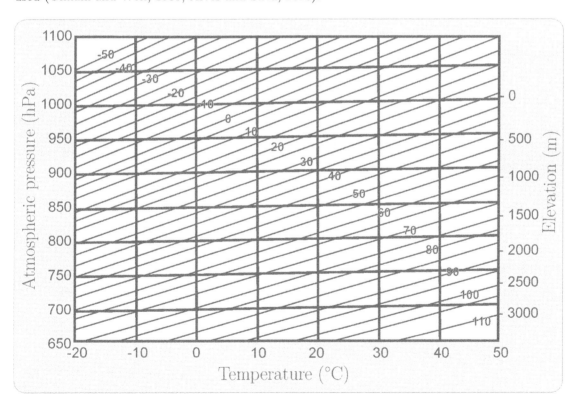

FIGURE 7.14: Errors in electronic instruments measuring distance as a function of temperature and pressure or altitude.

A horizontal geodetic field calibration network(GCN) with eight stations over an area of approximately 1 km^2 was constructed at Zanjan University (ZNU) in Iran with the objective of achieving submillimeter accuracy for horizontal positions in the field calibration of total stations and GNSS receivers. An accurate Leica TC2003 electro-optical total station was used to measure 56 distances and 48 angles. By applying the necessary corrections, reductions, and weightings of the observations, submillimeter ellipses of 95% absolute error were obtained at all stations after least squares adjustment. The performance of the GCN was evaluated with the Leica TC407 and Sanding STS-750 total stations with respect to distance and angle measurements. Four pairs of GNSS receivers of different types were also installed at the eight stations simultaneously for about 7 h for accuracy analysis (Saadati et al., 2019).

7.14   Computation

As a computation practice, we suggest to use field survey data of sampling grid points in a coffee plantation in Ijaci, MG, Cafua farm, obtained with electronic total station and GNSS. Survey data referring to WGS-84 ellipsoid is transformed to the Universal Transverse Mercator (UTM) coordinates, fuse 23 South, longitude (m), latitude (m) (x, y) of 67 points in the field.

Survey data is processed to obtain the distance matrix with the `dist` function. The Pythagorean theorem enables to check the calculated distance. Hierarchical dissimilarity cluster analysis and dendrogram of the distance matrix are evaluated in the surveyed points.

R package, `stats`is used to determine distance and statistical calculations. Other packages with tools for building, manipulating and using distance metrics can be used, such as the `distances` package (R Core Team, 2021). The `distances` package enables distance matrix calculation as well as functions for fast search for nearest and farthest neighbors (Savje, 2019).

7.14.1   Installing R packages

The `install.packages` function is used to install the `dist` package in the R console. If needed, use the same procedure to install the `stat` package. The `stat` package may already be enabled automatically when you install R.

```
## install.packages("stats")
```

7.14.2   Enabling R packages

If needed, enable the `stats` package using the `library` function in the R console.

```
library(stats)
```

7.14.3   Import UTM x, y coordinates of sample grid in coffee crop in Ijaci, MG

The measurement data of x and y coordinates are organized and stored in a `data.frame`, with rows and columns, with the `data.frame` function.

```
cafuaXY<-data.frame(x=c(502391.342, 502383.37, 502380.413, 502389.239,
502395.813, 502401.896, 502407.774, 502411.07, 502407.103, 502400.745,
502393.608, 502384.541,502369.001, 502374.903, 502381.018, 502387.105,
502389.678, 502388.269, 502382.771, 502376.238, 502366.355, 502352.024,
502344.385, 502331.455, 502341.925, 502334.139, 502334.432, 502341.844,
502347.31, 502353.328, 502359.306, 502365.288, 502367.732, 502365.758,
502360.488, 502352.202, 502342.176, 502330.834, 502318.687, 502319.158,
502318.427, 502317.288, 502318.147, 502321.125, 502329.738, 502331.77,
502331.281, 502316.118, 502299.447, 502297.982, 502296.672, 502298.676,
502301.545, 502278.527, 502276.113, 502275.835, 502278.395, 502282.37,
502290.152, 502294.431, 502279.182, 502260.54, 502257.093, 502254.031,
502254.991, 502258.558, 502263.301), y=c(7659044.65, 7658996.229,
7658946.355, 7658896.643, 7658872.969, 7658847.244, 7658822.878,
7658772.253, 7658747.478, 7658723.087, 7658699.469, 7658675.575,
7658888.791, 7658864.301, 7658839.999, 7658815.871, 7658791.005,
7658765.05, 7658739.87, 7658716.772, 7658692.687, 7658667.304,
7658647.918, 7658626.147,7659029.712, 7658980.678, 7658931.335,
7658911.677, 7658887.266, 7658861.975, 7658838.756, 7658814.212,
7658788.632, 7658763.706, 7658738.782, 7658714.826, 7658691.982,
7658668.731, 7658647.124, 7659021.919, 7658996.251, 7658972.494,
7658947.615, 7658922.771, 7658872.465, 7658822.38, 7658771.981,
7658724.261, 7659016.804, 7658991.434, 7658966.875, 7658941.765,
7658917.955, 7659009.287, 7658985.377, 7658960.456, 7658935.437,
7658909.855, 7658860.141, 7658810.317, 7658759.272, 7658711.44,
7659002.1, 7658977.254, 7658952.001, 7658927.487, 7658903.01))
```

7.14.4 Map the points x, y and the code number of each vertex

Next, a mapping of the points is performed, including the number of each vertex registered in the data.frame (Figure 7.15).

```
plot(cafuaXY)
text(cafuaXY,pos = TRUE, offset = 0.2)
```

7.14.5 Distance matrix calculation between points

The dist function is used to calculate a Euclidean distance matrix between the rows of the data frame.

```
dis <- dist(cafuaXY)
```

7.14.6 Check distance between points by Pythagorean theorem

The distance between points 1 and 2 is determined to check the results, obtaining the same result of 49.07286 m by both calculation methods.

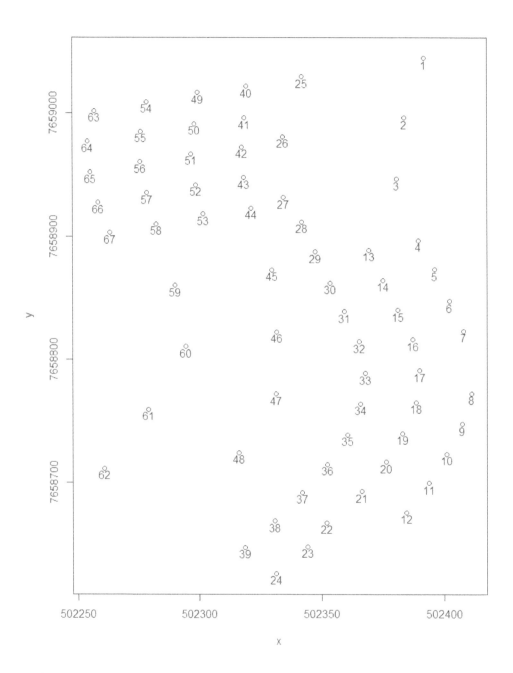

FIGURE 7.15: Mapping 67 georeferenced points with electronic total station at Cafua farm, Ijaci, Minas Gerais, Brazil.

```
H1_2<-sqrt((502391.342-502383.37)^2 + (7659044.65-7658996.229)^2)
```

7.14.7   Perform hierarchical cluster dissimilarity analysis and the distance matrix dendrogram

The hierarchical dissimilarity cluster analysis and the distance matrix dendrogram are used to evaluate the proximity between points defined with x and y coordinates, respectively (Figure 7.16).

```
hclust_centroid <- hclust(dis, "centroid")
plot(hclust_centroid, hang=-1)
```

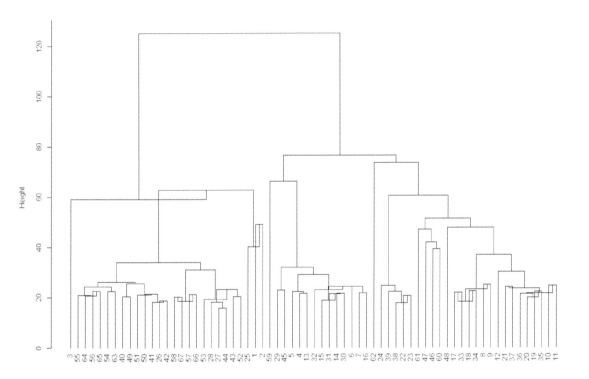

FIGURE 7.16: Distance cluster dendrogram between 67 points georeferenced with electronic total station at Cafua farm, Ijaci, Minas Gerais, Brazil.

The hclust function was used to perform a hierarchical cluster analysis by a set of dissimilarities across the 67 clustered objects. Initially, each object is assigned to its own cluster, and then the algorithm proceeded iteratively, at each stage joining the two most similar clusters, continuing until there is only a single cluster. At each stage, the distances between clusters are re-calculated by the Lance-Williams dissimilarity formula according to the centroid clustering method (Legendre and Legendre, 2012).

The study of clustering of distances between points can be useful to define sampling criteria in crops in order to optimize the data collection procedure.

7.15   Solved Exercises

7.15.1   What is the current wavelength and velocity of a beam of light from an electronic total station in the near infrared ($\lambda = 0.915$ m), modulated at the frequency of 320 MHz?

Required information:

Air temperature during the measurement of 34°C, relative humidity of 56% and atmospheric pressure of 1041.25 hPa.

A: The actual wavelength of the electronic total station is 0.9366012 m.

```
# Data
lambda<-0.915
t<-34
h<-56
P<-1041.25
c<-299792458
f<-320*10^6
# Refractive index of a group
Ng<-287.6155+(4.88660/(lambda^2))+(0.0680/(lambda^4))
# Water vapor partial pressure
a<-((7.5*t)/(237.3+t))+0.7858
E<-10^a
e<-E*(h/100)
# Refractive index in the atmosphere
na<-1+((273.15/1013.25)*((Ng*P)/(t+273.15))-
        ((11.27*e)/(t+273.15)))*10^-6
# Speed of light in the propagation medium
V<-c/na
# Current wavelength
lambdaa<-V/f
lambdaa
```

```
## [1] 0.9366012
```

7.15.2   A distance was measured with an electronic instrument in the field. Determine the value of the atmospheric correction for the measured distance.

Required information:

Atmosphere during measurement with air temperature of 41°C, relative humidity of 95% and atmospheric pressure of 745 mmHg = 993 mbar.

A: The atmospheric correction for the measured distance is 33.4514 ppm.

```
# Data
P<-993
t<-41
h<-95
# Calculate x
x<-((7.5*t)/(237.3+t))+0.7857
# Correction
deltappm<-281.5-((0.29035*P)/(1+0.00366*t))+
  (((11.27*h)/(100*(273.16+t)))*10^x)
deltappm
```

```
## [1] 33.4514
```

7.15.3 The wavelength used to measure a distance with an electronic instrument was precisely set to be 20 m. Assuming that the phase angle of the signal return was 115.7° and the number of integer wavelengths equals 9, determine the length of the line.

A: The line length is 93.2138 m.

```
# Data
lambda<-20
phase<-115.7
n<-9
# Wavelength fraction length
p<-(phase/360)*lambda
# Total length
L<-((n*lambda)+p)/2
L
```

```
## [1] 93.21389
```

7.15.4 A corrected slant distance of 165.360 m was measured from A to B, whose elevations were 447.401 m and 445.389 m above the datum, respectively. Determine the horizontal distance of the AB line considering the instrument and reflector heights equal to 1.417 m and 1.615 m, respectively.

A: The horizontal distance between points A and B is 165.35 m.

```
# Data
L<-165.360
eA<-447.401
eB<-445.389
hi<-1.417
hr<-1.615
```

```
# Level difference
d<-(eA+hi)-(eB+hr)
# Horizontal distance
HAB<-sqrt((L^2)-(d^2))
HAB
```

```
## [1] 165.35
```

7.15.5   A slant distance of **827.329 m** was measured between two points with electronic total station equipment with an error specification of $\pm$ (**2 mm + 2 ppm**). The instrument was centered with an estimated error of ± 3 mm. The estimated target centering error was $\pm$ **5 mm**. What is the estimated error in the observed distance and the relative precision?

A: The estimated error in the distance is ± 6.3826 m with relative precision of 1:129622.

```
# Data
L<-827329
ppm<-2*10^-6
ec<-2
ei<-3
er<-5
# Estimated error
Ed<-sqrt((ei^2)+(er^2)+(ec^2)+((ppm*L)^2))
Ed
```

```
## [1] 6.382624
```

```
# Precision
Pr<-(1/(Ed/L))
Pr
```

```
## [1] 129622.1
```

7.16   Homework

Choose one exercise presented by the teacher and solve the question with different input values. Compare the results obtained. Perform slant and horizontal distance calculation between several points in the field with electronic total station to create a irregular grid of sampling points.

7.17 Resources on the Internet

As a study guide, slides and illustrative videos are presented about the subject covered in the chapter in Table 7.1.

TABLE 7.1: Slide shows and video presentations on stadia indirect measurements.

Guide	Address for Access
1	Slides on electronic measurement of distances in geomatics[1]
2	Electronic distance measurement[2]
3	Installation and leveling of an electronic total station on a vertex[3]
4	Electronic distance measurement with robotic electronic total station[4]
5	Electronic distance measurement[5]
6	Setting up an electronic total station[6]
7	Atmospheric correction in measurement with electronic total station[7]

7.18 Research Suggestion

The development of scientific research on geomatics is stimulated by the activity proposals that can be used or adapted by the student to assess the applicability of the subject matter covered in the chapter (Table 7.2).

TABLE 7.2: Practical and research activities used or adapted by students using electronic distance measurement.

Activity	Description
1	In the content on electronic distance measurement, there may be interest in doing practical work based on the computational examples presented
2	Take angle, horizontal and slant distance measurements with an electronic total station in the field
3	Evaluate available electronic total station distance correction options as a function of atmospheric variation. Change the parameters and compare the results of measured distance with repetitions

[1] http://www.sergeo.deg.ufla.br/geomatica/book/c7/presentation.html#/
[2] https://youtu.be/hjcCrAejJI8
[3] https://youtu.be/lp824ZRIWQs
[4] https://youtu.be/L1w0ue_Teq8
[5] https://youtu.be/85UEwnyBdUI
[6] https://youtu.be/K8SEfTzpskg
[7] https://youtu.be/9AQrBKBu87E

7.19   Learning Outcome Assessment Strategy

Perform a summary of the chapter, "Electronic Distance and Level Measurements with Geomatics and R", on a single A4 page in order to show the student's abilities to summarize a subject presenting key points considered of greater importance today.

8

Radial Traverse Survey

8.1 Learning Questions

The emergent learning questions answered through reading the chapter are as follows:

- What are angle and direction measurements used for in radiating polygons?
- How to perform direct and indirect calculations on measurement data from radiated polygon.
- What are radiated polygon stations?
- How to calculate bearing, azimuth and horizontal distance between radiation vertices in R software, `circular` and `LearnGeom` packages.

8.2 Learning Outcomes

The learning outcomes expected from reading the chapter are as follows:

- Observe angles and directions in radiating polygons.
- Perform direct and indirect calculations on measurement data from a radiated polygon.
- Define radiated polygon stations.
- Calculate bearing, azimuth and horizontal distance between radiation vertices in R software, `circular` and `LearnGeom` packages.

8.3 Introduction

Polygonal or traverse is defined as a set of consecutive lines whose ends are demarcated in the field and whose length and direction are determined based on observations. When establishing polygonal lines in the field (traversing), the relative position of vertices must be determined with details necessary to describe the area (Ghilani and Wolf, 1989; Souza, 2003; Alves and Silva, 2016).

The following phases are performed in traversing a topographic survey (Comastri and Junior, 1998; Alves and Silva, 2016):

- Reconnaissance, which consisted of going around the region, selecting the starting point and the main vertices of the basic survey polygon, the necessary material and auxiliary team;

- Basic traverse, which defined the limits of the area, and can be used to evaluate uncertainties about the surveying quality;
- Detail survey, which consisted of determining the position of points in the surveyed area to define details needed for map representation.

Traverses can be classified into basic types, according to conformation, geometry and connection with higher-order traverses. The main types are open-path and closed-path traverses . Closed-path traverses are divided into the categories of polygon and link-path.

An open traverse, that is, geometrically and mathematically open, consists of a series of connected lines that did not return to the origin point. This type of traverse does not allow checking for observation errors and mistakes. Therefore, observations should be repeated carefully to avoid mistakes. Wooden or steel stakes and pickets are placed at each polygon station where there is a change of direction (Figure 8.1) (Ghilani and Wolf, 1989; Souza, 2003; Alves and Silva, 2016).

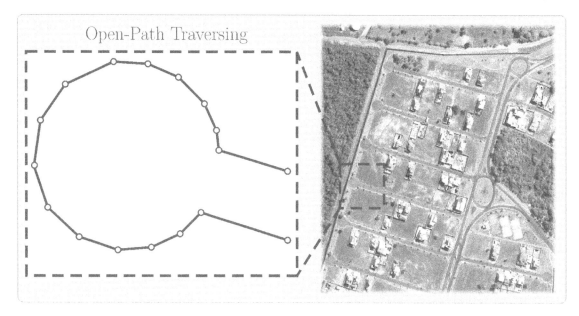

FIGURE 8.1: Open traverse used to demarcate access points in a residential condominium in Cuiabá, Mato Grosso, Brazil (left) and the location of the same area on a color composition 432 Ikonos satellite image (right).

Link-path traverses start at a point of known coordinates and end at another point of known coordinates. These polygons are geometrically open-path but mathematically closed-path, allowing the verification of angles and distances measured based on points with known coordinates, pre-existing at the beginning and end of the survey. The connecting polygon must end at another station with accuracy equal to or better than that of the starting point (Figure 8.2) (Ghilani and Wolf, 1989; Souza, 2003; Alves and Silva, 2016).

Polygon closed-path traverses returned to the starting point, forming a geometrically and mathematically closed figure. Closed-path traverses enable to check the observed angles and distances and have been used in control, construction, property, and topographic surveys (Figure 8.3) (Ghilani and Wolf, 1989; Alves and Silva, 2016).

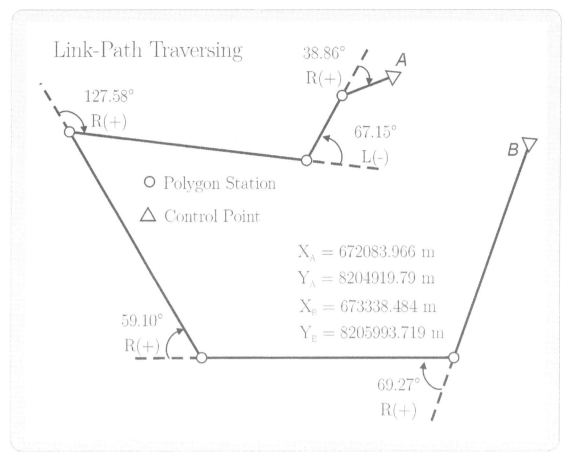

FIGURE 8.2: Traverse supported on control points A and B surveyed at Boa Vista farm, Mato Grosso, Brazil.

8.4 Observation of Traverse Angles and Directions

The angles and directions of polygon lines can be observed by the following methods (Ghilani and Wolf, 1989; Comastri and Junior, 1998):

- Internal angles;
- External angles;
- Deflection angles;
- Azimuths.

8.4.1 Traverses by internal angles

Traverses by internal angles are most appropriate in surveying property. Angles can be viewed clockwise or counterclockwise; however, to reduce mistakes in reading, recording and calculation, angles should always be read clockwise from the back station to the forward station. Internal angles can be improved by averaging forward and backward readings. External angles can also be read for checking (Ghilani and Wolf, 1989; Alves and Silva, 2016).

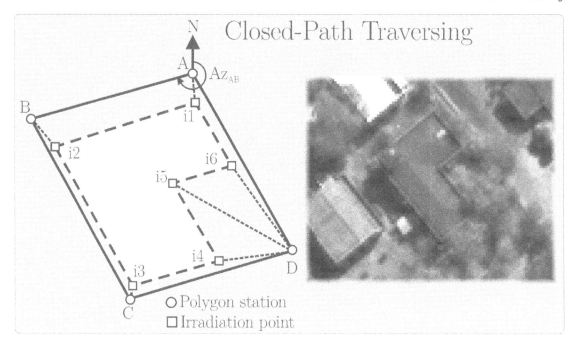

FIGURE 8.3: Closed polygon used for the survey of a building at the Federal University of Mato Grosso (left) and the location of the same area in color composition image 853 from the WorldView-2 satellite (right).

8.4.2   Traverses by external angles

Traverses by external angles are similar to traverses by internal angles; however, you walk the terrain clockwise, measuring the external angles of the polygon (Comastri and Junior, 1998; Alves and Silva, 2016).

8.4.3   Traverses by angles of deflection

Route surveys have typically been performed by observed deflection angles to the right or left of lines (Figure 8.2). A deflection angle is not complete without an indication to the right or left of the course and should not exceed 180°. Positive and negative values can be used to designate right and left deflection angles, respectively (Alves and Silva, 2016).

8.4.4   Traverses by azimuths

Traverses can be performed through azimuths. This process enabled reading the azimuths of all lines directly, eliminating the need to calculate them. In practice, we determine as accurately as possible the azimuth of the first alignment. At the next station, the azimuth of the previous alignment is recorded on the equipment and the backsight stadia rod is sighted with the telescope turned upside down. The telescope is placed in the normal position, and the forward mark is sighted, obtaining the new azimuth of the forward alignment. When closing the polygon, the equipment is reinstalled at the starting point, with the azimuth of the previous alignment set, and operated as at the other points to obtain the azimuth again. The difference between the first and last azimuth readings is the angular error of closure (Comastri and Junior, 1998) (Figure 8.4).

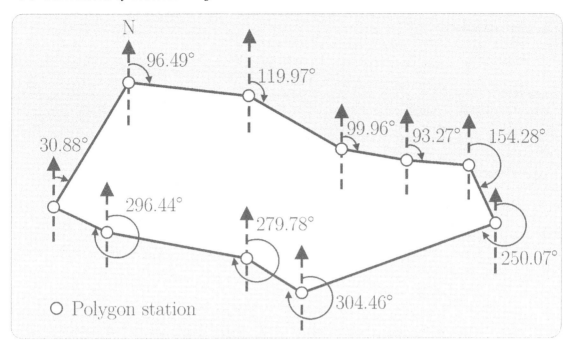

FIGURE 8.4: Azimuth traverse used to demarcate the legal reserve area of Boa Vista 2 farm in Santo Antônio do Leveger, Mato Grosso, Brazil.

8.5 Observation of Traverse Length

The length of each traverse line must be measured by the simplest and most economical method that satisfies the accuracy required by the project. One advantage of measuring with an electronic total station is that angles and distances can be observed with a single setting of each station. Averaging of fore and aft sight distances can increase the accuracy of the survey (Ghilani and Wolf, 1989; Alves and Silva, 2016).

8.6 Selection of Traverse Stations

Traverse stations should be located at easily accessible points. The number of stations can be reduced with careful reconnaissance of the area. Each type of survey has specific requirements for the location of stations. In the rural property survey, stations are placed at each vertex. In some cases it is necessary to place stakes near the vertices and take measurements, then consider the distance value. Distance stakes are also important in road surveys, as stakes can be destroyed during project execution. A polygon can also be used as a control when mapping topographical details such as roads, buildings and valleys (Ghilani and Wolf, 1989; Alves and Silva, 2016).

8.7 Identification of Traverse Stations

Traverse stations can be re-occupied after their construction. However, it is important to create tie-up observation points in case of destruction or new construction (Alves and Silva, 2016).

8.8 Radial Traverse

The relative position of some points can be determined by radial traverse or irradiation. Radial polygonal survey is ideal for quickly establishing a large number of points in the area (Figure 8.5) (Alves and Silva, 2016).

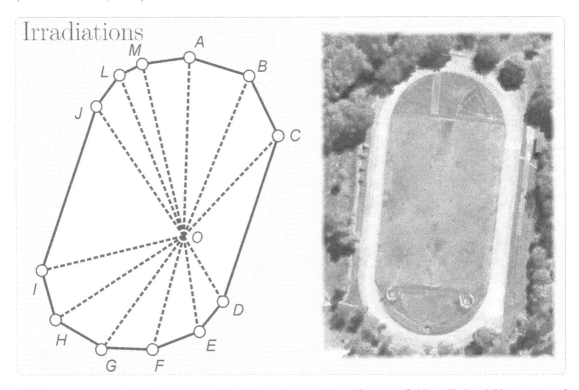

FIGURE 8.5: Radial traverse used to determine vertices of soccer field at Federal University of Mato Grosso, from an occupied station (left) and the location of the same area in color composition 853 from a WorldView-2 satellite image (right).

8.9 Radial Traversing of a Line

Longitude and latitude coordinates values determined changes in the X,Y components of a line in a rectangular grid system, defined by ΔX and ΔY. In traverse calculation, east longitude and north latitude are considered positive; west longitude and south latitude, negative. Azimuths used in the calculation of longitude and latitude range from 0 to 360° and algebraic sign of sine and cosine functions produces appropriate longitude and latitude values, according to the corresponding position in a rectangular quadrant (Alves and Silva, 2016) (Figure 8.6).

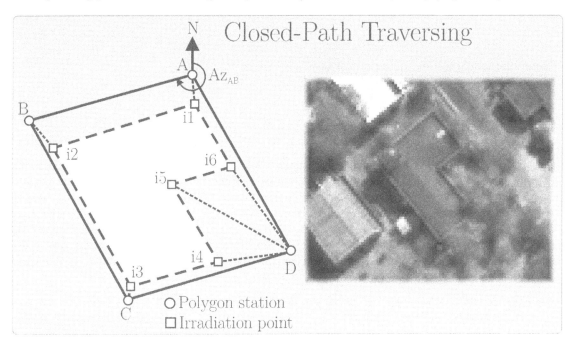

FIGURE 8.6: Longitude and latitude of a line where H consisted of the irradiation from vertices A to B and, α, the azimuth between these vertices.

The longitude (ΔX) and latitude (ΔY) of a line can be obtained by:

$$\Delta X = H\,sen\alpha \tag{8.1}$$

$$\Delta Y = H\,cos\alpha \tag{8.2}$$

where H is the horizontal length and, α, the azimuth of the alignment.

Therefore, a line with an azimuth of 125° has positive longitude and negative latitude; a line with an azimuth of 310° has negative longitude and positive latitude. When using bearings to calculate longitude and latitude, the angles always vary between 0 and 90°, so the sine and cosine values are always positive. In this case, the signs for longitude and latitude must be assigned to the longitude and latitude values. Thus, it is more convenient to use azimuth angles rather than bearing angles in computer traverse calculations (Ghilani and Wolf, 1989; Alves and Silva, 2016).

8.10 Computation of Rectangular Coordinates

The rectangular coordinates X and Y are used to reference the position of a point perpendicular to a reference axis. The X coordinate refers to the perpendicular distance in meters of the point on the Y axis. The Y coordinate refers to the perpendicular distance of the point on the X axis.

The reference of the axes characterized the position of the points, in the topographic survey. The Y axis is oriented in the south-north position, with north determining positive values in the Y direction. The X axis varies from east to west, with positive values on the east side (Ghilani and Wolf, 1989; Alves and Silva, 2016).

Rectangular coordinates are useful in several calculations listed below (Ghilani and Wolf, 1989; Alves and Silva, 2016):

- Determining lengths and directions of lines and angles;
- Calculating areas of land parcels;
- Performing calculations of curves;
- Locating inaccessible points;
- Plotting maps.

In practice, plane coordinate systems have usually been used as the basis for rectangular coordinates from plane surveys. However, any arbitrary system can be used for some types of calculations. For example, coordinates can be arbitrarily assigned to a polygon station. We can assume as the starting vertices of the polygon, X and Y values of 100 and 100, respectively. Using arbitrary coordinates with positive values prevents the occurrence of negative coordinate values, according to the magnitude defined in the arbitrary coordinates. In a closed polygon, assigning zero value to the farthest points to the south and west can make manual calculations easier (Ghilani and Wolf, 1989; Alves and Silva, 2016).

Given the X and Y coordinates of an initial A point, the X coordinate of the next B point is obtained by adding longitude (ΔX) of the AB alignment to X_A. Similarly, Y coordinate of B is the (ΔY) latitude of the AB alignment added to Y_A:

$$X_B = X_A + \Delta X_{AB} \tag{8.3}$$

$$Y_B = Y_A + \Delta Y_{AB} \tag{8.4}$$

Wheaton et al. (2012) developed an interactive geographic information system application for transforming non-projected total station data into real-world coordinates via three reference coordinates, which can be collected by a handheld global positioning system (GPS). With the application, we can inspect transformation options, while comparing residual error estimates to interactively choose the best transformation. This provides a cost-effective and easy-to-use workflow with facilities to share and view accurate total station survey data in real-world coordinates through a webGIS or virtual globes.

8.11   Inversing Computation

If the longitude and latitude of a line segment AB are known, the length (H), bearing (B) and azimuth (Az) can be obtained by:

$$H_{AB} = \frac{\Delta X_{AB}}{sen Az_{AB}} = \frac{\Delta Y_{AB}}{cos Az_{AB}} = \sqrt{\Delta X_{AB}^2 + \Delta Y_{AB}^2} \tag{8.5}$$

$$B_{AB} = tg^{-1}(\frac{|X_B - X_A|}{|Y_B - Y_A|}) = tg^{-1}(\frac{|\Delta X_{AB}|}{|\Delta Y_{AB}|}) \tag{8.6}$$

where the sign of the partial longitude and latitude determines the bearing quadrant. The quadrant is Northeast (NE), if ΔX and $\Delta Y > 0$, Southeast (SE), if $\Delta Y < 0$ and $\Delta X > 0$, Southwest (SW), if $\Delta X < 0$ and $\Delta Y < 0$, and Northwest (NW), if $\Delta X < 0$ and $\Delta Y > 0$.

$$Az_{AB} = tg^{-1}(\frac{\Delta X_{AB}}{\Delta Y_{AB}}) + C \tag{8.7}$$

where C is $0°$ if ΔX and $\Delta Y > 0$; $180°$ if $\Delta Y < 0$; and $360°$ if $\Delta X < 0$ and $\Delta Y > 0$.

The equations for longitude and latitude can be expressed in terms of rectangular coordinate differences:

$$\Delta X_{AB} = X_B - X_A \tag{8.8}$$

$$\Delta Y_{AB} = Y_B - Y_A \tag{8.9}$$

8.12   Area and Perimeter Assessment

The equation for obtaining the distance between two vertices that defined the boundary of the property can be used to determine the perimeter of a closed-path traverse, from the northernmost vertex, to the right, to cover the entire perimeter, arriving back at the first vertex. The sum of all the sides that defined the divisions of the property is equivalent to the perimeter (P):

$$P = \sum_{n=1}^{i} \sqrt{\Delta X_i^2 + \Delta Y_i^2} \tag{8.10}$$

The area calculation can be performed by the Cartesian coordinates of vertices matrix determinant method, in a practical way, by arranging the coordinates of the points in two columns X and Y, repeating the coordinates of the first point at the last point in the database. If the sequence of points is arranged clockwise, the products indicated by upward arrows (solid line) are given the sign (+), and those indicated by downward arrows (dashed line) are given the sign (-). If the sequence of points is organized counterclockwise, the products indicated by ascending arrows (solid line) will receive the sign (-), and those indicated by descending arrows (dashed line) will receive the sign (+). With the algebraic sum of the ascending and descending products divided by two, the area of the polygon 1 m^2 is obtained (Figure 8.7) (Alves and Silva, 2016).

8.13   Computation

As a computing practice, we proposed to use radial survey data of a polygon made in a soccer field by the radiation method and propose solutions for mapping the points, determining Cartesian coordinates of the vertices, perimeter and area of the polygon surveyed. Next, it is demonstrated how to store the data in a table and export in a file for later use.

The `circular` package (Lund et al., 2017a) is used to convert angular measurements from degrees to radians associated with trigonometry operations to determine the Cartesian coordinates of the

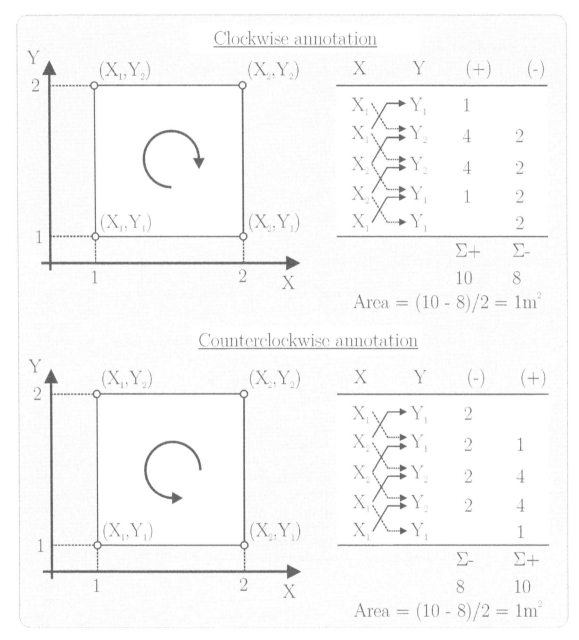

FIGURE 8.7: Specificities of area calculation by the determinant method of the Cartesian coordinates matrix of vertices with data registration to perform the calculation considering clockwise or counterclockwise notation.

vertices at the soccer field boundaries. The `LearnGeom` package (Jammalamadaka and SenGupta, 2001; Briz-Redon and Serrano-Aroca, 2020) is used for mapping the vertices in the surveyed area. The area calculation is performed by matrix determinants of order two. The perimeter is determined as a function of the calculation of alignments between vertices by the Pythagorean theorem using `LearnGeom` package functions.

8.13.1   Installing R packages

The `install.packages` function is used to install the `circular` and `LearnGeom` packages in the R console.

```
## install.packages("circular")
## install.packages("LearnGeom")
```

8.13.2   Enabling R packages

The `library` function is used to enable the `circular` and `LearnGeom` packages in the R console.

```
library(circular)
library(LearnGeom)
```

8.13.3   Import field notes from survey

Field notes from theodolite radiation survey on a soccer field from the center of the field to the edges are imported with the headings ID, Az, AH, F_I, F_S and H, where ID is the vertex identification number; Az, the zenith angle in decimal degrees; AH, the horizontal angle in decimal degrees; F_I, the lower stadia line (mm); F_S, the upper stadia line (mm); and H, the horizontal distance (m).

```
irr<-read.table("files/irr.txt", header = TRUE, sep = " ", dec = ".")
```

8.13.4   Determine the partial X, Y coordinates of the polygon vertices

The partial coordinates x and y of the vertices $A, B, C, D, E, F, G, H, I, J, L, M$ are determined by the latitude and longitude calculation equation based on angular and distance measurements.

```
x<- irr$H*(sin(rad(irr$AH)))
y<- irr$H*(cos(rad(irr$AH)))
```

8.13.5   Determine the rectangular X, Y coordinates of the polygon vertices

The rectangular X, Y coordinates of vertices $A, B, C, D, E, F, G, H, I, J, L, M$ are determined by adding an arbitrary value 100 to the base instrument installation point (O), $X_O = 599870.71$

m; $Y_O = 8273913.86$ m. This is done so that the final coordinates of each vertex are established in the plane UTM coordinate system, UTM zone 21S, transformed from WGS-84 ellipsoid.

```
X<- 599870.71+x
Y<- 8273913.86+y
```

8.13.6   Merge calculated coordinate results in the irradiation table

The partial and rectangular coordinate results are merged as new columns in the irradiation table.

```
irr<-cbind(irr, x, y, X, Y)
```

8.13.7   Map the soccer field based on the rectangular X, Y coordinates

The coordinate axis with minimum and maximum values of X, Y is defined based on the minimum and maximum values of the measurements. The CoordinatePlane function is used to draw the axis on which the polygon is mapped. The Draw function is used with the CreatePolygon to map the vertices of the area and the polygon is colored gray (Figure 8.8).

```
# Assign coordinates to vertices A through M
A <- c(irr$X[1],irr$Y[1])
B <- c(irr$X[2],irr$Y[2])
C <- c(irr$X[3],irr$Y[3])
D <- c(irr$X[4],irr$Y[4])
E <- c(irr$X[5],irr$Y[5])
F <- c(irr$X[6],irr$Y[6])
G <- c(irr$X[7],irr$Y[7])
H <- c(irr$X[8],irr$Y[8])
I <- c(irr$X[9],irr$Y[9])
J <- c(irr$X[10],irr$Y[10])
L <- c(irr$X[11],irr$Y[11])
M <- c(irr$X[12],irr$Y[12])


# Set X,Y axis dimensions
x_min <- 599816.0
x_max <- 599908.0
y_min <- 8273830.0
y_max <- 8273990.9
# Draw the X,Y axis
CoordinatePlane(x_min, x_max, y_min, y_max)
# Draw a polygon from the vertices
Draw(CreatePolygon(A, B, C, D, E, F, G, H, I, J, L, M), "gray", label=TRUE)
```

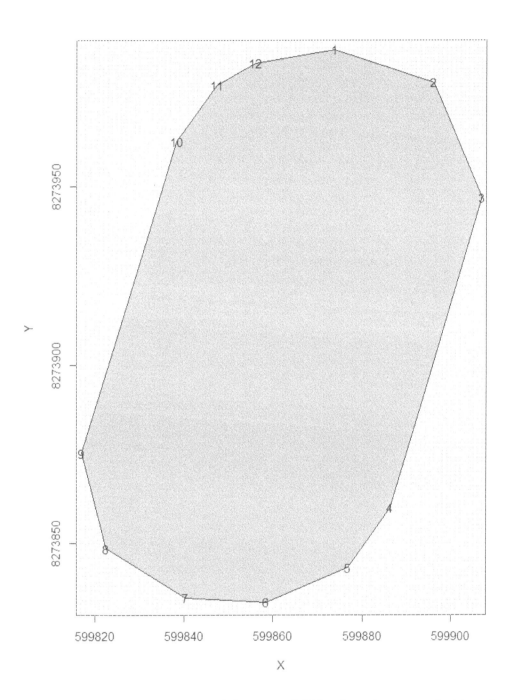

FIGURE 8.8: Mapping vertices of the soccer field by the radiation method in order to define a closed polygon.

8.13.8 Determine the perimeter of the soccer field as a function of the rectangular coordinates X, Y

The perimeter is determined using the Pythagorean theorem, and then the result was checked by the DistancePoints function. The perimeter of the soccer field calculated by both methods is 392.2032 m.

```
# Calculate the perimeter
AB<-sqrt(((irr$X[2]-irr$X[1])^2)+((irr$Y[2]-irr$Y[1])^2))
BC<-sqrt(((irr$X[3]-irr$X[2])^2)+((irr$Y[3]-irr$Y[2])^2))
CD<-sqrt(((irr$X[4]-irr$X[3])^2)+((irr$Y[4]-irr$Y[3])^2))
DE<-sqrt(((irr$X[5]-irr$X[4])^2)+((irr$Y[5]-irr$Y[4])^2))
EF<-sqrt(((irr$X[6]-irr$X[5])^2)+((irr$Y[6]-irr$Y[5])^2))
FG<-sqrt(((irr$X[7]-irr$X[6])^2)+((irr$Y[7]-irr$Y[6])^2))
GH<-sqrt(((irr$X[8]-irr$X[7])^2)+((irr$Y[8]-irr$Y[7])^2))
HI<-sqrt(((irr$X[9]-irr$X[8])^2)+((irr$Y[9]-irr$Y[8])^2))
IJ<-sqrt(((irr$X[10]-irr$X[9])^2)+((irr$Y[10]-irr$Y[9])^2))
JL<-sqrt(((irr$X[11]-irr$X[10])^2)+((irr$Y[11]-irr$Y[10])^2))
LM<-sqrt(((irr$X[12]-irr$X[11])^2)+((irr$Y[12]-irr$Y[11])^2))
MA<-sqrt(((irr$X[12]-irr$X[1])^2)+((irr$Y[12]-irr$Y[1])^2))
# Determine the perimeter
P<-sum(AB, BC, CD, DE, EF, FG, GH, HI, IJ, JL, LM, MA)
P
```

```
## [1] 392.2032
```

```
# Or
# Determine the length of each line segment
AB <- DistancePoints(A, B)
BC <- DistancePoints(B, C)
CD <- DistancePoints(C, D)
DE <- DistancePoints(D, E)
EF <- DistancePoints(E, F)
FG <- DistancePoints(F, G)
GH <- DistancePoints(G, H)
HI <- DistancePoints(H, I)
IJ <- DistancePoints(I, J)
JL <- DistancePoints(J, L)
LM <- DistancePoints(L, M)
MA <- DistancePoints(M, A)
# Determine the perimeter
P1<-sum(AB, BC, CD, DE, EF, FG, GH, HI, IJ, JL, LM, MA)
P1
```

```
## [1] 392.2032
```

```
# Check
isTRUE(all.equal(P, P1))
```

```
## [1] TRUE
```

8.13.9   Calculate the area of the soccer field by matrix determinants

Afterward, the area of the polygon was determined by the method of matrix determinants of order two. The area obtained as a function of the radiation survey was 9779.434 m^2.

```
# Determine the sum of the ascending products
XOB_YOA <-irr$X[2]*irr$Y[1]
XOC_YOB <-irr$X[3]*irr$Y[2]
XOD_YOC <-irr$X[4]*irr$Y[3]
XOE_YOD <-irr$X[5]*irr$Y[4]
XOF_YOE <-irr$X[6]*irr$Y[5]
XOG_YOF <-irr$X[7]*irr$Y[6]
XOH_YOG <-irr$X[8]*irr$Y[7]
XOI_YOH <-irr$X[9]*irr$Y[8]
XOJ_YOI <-irr$X[10]*irr$Y[9]
XOL_YOJ <-irr$X[11]*irr$Y[10]
XOM_YOL <-irr$X[12]*irr$Y[11]
XOA_YOM <-irr$X[1]*irr$Y[12]
sumAsc<-sum(XOB_YOA, XOC_YOB, XOD_YOC, XOE_YOD, XOF_YOE, XOG_YOF,
            XOH_YOG, XOI_YOH, XOJ_YOI, XOL_YOJ, XOM_YOL, XOA_YOM)
sumAsc
```

```
## [1] 5.955826e+13
```

```
# Determine the sum of descendant products
XOA_YOB <-irr$X[1]*irr$Y[2]
XOB_YOC <-irr$X[2]*irr$Y[3]
XOC_YOD <-irr$X[3]*irr$Y[4]
XOD_YOE <-irr$X[4]*irr$Y[5]
XOE_YOF <-irr$X[5]*irr$Y[6]
XOF_YOG <-irr$X[6]*irr$Y[7]
XOG_YOH <-irr$X[7]*irr$Y[8]
XOH_YOI <-irr$X[8]*irr$Y[9]
XOI_YOJ <-irr$X[9]*irr$Y[10]
XOJ_YOL <-irr$X[10]*irr$Y[11]
XOL_YOM <-irr$X[11]*irr$Y[12]
XOM_YOA <-irr$X[12]*irr$Y[1]
sumDesc<-sum(XOA_YOB, XOB_YOC, XOC_YOD, XOD_YOE, XOE_YOF, XOF_YOG,
             XOG_YOH, XOH_YOI, XOI_YOJ, XOJ_YOL, XOL_YOM, XOM_YOA)
sumDesc
```

```
## [1] 5.955826e+13
```

```
area<-( sumAsc-sumDesc)/2
area
```

```
## [1] 9779.434
```

8.13.10 Export the irradiation table in .txt extension

The results are exported in text file format (.txt) for later use.

```
write.table(irr, file = "E:/Aulas/Topografia/Aula7/irr.txt",
            sep = " ", row.names = TRUE, col.names = TRUE)
```

8.14 Solved Exercises

8.14.1 Name the main types of traverses.

A: Open-path, closed-path, link-path and polygon traverses.

8.14.2 Cite one advantage and disadvantage of the radiation survey method.

A: Advantage: It is a quick method considering the irradiation of all vertices of a single base. Disadvantage: It does not make it possible to circumvent obstacles in surveying areas where all points are not intervisible between the base and the forward reading.

8.14.3 Determine the bearing and azimuth between vertices OA and OB of the computation practice.

A: The bearing between vertices OA and OB is 67.5° SE.

```
# Obtain X
XA<-irr$X[1]
XB<-irr$X[2]
# Obtain Y
YA<-irr$Y[1]
YB<-irr$Y[2]
# Determine the longitude and latitude
dX<-XB-XA
dX
```

```
## [1] 22.13009
```

```
dY<-YB-YA
dY
```

```
## [1] -9.164768
```

```
# Determine the bearing
r<-deg(atan(dX/dY))
r
```

```
## [1] -67.50401
```

8.14.4 Determine the azimuth between vertices OA and OB of the computation practice.

A: The azimuth between vertices OA and OB is 112.496°.

```
az<-deg(atan(dX/dY))+180
az
```

```
## [1] 112.496
```

8.14.5 Determine the horizontal distance between vertices OA and OB of the computation practice.

A: The horizontal distance between vertices OA and OB is 23.9527 m.

```
H<-sqrt(dX^2+dY^2)
H
```

```
## [1] 23.95274
```

8.15 Homework

Choose one exercise presented by the teacher and solve the question with different input values. Compare the results obtained. Perform the radiation method in the field. Propose solutions for mapping the points using Cartesian coordinates of the vertices, perimeter and area of the polygon surveyed.

8.16   Resources on the Internet

As a study guide, slides and illustrative videos are presented about the subject covered in the chapter in Table 8.1.

TABLE 8.1: Slide shows and video presentations on radial traverse survey.

Guide	Address for Access
1	Slides on traversing and coordinate geometry in radial surveying[1]
2	Radial surveying[2]
3	Radial surveying using compass[3]
4	Offset and radial surveys[4]
5	Radial surveys[5]

8.17   Research Suggestion

The development of scientific research on geomatics is stimulated by the activity proposals that can be used or adapted by the student to assess the applicability of the subject matter covered in the chapter (Table 8.2).

TABLE 8.2: Practical and research activities used or adapted by students using radial traverse survey.

Activity	Description
1	In the content on radial traversing and coordinate geometry, interest may arise in doing the work based on computational examples presented
2	Perform field measurements with the radial method. Determine the horizontal distance to each measured point
3	Perform perimeter and area evaluation of a closed polygon from field measurements with the radial method

8.18   Learning Outcome Assessment Strategy

Perform a summary of the chapter, "Radial Traverse Survey with Geomatics and R", on a single A4 page in order to show the student's abilities to summarize a subject presenting key points considered of greater importance today.

[1] http://www.sergeo.deg.ufla.br/geomatica/book/c8/presentation.html#/
[2] https://youtu.be/s8nihBwuVYc
[3] https://youtu.be/rwFyJ3OvXo8
[4] https://youtu.be/5fPIGtac7bs
[5] https://youtu.be/wufVJ5invJY

9

Coordinate Geometry of Closed-Path Traverse Surveying

9.1 Learning Questions

The emergent learning questions answered through reading the chapter are as follows:

- How to perform the calculation of Closed-path traverses with irradiations in R.
- How to calculate bearings, azimuths, latitudes and longitudes in a closed-path traverse survey with irradiations.
- What are the sources and errors in the calculation of closed-path traverses with irradiations?
- How to determine the errors of linear closure, relative accuracy of closed-path traverse.
- How to perform error adjustment on a closed-path traverse with irradiations by the Bowditch method.
- How can R software and packages `circular`, `LearnGeom`, `rgdal` and `plotKML` be used in the analysis and mapping of closed-path traverse survey?

9.2 Learning Outcomes

The learning outcomes expected from reading the chapter are as follows:

- Perform the calculation of closed-path traverse with irradiations in R.
- Calculate bearings, azimuths, latitudes and longitudes in closed-path traverse surveys by walking with radiations.
- Know the sources and errors in the calculation of closed-path traverses with irradiations.
- Calculate the errors of linear closure, relative accuracy of closed-path polygonal survey by walking with irradiations.
- Perform error adjustment on a closed-path traverse by the Bowditch method.
- Use R software and packages `circular`, `LearnGeom`, `rgdal` and `plotKML` in the analysis and mapping of closed-path traverse survey data with irradiations.

9.3 Introduction

Depending on the type of project and the size of the area, a polygon with georeferenced vertices can be characterized by local plane topographic survey or with the plane of cartographic projection, such as Universal Transverse Mercator (UTM). Generally, a set of vertices in topographic surveys are performed for specific purposes, such as (Silva and Segantine, 2015):

- Topographic mapping;
- Cadastral mapping;
- Control points for aerophotogrammetry;
- Support network for surveying and implementation of works.

The closed-path polygonal traverse has been used in civil engineering works and for rural cadastral and surveying mapping. When surveying a polygon, we must observe through field reconnaissance whether the vertices to be measured on the terrain are intervisible two-by-two. The polygon should have as few vertices as possible to minimize the effect of the centering error of the instrument at station changes. The installation site of the measuring instrument should be firm, flat, and free of vibrations. Wooden pickets can be used instead of concrete blocks in surveying work where the supporting vertices have a short temporal function (Silva and Segantine, 2015).

9.4   Closed-Path Traverse Surveying Calculation

In closed polygons, the vertices are materialized in the field, returning to the initial point at the end of the survey, forming a figure with closed geometry. This enables to check the angles and distances observed in relation to the accuracy required in different types of topographic surveys (Figure 9.1)(Alves and Silva, 2016).

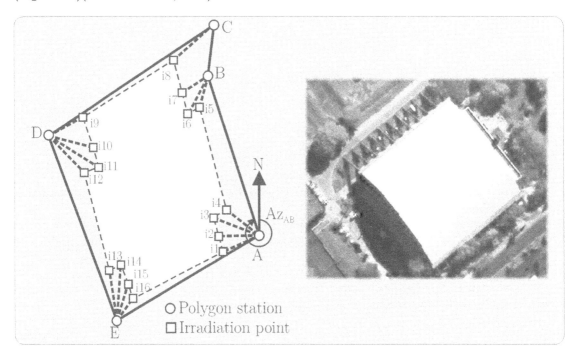

FIGURE 9.1: Closed-path polygonal traverse used for as-built survey of a building and the location of the same area in a color composition 853 from WorldView-2 satellite image (bottom).

The angles and directions obtained when measuring closed-path polygonal traverses can be evaluated after the fieldwork has been completed. After determining the errors of the linear and angular measurements, the polygon can be adjusted to determine a geometrically consistent closure between angles and lengths. Depending on the magnitude of the error, field observations must be repeated until adequate results are obtained. Determining the accuracy of the traverse adjustment has been extremely important in assessing whether the property mapping is within the specifications determined by law (Ghilani and Wolf, 1989).

Different procedures can be used to calculate and adjust polygon traverses. These procedures range from elementary methods to more advanced techniques based on the least squares method. Only elementary procedures will be considered in this chapter, with coordinate adjustment based on the Bowditch method.

The basic steps for calculating closed-path polygonal traverses are (Alves and Silva, 2016):

- Adjusting angular errors to fix geometric conditions;
- Determination of azimuths (or directions) of alignments;
- Calculation of longitude and latitude and adjustment of linear closure error;
- Calculation of rectangular coordinates of polygon stations;
- Calculation of lengths and azimuths of polygon lines after adjustment;
- Calculation of rectangular coordinates of irradiations;
- Mapping the surveyed area with polygon and irradiations (drawing);
- Determination of the area of interest on the map;
- Performing the inverse calculation and descriptive memorial of the area of interest;
- Mapping the area referring to the descriptive memorial (drawing).

9.5 Balancing Angles

In traditional adjustment methods, the first step is to adjust the angles according to geometric properties. The balancing of angles is performed, starting from the knowledge of the total error of the sum of angles of the polygon. Angular correction can be performed by applying an average correction for each angle measured at the stations by dividing the total angular closure error by the number of angles (Ghilani and Wolf, 1989).

The angular error can be determined as a function of the expected or theoretical value of the sum of angles that the polygon should have. The expected sum of the internal angles $\sum ai_T$ and external angles $\sum ae_T$ of a closed-path polygon traverse, as well as the sum of the azimuths of an open-path traverse $\sum Az_T$ and the algebraic sum of the deflections $\sum d_T$ should present the angular geometric value based on the following summations:

$$\sum ai_T = (n-2)180° \tag{9.1}$$

$$\sum ae_T = (n+2)180° \tag{9.2}$$

$$\sum Az_T = Az_F - Az_I \tag{9.3}$$

$$\sum d_T = \pm360° \tag{9.4}$$

where in the case of deflection angles, right deflections are given the sign (+) and left deflections are given the sign (-). The sum of the right deflections minus the sum of the left deflections must equal 360°.

The sum of the observed angles minus the theoretical angles will determine the angular closure error (Efa) in the survey methods used.

The angular closure error to be distributed to each survey station vertex by internal angles ($EFai$), external angles ($EFae$), azimuths ($EFAz$) and deflection angles (EFd) is calculated by:

$$EFai = \frac{(\sum ai_O - \sum ai_T)}{n} \tag{9.5}$$

$$EFae = \frac{(\sum ae_O - \sum ae_T)}{n} \tag{9.6}$$

$$EFAz = \frac{(\sum Az_O - \sum Az_T)}{n} \tag{9.7}$$

$$EFd = \frac{(\sum ad_O - \sum d_T)}{n} \tag{9.8}$$

where $\sum ai_O$ is the sum of the observed internal angles, $\sum ae_O$, the sum of the observed external angles, $\sum Az_O$, the sum of the observed open polygon angles, $\sum ad_O$, the sum of the observed deflection angles n, the number of polygon vertices.

The error must be distributed in equal portions for the total number of measured angles, considering that the correction will always have a sign opposite to the sign of the error. For example, if the sum of the measured angles is greater than the expected value, the error will be negative and, therefore, the correction for each vertex must be added (Souza, 2003).

It should be noted that, although the adjusted angles satisfy the geometric condition of a closed geometry, the resulting values may not be close to the observed angles, because adjustments applied to angles are independent of the angle size (Ghilani and Wolf, 1989).

The angular closure error tolerance (T) for theodolite surveys can be obtained by (Comastri and Junior, 1998):

$$T = 5'\sqrt{n} \tag{9.9}$$

In surveys using equipment with higher measurement accuracy, an evaluation of the survey quality is proposed as a function of the relative precision obtained based on the linear error of the survey, demonstrated later (Table 9.1).

9.6   Azimuth Calculation

After balancing angles, the next step is to calculate preliminary azimuths or bearings. At this stage, at least one direction value is required from a polygon vertex to assign the north direction. In some surveys, the magnetic bearing of alignment can be determined and used as reference direction; however, for boundary surveys, true directions are required.

The true north direction can be obtained by (Ghilani and Wolf, 1989):

- Incorporating into the polygon a line with the true direction;
- Performing astronomical observations to determine the magnetic declination;
- Using satellite positioning system measurements.

Considering an initial azimuth obtained from a geodetic reference network or determined in the field, the azimuths of the other vertices can be calculated as a function of the known internal angles (Souza, 2003):

$$Az_{i+1} = Az_i + 180° \pm A_i \tag{9.10}$$

where Az_{i+1} is the azimuth to be calculated (forward), Az_i, the known azimuth at the previous vertex (backward), and A_i, the internal angle. If the result of the equation is negative, we must add 360°, and if it is greater than 360°, we must subtract 360° as many times as necessary. The (+) sign must be used when the angle between the alignments is measured clockwise from the aft alignment, otherwise the (-) sign must be used.

In the case of a polygonal survey with detail points included, the azimuth of the irradiated point must be calculated, so that the mapping is presented in the same direction. The azimuth of the irradiation can be calculated by:

$$Az_{ir} = Az_i + 180° \pm Air \tag{9.11}$$

where Az_{ir} is the azimuth of irradiation measured at the forward vertex of the polygon to be calculated, Az_i, the known azimuth at the anterior vertex (backward) of the main polygon and, Air, the irradiated angle. Other procedures are similar to those described for polygon azimuth calculation.

9.7 Latitude and Longitude Calculation

After angles balancing and calculating preliminary azimuths, the closure of the polygon can be evaluated by calculating the longitude and latitude of each line. The longitude of an alignment referred to the orthographic projection in the direction of the east-west axis of the survey and is equal to the length of the alignment multiplied by the sine of the azimuth or bearing angle. The latitude of a line consisted of the orthographic projection in the direction of the south-north axis of the survey and is equal to the length of the line multiplied by the cosine (Ghilani and Wolf, 1989; Alves and Silva, 2016):

$$\Delta X = H sen\alpha \tag{9.12}$$

$$\Delta Y = H cos\alpha \tag{9.13}$$

9.8 Traverse Linear Misclosure Error and Relative Precision

Considering a closed-path polygon, if all angles and distances are measured perfectly, the algebraic sum of the longitudes of all polygon alignments should be zero. The same condition applies to the algebraic sum of all latitudes. Since the conditions are not perfect, with errors in the measurement of angles and distances, the difference between the observed and expected value determines the longitude closure error and the latitude closure error. These values are determined by the algebraic sum of the longitudes and latitudes, which are compared to a required standard. The magnitude of the longitude and latitude closure error indicates the precision of the observation of angles and distances. High error values indicate the existence of significant errors or even mistakes. In the occurrence of small errors, the observed data are accurate and free of mistakes, but it is not guaranteed that no systematic or compensation errors occur (Ghilani and Wolf, 1989).

The longitude and latitude closure errors determine the linear polygon closure error (E_L) based on the following equation:

$$E_L = \sqrt{\left(\sum X\right)^2 + \left(\sum Y\right)^2} \tag{9.14}$$

where $\sum X$ is the longitude closure error and, $\sum Y$, the latitude closure error.

The closed-path polygon relative precision (P_R) is expressed by the fraction of the linear closure error (E_L) as numerator and the polygon perimeter (P) as denominator (Alves and Silva, 2016):

$$P_R = \frac{E_L}{P} \tag{9.15}$$

The relative precision result must be divided by $1/P_R$ to express the 1 m error for the total area surveyed. For example, in a survey with relative precision of 1:10000, there is 1 m of error for every 10000 m of perimeter surveyed.

Linear closure error tolerance can be checked against relative precision (Table 9.1) (Silva and Segantine, 2015).

TABLE 9.1: Linear closure error tolerance values based on relative precision.

Quality	Precision	Application	Observation with Total Station
High	≥ 1:50000	High-precision engineering surveying	1 mm + 1 ppm accuracy
Good	1:10000 to 1:50000	General high precision engineering surveying	2 mm + 2 ppm accuracy
Regular	1:5000 to 1:10000	Rural surveying	3 mm + 3 ppm precision
Low	≤ 1:5000	Rural surveying	5 mm + 5 ppm precision

NBR 13133/1994 can be consulted for more details in specific situations (ABNT, 1994).

9.9   Traverse Adjustment

The linear closure error must be adjusted or distributed to balance the polygon for any closed polygon. The adjustment must be performed even if the closure error is negligible, when representing the polygon at map scale. There are several elementary methods available for adjusting the polygon, but the most common is the compass rule (Bowditch method). Least-squares fitting is an advanced technique that can also be used in traverse adjustment and is discussed with more details in Ghilani (2017).

9.10 Compass Rule

In the compass rule, the longitude and latitude values are adjusted in relation to their length. Although this method is not as rigorous as the least squares method, there is a logical distribution of the closure errors proportional to the sides of the polygon (Souza, 2003). The longitude $C(X_i)$ and latitude $C(Y_i)$ corrections are performed according to the following rules (Alves and Silva, 2016):

$$C(X_i) = -\frac{\sum x}{P} H_i \tag{9.16}$$

$$C(Y_i) = -\frac{\sum y}{P} H_i \tag{9.17}$$

where H_i is the horizontal distance of the alignment to be corrected.

The correction sign is opposite the closure error (Ghilani and Wolf, 1989). The corrections should be summed algebraically at each projection. A check can be made so that the column sum of the corrected longitude and latitude values must be zero; there may be small differences resulting from approximations that must be eliminated by revising one of the corrections (Ghilani and Wolf, 1989; Souza, 2003).

9.11 Rectangular Coordinate Calculation

Plane coordinate systems or arbitrary values can normally be used as a basis for determining rectangular coordinates from the longitude and latitude values calculated from the surveying data (Alves and Silva, 2016):

$$X_B = X_A + \Delta X_{AB} \tag{9.18}$$

$$Y_B = Y_A + \Delta Y_{AB} \tag{9.19}$$

We can assume as initial vertex of the polygon, X and Y values of 1000 and 1000, respectively, depending on the magnitude of the negative longitude and latitude values, in order to avoid the occurrence of negative values in the final plane coordinates.

9.12 Computing Final Adjusted Traverse Lengths and Directions

In the polygonal adjustment, corrections are applied to longitudes and latitudes to obtain adjusted values. These values are used to calculate X, Y coordinates of polygon stations. As longitude and latitude lines values are modified in the adjustment process, length and azimuth values are also modified. Thus, it is necessary to calculate the final or adjusted lengths and directions. The equations presented above can be used to obtain the final length and direction of adjusted polygon lines based on adjusted longitude and latitude coordinates (Alves and Silva, 2016).

$$H_{AB} = \sqrt{\Delta X_{AB}^2 + \Delta Y_{AB}^2} \tag{9.20}$$

$$Az_{AB} = tg^{-1}(\frac{\Delta X_{AB}}{\Delta Y_{AB}}) + C \tag{9.21}$$

where C is $0°$ if ΔX and $\Delta Y > 0$; $180°$ if $\Delta Y < 0$; and $360°$ if $\Delta X < 0$ and $\Delta Y > 0$.

9.13   Traverse Perimeter and Area Calculation

The traverse perimeter of a closed polygon can be calculated by the sum of distances determined by inverse calculation between each vertex (H_i) of the surveyed area, starting from the northern most point of the polygon, going to the right, until it runs around the entire perimeter, arriving back at the first vertex (Alves and Silva, 2016).

$$P = \sum_{n=1}^{i} H_i \tag{9.22}$$

The area calculation can be performed by the method of determinants of the matrix of Cartesian coordinates of vertices of the closed polygon in the surveyed region. The area of the polygon can be obtained as the algebraic sum of the ascending and descending products divided by two, of the x, y coordinates that defined the area of the polygon. Other methods can be used for area calculation and will be described later.

9.14   Error Sources in Closed Traverse Surveying

Some sources of error in traversing can be caused by (Ghilani and Wolf, 1989; Alves and Silva, 2016):

- Improper selection of station points, resulting in improper sighting conditions due to alternating sun and shade, visibility of only one end of the line, line of sight close to the ground, lines that are too short and, sighted in the direction of the sun;
- Errors in observing angles and distances;
- Failure to observe angles in equal numbers of times in the direct and reverse directions.

9.15   Mistakes in Traverse Computations

The most common mistakes in traverse surveying are (Ghilani and Wolf, 1989; Alves and Silva, 2016):

- Occupying the wrong station and performing sighting on the wrong station;
- Incorrect orientation;
- Confusing left and right angles;
- Forgetting to register the point;
- Not identifying the station targeted.

The most common mistakes in traverse calculations are (Ghilani and Wolf, 1989; Alves and Silva, 2016):

- Failing to adjust angles before calculating azimuths or bearings;
- Applying angle adjustment in the wrong direction and not getting to the sum of all angles;
- Swapping longitude and latitude values;
- Confusing coordinate signs.

9.16 Computation

As a computation practice, we proposed to use data from anticlockwise internal angle walk survey of a closed polygon traversing around a building on a university campus, performed with mechanical theodolite. The corners of the building are surveyed as detail points by irradiation. Solutions for mapping the points are proposed, determining linear closure error, relative precision, partial and final coordinates, area calculation and the surveyed perimeter. Next, we demonstrate how to create and store attribute data in spatial points shapefile and (KML) formats.

The `circular` package (Lund et al., 2017a) is used to convert angular measurements from degrees to radians associated with trigonometry operations to determine coordinates of polygon vertices and radiations. The `LearnGeom` package (Jammalamadaka and SenGupta, 2001; Briz-Redon and Serrano-Aroca, 2020) is used to map vertices in the surveyed area. Area calculation is performed by matrix determinants of order two and the perimeter is determined as a function of the summation of distances of alignments between building vertices. The `rgdal` R package enables access to the Geospatial Data Abstraction Library (GDAL), projection and transformation operations of the library `PROJ` (Bivand et al., 2021). This enabled to transform coordinate vertices in spatial points with attributes and map them. In the `plotKML` R package, objects from `sp`, `space-time` and `raster` classes can be converted to KML with basic cartographic rules (Hengl et al., 2020). Vertices are mapped in Google Earth using WGS-84 datum.

9.16.1 Installing R packages

The `install.packages` function is used to install `circular` and `LearnGeom`, `rgdal` and `plotKML` packages in the R console.

```
## install.packages("circular")
## install.packages("LearnGeom")
## install.packages("rgdal")
## install.packages("plotKML")
```

9.16.2   Enabling R packages

The `library` function is used to enable the `circular`, `LearnGeom`, `rgdal` and `plotKML` packages in the R console.

```
library(circular)
library(LearnGeom)
library(rgdal)
library(plotKML)
```

9.16.3   Adjusting a closed-path polygon traverse and radiations by the compass rule (Bowditch)

Polygon and irradiation adjustment are performed by the compass rule (Bowditch method). Data obtained from a walking survey with theodolite is recorded in a spreadsheet with the following coding. Occupied station (E); Point targeted (PV); Horizontal angle in degrees (AH); Horizontal distance (H) in m; Magnetic azimuth (AB $Az_{AB} = 237°43'10"$).

```
# Import field notes
cam<-data.frame(E=c('A', 'A', 'B', 'B', 'C', 'C', 'D', 'D', 'D', 'D'),
PV=c('RDVB', 'i1', 'RAVC', 'i2', 'RBVD', 'i3', 'RCVA', 'i4', 'i5',
     'i6'),
AH=c((90+31/60+10/3600), (17+25/60+30/3600), (88+58/60+20/3600),
(78+30/60+10/3600), (90+10/60+30/3600), (33+20/60+50/3600),
(90+20/60+20/3600), (10+52/60+10/3600), (60+13/60+40/3600),
(85+18/60+50/3600)), H=c(16.97, 3.99, 25.41, 4.31, 16.99, 1.60,
                         25.89, 7.27, 14.69, 12.71))
```

The data is digitized and exported in `.txt` for later use.

```
## write.table(cam, file = "E:/Aulas/Topografia/Aula8/cam.txt",
##             sep = " ", row.names = TRUE, col.names = TRUE)
```

A subset of rows and columns of the dataset is made only with measurements performed in the traverse, in order to separate closed-path polygon data and irradiations data. The calculations are performed and organized in a sequential and logical manner to facilitate understanding of adjustment steps for initial polygon mapping and, subsequently, irradiations.

```
pol<-cam[c(1, 3, 5, 7), c(1:4)]
```

The angular error of closure is determined based on the equation of internal angles, since the survey is performed counterclockwise when traversing the area along the station vertices. If of interest, the quality of the angular error can be evaluated based on a tolerance value. For theodolite surveys, an angular error tolerance of $5'\sqrt{n}$ (Comastri and Junior, 1998).

```
# Determining angular closure error
# Sum the interior angles of the polygon
sumAH<-sum(pol$AH)
sumAH
```

```
## [1] 360.0056
```

```
# Determine the theoretical sum of the polygon angles
sumAHt<-180*(nrow(pol)-2)
sumAHt
```

```
## [1] 360
```

```
# Determine the angular closure error
EFA<-sumAH-sumAHt
EFA
```

```
## [1] 0.005555556
```

Based on the magnitude of the angular error of closure, error correction is performed in equal proportions for all vertices and the corrected angles are added as a new column to the polygon dataset.

```
# Determine the angular correction per vertex
EFAv<-EFA/nrow(pol)
EFAv
```

```
## [1] 0.001388889
```

```
# Perform correction of angular closure error per polygon vertex
# As sumAH was greater than sumAHt, correct by subtracting
AHc<-pol$AH-EFAv
AHc
```

```
## [1] 90.51806 88.97083 90.17361 90.33750
```

```
# Join the corrected angle results in the polygonal table
pol<-cbind(pol, AHc)
```

Based on the corrected horizontal angle values, azimuth calculation is performed between the polygon vertices. The results of azimuth calculation are checked by calculating the azimuth that is read with the theodolite in order to evaluate if the same value of the observed reading and the azimuth compensation calculation is obtained. The calculated azimuths are organized in a new column of the polygon dataset.

```
# Perform the azimuth compensation
AZAB<-237+43/60+10/3600
AZAB
```

```
## [1] 237.7194
```

```
AZBC<-AZAB+180+pol$AHc[2]-360
AZBC
```

```
## [1] 146.6903
```

```
AZCD<-AZBC+180+pol$AHc[3]-360
AZCD
```

```
## [1] 56.86389
```

```
AZDA<-AZCD+180+pol$AHc[4]
AZDA
```

```
## [1] 327.2014
```

```
# Check the results by calculating AZAB
AZAB<-AZDA+180+pol$AHc[1]-360
AZAB
```

```
## [1] 237.7194
```

```
AZ<-c(AZAB, AZBC, AZCD, AZDA)
AZ
```

```
## [1] 237.71944 146.69028  56.86389 327.20139
```

```
# Join the azimuth results calculated in the polygon table
pol<-cbind(pol, AZ)
```

Based on the azimuth and horizontal distance values between each alignment, the longitude and partial latitude values of each vertex of the polygon are determined. The results of the partial projections are added in a new column of the polygonal dataset.

```
# Determine the partial projections x and y
xA<-(sin(rad(pol$AZ[1])))*pol$H[1]
xA
```

```
## [1] -14.34717
```

```
xB<-(sin(rad(pol$AZ[2])))*pol$H[2]
xB
```

```
## [1] 13.95427
```

```
xC<-(sin(rad(pol$AZ[3])))*pol$H[3]
xC
```

```
## [1] 14.22699
```

```
xD<-(sin(rad(pol$AZ[4])))*pol$H[4]
xD
```

```
## [1] -14.0243
```

```
x<-c(xA, xB, xC, xD)
x
```

```
## [1] -14.34717  13.95427  14.22699 -14.02430
```

```
yA<-(cos(rad(pol$AZ[1])))*pol$H[1]
yA
```

```
## [1] -9.063091
```

```
yB<-(cos(rad(pol$AZ[2])))*pol$H[2]
yB
```

```
## [1] -21.2355
```

```
yC<-(cos(rad(pol$AZ[3])))*pol$H[3]
yC
```

```
## [1] 9.287241
```

```
yD<-(cos(rad(pol$AZ[4])))*pol$H[4]
yD
```

```
## [1] 21.76261
```

```
y<-c(yA, yB, yC, yD)
y
```

```
## [1]  -9.063091 -21.235498   9.287241  21.762609
```

```
# Merge the results of partial projections into the polygon table
pol<-cbind(pol, x, y)
```

The linear closure error and the relative accuracy of the polygon are determined by the partial projection. Quality classification with low accuracy of the survey is observed (Silva and Segantine, 2015); however, it should be noted that this survey is conducted in practice with mechanical theodolite and stadia rod reading and not with total station equipment and with students in the training phase, resulting in a greater magnitude of error in the survey.

```
# Determine the linear closure error
EL<-sqrt((sum(pol$x)^2)+(sum(pol$y)^2))
EL
```

```
## [1] 0.7749657
```

```
# Determine the relative precision
PR<-1/(EL/sum(pol$H))
PR
```

```
## [1] 110.0178
```

The linear error correction is determined for each vertex of the polygon in relation to the distance of each measured alignment. The linear error correction values of the X and Y projections are merged into the polygonal dataset.

```
# Determine the linear error correction at x and y
Cx<--(sum(pol$x)/sum(pol$H))*pol$H
Cx
```

```
## [1] 0.03785789 0.05668644 0.03790250 0.05775726
```

```
Cy<--(sum(pol$y)/sum(pol$H))*pol$H
Cy
```

```
## [1] -0.1495298 -0.2238982 -0.1497060 -0.2281277
```

```
# Join the linear error correction values in the polygon table
pol<-cbind(pol, Cx, Cy)
```

Then, linear error correction is performed in order to determine corrected longitude and latitude values to obtain accurate polygon closure. The corrected longitude and latitude results are joined to the polygonal dataset.

```
# Perform the linear error correction of x and y
xc<- x+Cx
xc
```

```
## [1] -14.30931  14.01096  14.26489 -13.96654
```

```
yc<- y+Cy
yc
```

```
## [1]  -9.212621 -21.459396   9.137535  21.534482
```

```
# Join the corrected longitude and latitude results in the polygon table
pol<-cbind(pol, xc, yc)
```

Absolute or final coordinates of the X_A, Y_A vertices are obtained from imaging of the area available on Google Earth with the values of 600104.9400 and 8273805.3100 m, respectively. In this case, the coordinates are obtained in UTM projection, UTM zone 21S, transformed from WGS-84 datum. From the coordinates of vertex A, the coordinates of B, C and D vertices are determined. The checking of the arbitrated values for vertex A is accomplished, performing the same calculation procedure of the other vertices. The results of X, Y UTM coordinates are joined to the polygonal dataset.

In situations where more survey accuracy is required, plane coordinates can be obtained by surveying with GNSS technology near the location of interest. A view at the point where GNSS is tracked can be used as a reference to determine plane coordinates of the other vertices.

```
# Determine the absolute coordinates
XA<-600104.9400
XB<-XA+pol$xc[1]
XB
```

```
## [1] 600090.6
```

```
XC<-XB+pol$xc[2]
XC
```

```
## [1] 600104.6
```

```
XD<-XC+pol$xc[3]
XD
```

```
## [1] 600118.9
```

```
XA<-XD+pol$xc[4] # check XA vertex
XA
```

```
## [1] 600104.9
```

```
X<-c(XA, XB, XC, XD)
YA<-8273805.3100
YB<-YA+pol$yc[1]
YB
```

```
## [1] 8273796
```

```
YC<-YB+pol$yc[2]
YC
```

```
## [1] 8273775
```

```
YD<-YC+pol$yc[3]
YD
```

```
## [1] 8273784
```

```
YA<-YD+pol$yc[4] # check YA vertex
YA
```

```
## [1] 8273805
```

```
Y<-c(YA, YB, YC, YD)
# join X and Y coordinates in the polygon table
pol<-cbind(pol, X, Y)
```

The polygon dataset with all calculations is exported in a `.txt` file.

```
## write.table(pol, file = "E:/Aulas/Topografia/Aula8/pol.txt",
##             sep = " ", row.names = TRUE, col.names = TRUE)
```

The `format` function is used to obtain X, Y coordinate values to 4 decimal places. The coordinate values with 4 decimal places can be useful for the construction of the specifications or other future demands.

```
# Display X and Y coordinates with 4 decimal places
Xa<-format(XA, digits=4, nsmall=4)
Xa
```

```
## [1] "600104.9400"
```

```
Xb<-format(XB, digits=4, nsmall=4)
Xb
```

```
## [1] "600090.6307"
```

```
Xc<-format(XC, digits=4, nsmall=4)
Xc
```

```
## [1] "600104.6416"
```

```
Xd<-format(XD, digits=4, nsmall=4)
Xd
```

```
## [1] "600118.9065"
```

```
Ya<-format(YA, digits=4, nsmall=4)
Ya
```

```
## [1] "8273805.3100"
```

```
Yb<-format(YB, digits=4, nsmall=4)
Yb
```

```
## [1] "8273796.0974"
```

```
Yc<-format(YC, digits=4, nsmall=4)
Yc
```

```
## [1] "8273774.6380"
```

```
Yd<-format(YD, digits=4, nsmall=4)
Yd
```

```
## [1] "8273783.7755"
```

The polygon vertices are mapped through LearnGeom package functions. For this, dimensions of the X, Y axis are defined, UTM coordinates are assigned at vertices A to D and the polygon is drawn with a gray color (Figure 9.2).

```
## # Define dimensions of the X,Y axis
## x_min <- 600088
## x_max <- 600120
## y_min <- 8273768
## y_max <- 8273810
## # Draw the x,y axis
## CoordinatePlane(x_min, x_max, y_min, y_max)
## # Assign coordinates to vertices A through D
## A <- c(pol$X[1],pol$Y[1])
## B <- c(pol$X[2],pol$Y[2])
## C <- c(pol$X[3],pol$Y[3])
## D <- c(pol$X[4],pol$Y[4])
## # Draw a polygon from vertices A through D
## Draw(CreatePolygon(A, B, C, D), "gray", label=T)
```

After the polygonal calculation and correction, the irradiations calculation is performed. A subset of rows and columns of the dataset is made with irradiations measurements only. The calculation execution is organized in a sequential and logical way to facilitate the understanding of irradiation adjustment and mapping steps.

```
# Perform a subset with irradiation data only
irr<-cam[c(2, 4, 6, 8, 9, 10), c(1:4)]
```

Considering that the angular correction is not performed for the irradiation data, the azimuth calculation is performed with original angle values of each irradiation. The polygon azimuth and horizontal angle values of irradiations are used to determine the azimuth of irradiations performed from the polygon vertices. The calculated azimuths are organized in a new column of the irradiation dataset.

```
# Perform the azimuth compensation of irradiations
i1<-pol$AZ[4]+180+irr$AH[1]-360
i1
```

```
## [1] 164.6264
```

```
i2<-pol$AZ[1]+180+irr$AH[2]-360
i2
```

```
## [1] 136.2222
```

```
i3<-pol$AZ[2]+180+irr$AH[3]-360
i3
```

```
## [1] 0.0375
```

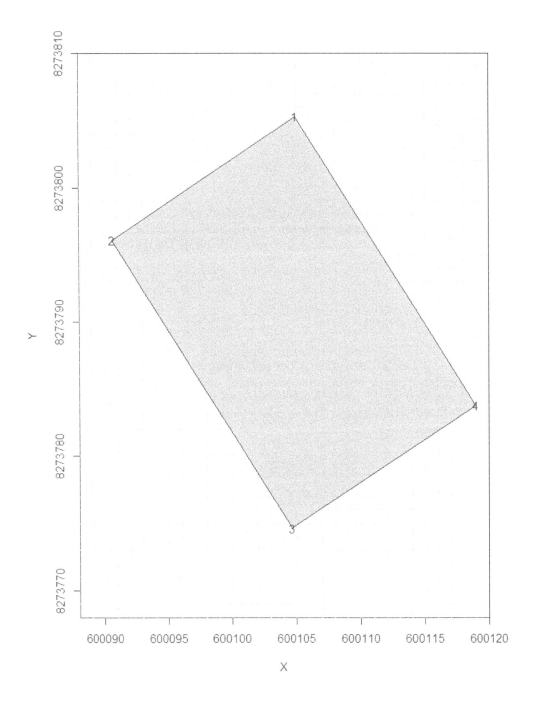

FIGURE 9.2: Vertex mapping of a closed polygon conducted around the animal anatomy building on the Federal University of Mato Grosso (UFMT) campus, with the `LearnGeom` package.

```
i4<-pol$AZ[3]+180+irr$AH[4]
i4
```

```
## [1] 247.7333
```

```
i5<-pol$AZ[3]+180+irr$AH[5]
i5
```

```
## [1] 297.0917
```

```
i6<-pol$AZ[3]+180+irr$AH[6]
i6
```

```
## [1] 322.1778
```

```
AZi<-c(i1, i2, i3, i4, i5, i6)
# Join the results of calculated azimuths of radiations in the polygon table
irr<-cbind(irr, AZi)
```

Based on the azimuth and horizontal distance values between each irradiation alignment, the partial longitude and latitude values of each irradiation point are determined. The results of the partial projections are added in a new column of the irradiation dataset.

```
# Determine x,y partial projections of irradiations
x1<-(sin(rad(irr$AZi[1])))*irr$H[1]
x1
```

```
## [1] 1.057797
```

```
x2<-(sin(rad(irr$AZi[2])))*irr$H[2]
x2
```

```
## [1] 2.98193
```

```
x3<-(sin(rad(irr$AZi[3])))*irr$H[3]
x3
```

```
## [1] 0.001047197
```

```
x4<-(sin(rad(irr$AZi[4])))*irr$H[4]
x4
```

```
## [1] -6.727878
```

```
x5<-(sin(rad(irr$AZi[5])))*irr$H[5]
x5
```

```
## [1] -13.0782
```

```
x6<-(sin(rad(irr$AZi[6])))*irr$H[6]
x6
```

```
## [1] -7.793943
```

```
xi<-c(x1, x2, x3, x4, x5, x6)
y1<-(cos(rad(irr$AZi[1])))*irr$H[1]
y1
```

```
## [1] -3.847228
```

```
y2<-(cos(rad(irr$AZi[2])))*irr$H[2]
y2
```

```
## [1] -3.111943
```

```
y3<-(cos(rad(irr$AZi[3])))*irr$H[3]
y3
```

```
## [1] 1.6
```

```
y4<-(cos(rad(irr$AZi[4])))*irr$H[4]
y4
```

```
## [1] -2.754733
```

```
y5<-(cos(rad(irr$AZi[5])))*irr$H[5]
y5
```

```
## [1] 6.690053
```

```
y6<-(cos(rad(irr$AZi[6])))*irr$H[6]
y6
```

```
## [1] 10.03985
```

```
yi<-c(y1, y2, y3, y4, y5, y6)
# Merge the results of partial projections into the irradiation table
irr<-cbind(irr, xi, yi)
```

Considering that the linear error correction is not performed for the irradiations, X, Y coordinates of each irradiation are determined by the UTM coordinates of the polygon vertices adjusted previously. In this case, the coordinates are obtained in UTM projection, UTM zone 21S, from WGS-84 datum transformation. The results of UTM X, Y coordinates of irradiations are joined to the irradiations dataset.

```
# Determine absolute coordinates of irradiations
X1<-XA+irr$xi[1]
X1
```

```
## [1] 600106
```

```
X2<-XB+irr$xi[2]
X2
```

```
## [1] 600093.6
```

```
X3<-XC+irr$xi[3]
X3
```

```
## [1] 600104.6
```

```
X4<-XD+irr$xi[4]
X4
```

```
## [1] 600112.2
```

```
X5<-XD+irr$xi[5]
X5
```

```
## [1] 600105.8
```

```
X6<-XD+irr$xi[6]
X6
```

```
## [1] 600111.1
```

```
Xi<-c(X1, X2, X3, X4, X5, X6)
Y1<-YA+irr$yi[1]
Y1
```

```
## [1] 8273801
```

```
Y2<-YB+irr$yi[2]
Y2
```

```
## [1] 8273793
```

```
Y3<-YC+irr$yi[3]
Y3
```

```
## [1] 8273776
```

```
Y4<-YD+irr$yi[4]
Y4
```

```
## [1] 8273781
```

```
Y5<-YD+irr$yi[5]
Y5
```

```
## [1] 8273790
```

```
Y6<-YD+irr$yi[6]
Y6
```

```
## [1] 8273794
```

```
Yi<-c(Y1, Y2, Y3, Y4, Y5, Y6)
# Join Xi and Yi coordinates in the irradiation table
irr<-cbind(irr, Xi, Yi)
```

The irradiation dataset with all calculations is exported in a .txt file.

```
## write.table(irr, file = "E:/Aulas/Topografia/Aula8/irr.txt",
##             sep = " ", row.names = TRUE, col.names = TRUE)
```

The format function is used to obtain X, Y coordinate values of irradiations with 4 decimal places.

```
# Show X and Y coordinates with 4 decimal places
Xi1<-format(X1, digits=4, nsmall=4)
Xi1
```

```
## [1] "600105.9978"
```

```
Xi2<-format(X2, digits=4, nsmall=4)
Xi2
```

```
## [1] "600093.6126"
```

```
Xi3<-format(X3, digits=4, nsmall=4)
Xi3
```

```
## [1] "600104.6427"
```

```
Xi4<-format(X4, digits=4, nsmall=4)
Xi4
```

```
## [1] "600112.1787"
```

```
Xi5<-format(X5, digits=4, nsmall=4)
Xi5
```

```
## [1] "600105.8283"
```

```
Xi6<-format(X6, digits=4, nsmall=4)
Xi6
```

```
## [1] "600111.1126"
```

```
Yi1<-format(Y1, digits=4, nsmall=4)
Yi1
```

```
## [1] "8273801.4628"
```

```
Yi2<-format(Y2, digits=4, nsmall=4)
Yi2
```

```
## [1] "8273792.9854"
```

```
Yi3<-format(Y3, digits=4, nsmall=4)
Yi3
```

```
## [1] "8273776.2380"
```

```
Yi4<-format(Y4, digits=4, nsmall=4)
Yi4
```

```
## [1] "8273781.0208"
```

```
Yi5<-format(Y5, digits=4, nsmall=4)
Yi5
```

```
## [1] "8273790.4656"
```

```
Yi6<-format(Y6, digits=4, nsmall=4)
Yi6
```

```
## [1] "8273793.8154"
```

Vertices of irradiations are mapped inside the polygon through LearnGeom package functions. The X, Y axis dimensions are defined, UTM coordinates are assigned at vertices A to D and the polygon is drawn with a gray color and irradiations with a white color (Figure 9.3).

```
## # Mapping Xi and Yi vertices of irradiations inside the polygon
## # Assign coordinates of radiations 1 to 6
## i1 <- c(irr$Xi[1],irr$Yi[1])
## i2 <- c(irr$Xi[2],irr$Yi[2])
## i3 <- c(irr$Xi[3],irr$Yi[3])
## i4 <- c(irr$Xi[4],irr$Yi[4])
## i5 <- c(irr$Xi[5],irr$Yi[5])
## i6 <- c(irr$Xi[6],irr$Yi[6])
## # Draw the x,y-axis
## CoordinatePlane(x_min, x_max, y_min, y_max)
## # Draw the polygon
## Draw(CreatePolygon(A, B, C, D), "gray", label=T)
## # Draw irradiations
## Draw(CreatePolygon(i1, i2, i3, i4, i5, i6), "white", label=T)
```

The area defined by the perimeter of irradiations is determined using the method of determinants by matrix of order 2. The area of the building obtained from the survey is 235.3 m^2.

```
# Calculate the area of the building by determinants
# Determine the sum of the ascending products
i6_i1 <-irr$Xi[6]*irr$Yi[1]
i5_i6 <-irr$Xi[5]*irr$Yi[6]
i4_i5 <-irr$Xi[4]*irr$Yi[5]
i3_i4 <-irr$Xi[3]*irr$Yi[4]
i2_i3 <-irr$Xi[2]*irr$Yi[3]
i1_i2 <-irr$Xi[1]*irr$Yi[2]
sumAsc<-sum(i6_i1, i5_i6, i4_i5, i3_i4, i2_i3, i1_i2)
# Determine the sum of descendant products
i1_i6 <-irr$X[1]*irr$Y[6]
i6_i5 <-irr$X[6]*irr$Y[5]
i5_i4 <-irr$X[5]*irr$Y[4]
```

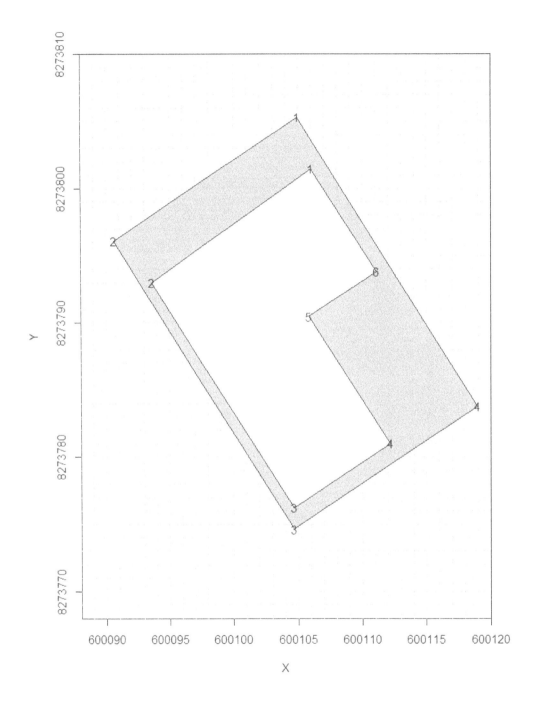

FIGURE 9.3: Mapping closed polygon vertices with irradiations performed at the boundaries of the animal anatomy building on the UFMT campus.

```
i4_i3 <-irr$X[4]*irr$Y[3]
i3_i2 <-irr$X[3]*irr$Y[2]
i2_i1 <-irr$X[2]*irr$Y[1]
sumDesc<-sum(i1_i6, i6_i5, i5_i4, i4_i3, i3_i2, i2_i1)
area<-(sumAsc-sumDesc)/2
area
```

```
## [1] 235.3223
```

9.16.4 Investigate irradiation boundaries in Google Earth and R

Irradiations boundaries are mapped in R and Google Earth. To perform the mapping we define data columns with coordinates in the irradiations dataset and assign UTM map projection system to the vertices before performing the mapping. The coordinates of the vertices are re-projected to WGS-84 geographic projection with the spTransform function. Subsequent mapping is performed with the plot and plotKML functions (Figure 9.4).

```
# Define columns with coordinates
irrGeo<-irr
coordinates(irrGeo) <- c("Xi","Yi")
# Define and apply UTM map projection system
prj_mt <- CRS("+init=epsg:32721")##Mato Grosso
proj4string(irrGeo) <- prj_mt
# Reproject to WGS84 geographical projection
prj_wgs84 <- CRS("+proj=longlat +ellps=WGS84 +datum=WGS84")
irr_wgs84 <- spTransform(irrGeo, CRS= prj_wgs84)
```

```
plot(irrGeo, axes=T) # Mapping point shapefile
```

The file irr_wgs84.kml is opened in the source directory, requiring the Google Earth application installed on the computer to be displayed (Figure 9.5).

```
## plotKML(irr_wgs84, colour_scale = SAGA_pal[[1]]) # Mapping KML
```

9.16.5 Export geographic coordinates of irradiations as ESRI Shapefile and KML file

The vertices with the geographic coordinates of irradiations are exported as two distinct file types: ESRI Shapefile and KML.

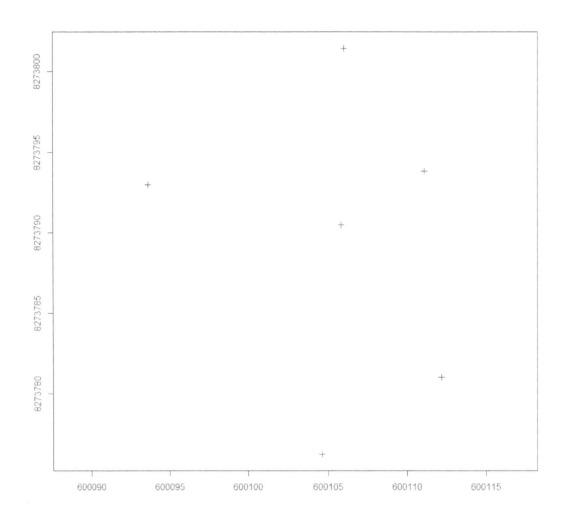

FIGURE 9.4: Mapping irradiations of building on the UFMT campus transformed in `SpatialPointsDataFrame` class object.

```
## # Export data as SpatialPointsDataFrame
## writeOGR(obj=irrGeo, dsn="E:/Aulas/Topografia/Aula8/irrGeo.shp",
##          layer="irrGeo", driver="ESRI Shapefile")
## # Export data as KML
## writeOGR(irr_wgs84, dsn="E:/Aulas/Topografia/Aula8/irr_wgs84.kml",
##          layer= "irr_wgs84", driver ="KML", dataset_options=c("NameField=name"))
```

According to Bird (1970), comparing polygonal fitting between Bowditch and least squares methods, the best fitting results are obtained by the least squares method. Therefore, the least squares method should be implemented in the future to obtain better polygon traverse adjustment using the R software.

FIGURE 9.5: Mapping of irradiations performed at building boundaries at the UFMT campus on Google Earth.

9.17 Solved Exercises

9.17.1 In a survey conducted by walking by internal angles of a closed-path polygon traverse, the sum of the polygon angles is:

a. $\sum a = (n+2)\ 180°$.
b. $\sum a = (n-2)\ 180°$. [X]
c. $\sum a = Az_F - Az_I$.
d. None of the alternatives.

9.17.2 List two advantages of the walkover survey method of closed-path polygon traverse.

A: The closed-path polygonal traverse survey method enables to circumvent obstacles in surveys. Thus, it is possible to check the angles and distances observed in relation to the accuracy required, considering the survey of a geometrically closed polygon.

9.17.3   In surveying a triangular area by the walkover survey method of closed-path polygon traverse, the sum of the interior angles of the polygon is:

    a.   900°. [X]
    b.   540°.
    c.   720°.
    d.   360°.

9.17.4   In a walkover survey of closed-path polygon, a perimeter of 86 m was observed. The sum of the X and Y errors of partial longitude and latitude were -0.045 and +0.774 m, respectively. Determine the linear closure error and relative precision.

A: The linear closure error is 0.77 m. The relative precision is 1:110.

```
# Determine the linear closure error
EL<-sqrt((sum(-0.045)^2)+(sum(0.774)^2))
EL
```

```
## [1] 0.775307
```

```
# Determine the relative precision
PR<-1/(EL/86)
PR
```

```
## [1] 110.9238
```

9.17.5   In a walkover survey with an electronic total station, the relative precision was 1:9000. Determine the quality rating of the survey according to Silva and Segantine (2015).

    a.   Low.
    b.   Good.
    c.   High.
    d.   Regular.[X]

9.18   Homework

Choose one exercise presented by the teacher and solve the question with different input values. Compare the results obtained. Performe closed-path polygon traverse surveying method in the field. Propose solutions to provide uncertainty information on the results and mapping the surveyed points.

9.19   Resources on the Internet

As a study guide, slides and illustrative videos are presented about the subject covered in the chapter in Table 9.2.

TABLE 9.2: Slide shows and video presentations on coordinate geometry of closed-path polygonal traverse surveying.

Guide	Address for Access
1	Slides on traversing and coordinate geometry in closed-path polygonal surveying[1]
2	Surveying closed-path traversing[2]
3	Closed-path traversing with animation[3]
4	Closure error in surveying and correction[4]
5	Closed-path polygon surveying in the field[5]
6	Traverse adjustment, purpose and observations needed[6]
7	Traverse adjustment[7]

9.20   Research Suggestion

The development of scientific research on geomatics is stimulated by the activity proposals that can be used or adapted by the student to assess the applicability of the subject matter covered in the chapter (Table 9.3).

TABLE 9.3: Practical and research activities used or adapted by students using coordinate geometry of closed-path polygonal traverse surveying.

Activity	Description
1	There is demand to establish a function for azimuth calculation and implement least squares methods for traversing adjustment
2	Survey a closed-path polygon by the walking method. Determine the angular closure error and the relative precision
3	Perform coordinate corrections obtained in field survey using the Bowditch method. Determine polygon area and perimeter after corrections

[1] http://www.sergeo.deg.ufla.br/geomatica/book/c9/presentation.html#/
[2] https://youtu.be/r12_bAuce5Y
[3] https://youtu.be/pGS2YX30nI8
[4] https://youtu.be/Ww7EcE3w_x4
[5] https://youtu.be/7slV7bl3Dds
[6] https://youtu.be/bCwoonQgkIs
[7] https://youtu.be/gtv-1GjHqVE

9.21 Learning Outcome Assessment Strategy

Perform a summary of the chapter, "Coordinate Geometry of Closed-Path Traverse Surveying with Geomatics and R", on a single A4 page in order to show the student's abilities to summarize a subject presenting key points considered of greater importance today.

10

Coordinate Geometry of Intersection Surveying

10.1 Learning Questions

The emergent learning questions answered through reading the chapter are as follows:

- What is the purpose of intersection surveying?
- How to determine rectangular coordinates in intersection surveying.
- How to map intersection survey vertices with R packages `LearnGeom` and `maptools`, and Google Earth.
- How to convert survey points to spatial points with attributes.

10.2 Learning Outcomes

The learning outcomes expected from reading the chapter are as follows:

- Evaluate intersection surveying applications.
- Calculate rectangular coordinates in intersection surveying.
- Map intersection survey vertices in R with the packages `LearnGeom`, `maptools` and Google Earth.
- Convert survey points to spatial points with attributes.

10.3 Introduction

Intersection surveys have been used in geomatics for difference in level determination between two accessible points separated by a large distance, or when some points are accessible and others inaccessible. In this case, trigonometric leveling can be used in conjunction with the intersection process, given the plane coordinates and azimuth between the two visible vertices.

Rectangular plane coordinate systems have been widely used in geomatics, except for geodetic control surveys over large areas. The advantages of referencing points in a rectangular coordinate system are (Alves and Silva, 2016):

- Relative positions of vertices are defined individually;
- Vertices can easily be represented using graphs;
- In case of lost points in the field, values can be retrieved from other available points in the same reference system;

DOI: 10.1201/9781003184263-10

- Computation calculations have been made easier.

The situations using coordinate geometry problems are intersection of points with two lines, a line and a circle, and two circles. The method applied to determine the intersection of points between line and circle can be solved by knowing the azimuth of one line and the length of another line. The intersection between two circles can be solved by knowing the length of two lines. These types of problems are commonly encountered on horizontal survey alignments where it is necessary to calculate intersections of tangents and circular curves to divide land parcels with straight lines and circular arcs (Ghilani and Wolf, 1989).

10.4   Intersection Surveying

Intersection problems can be solved by realizing triangles between two stations with known positions from where the observations are made. Two important functions used to solve oblique triangles are the law of sines and the law of cosines.

With the law of sines, the length of the sides of a triangle is related to the sine of opposite angles (Figure 10.1) (Alves and Silva, 2016).

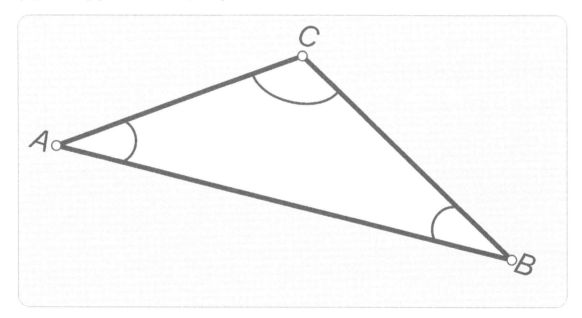

FIGURE 10.1: Oblique triangle used in intersection surveying.

$$\frac{BC}{sen A} = \frac{AC}{sen B} = \frac{AB}{sen C} \tag{10.1}$$

where AB, BC, and AC are the lengths of three sides of the triangle ABC, and A, B, C are the angles.

In the cosine law, two sides and the internal angle are related to the length of the side opposite the angle (Alves and Silva, 2016):

$$BC^2 = AC^2 + AB^2 - 2(AC)(AB)cos A \tag{10.2}$$

$$AC^2 = AB^2 + BC^2 - 2(AB)(BC)cosB \qquad (10.3)$$

$$AB^2 = BC^2 + CA^2 - 2(BC)(AC)cosC \qquad (10.4)$$

10.5   Intersection of Two Lines Knowing Directions

The situation where the intersection of two lines AP and BP have known coordinates of endpoints and each line has known direction is called the "direction-direction problem". The intersection of a point P can be calculated by simply determining the parts of the oblique triangle ABP. Since the coordinates of A and B are known, the length and azimuth AB (dotted line) can be obtained by means of the equations (Figure 10.2) (Ghilani and Wolf, 1989; Alves and Silva, 2016).

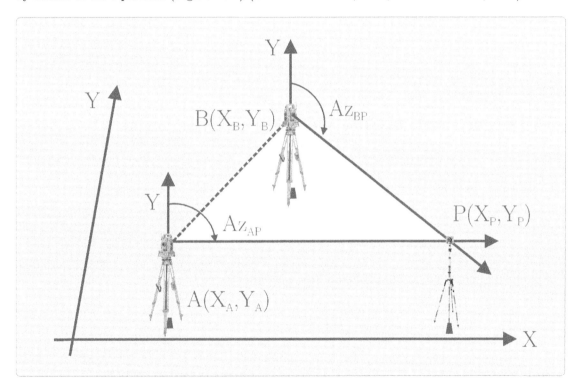

FIGURE 10.2: Intersection of two lines with known directions.

We observed that the angle A is obtained by the difference between the azimuths AB and AP:

$$A = Az_{AP} - Az_{AB} \qquad (10.5)$$

Similarly, the angle B is obtained by the difference of the azimuths BA and BP:

$$B = Az_{BA} - Az_{BP} \qquad (10.6)$$

After determining the two angles of the triangle ABP, the angle P is:

$$P = 180° - A - B \tag{10.7}$$

The length of the AP side is:

$$AP = AB\frac{sen(B)}{sen(P)} \tag{10.8}$$

With the known length and azimuth of AP, the coordinates of P are:

$$X_P = X_A + APsenAz_{AP} \tag{10.9}$$

$$Y_P = Y_A + APcosAz_{AP} \tag{10.10}$$

The solution can be checked against the length of BP, and using the azimuth of BP to calculate the coordinates of P. Both solutions must be equal. It should be noted that if the azimuths of the lines AP and BP are equal, the lines are parallel and there is no intersection (Alves and Silva, 2016).

10.6   Trigonometric Leveling with Intersection Surveying

When the problem is to determine the difference in level between two accessible points separated by a large distance, or when some points are accessible and others inaccessible, trigonometric leveling can be used in conjunction with the intersection process (Alves and Silva, 2016).

The difference in level between the topographical points A and C on the terrain is determined as follows (Comastri and Junior, 1998) (**Figure** 10.3):

- The instrument is installed at point A. The height of the instrument (Ai) at point A is measured. A base AB is measured, in order to be able to sight, from the ends, the point P, in an inaccessible place;
- With the instrument centered at endpoint A, we sight the endpoint B. Then, the horizontal movement of the limb is performed until the point P is focused, stopping the movement of the limb. The angle $BAP = \alpha$ is recorded, as well as the vertical angle Φ_A;
- The instrument is changed from point A to point B. From point B, point A is sighted and then, the point P, recording the horizontal angle $ABC = \beta$ and the vertical angle Φ_B.

The difference in level between A and P is obtained by trigonometric leveling (Comastri and Junior, 1998):

$$d_{AP} = H_{AC}tg\Phi_A \pm Ai \tag{10.11}$$

where, if the sighted point P is above the telescope's axis, the instrument's height must be added, and otherwise subtracted from the obtained result.

The distance H_{AC} can be determined by:

$$\frac{H_{AB}}{sen\delta} = \frac{H_{AC}}{sen\beta} = \frac{H_{BC}}{sen\alpha} \tag{10.12}$$

FIGURE 10.3: Trigonometric leveling of inaccessible points from accessible points.

where $\delta = 180° - (\alpha + \beta)$,

$$H_{AC} = \frac{H_{AB} sen\beta}{sen[180° - (\alpha + \beta)]} \tag{10.13}$$

Replacing H_{AC} into the equation used to determine the difference in level between A and P:

$$d_{AP} = \frac{H_{AB} sen\beta}{sen[180° - (\alpha + \beta)]} tg\Phi_A + Ai \tag{10.14}$$

where the instrument height Ai is added to the result obtained because the targeted point is above the telescope's axis.

The difference in level between B and P, (d_{BP}), is obtained by:

$$d_{BP} = H_{BC} tg\Phi_B \pm Ai \tag{10.15}$$

As the sighted point P is above the telescope's axis,

$$d_{BP} = \frac{H_{BC} sen\alpha}{sen[180° - (\alpha + \beta)]} tg\Phi_B + Ai \tag{10.16}$$

The level difference between A and B, (d_{AB}), is obtained by:

$$d_{AB} = d_{AC} - d_{BC} \tag{10.17}$$

This chapter presented methodologies of coordinate geometry that can be used in planar surveys, without considering the curvature of the Earth. For geodetic calculations, the n-vector method (Gade, 2010) presented simple and non-singular solutions, for any global position for different problems of geographical position calculations such as (Spinielli, 2020):

- Given path A going through A_1 and A_2, and path B going through B_1 and B_2, find the intersection of the two paths.

Geodetic calculations involving coordinate systems are presented in more detail in Chapter 12.

10.7   Computation

As a computation practice, the objective is to use the intersection method where two lines AP and BP have known endpoints coordinates and each line has known direction, called the "direction-direction problem". The intersection of point P is calculated by a simple method by determining the parts of the oblique triangle ABP. Since the coordinates of A and B are known, the length and azimuth AB is determined by applying equations and the law of sine. The coordinates of vertex P are determined by intersection, and then the results are checked by determining the same coordinates from another vertex.

The `circular` R package (Lund et al., 2017a) is used to convert angular measurements from degrees to radians and trigonometry operations to determine the coordinates of the vertex P by the intersection method. The `LearnGeom` package (Jammalamadaka and SenGupta, 2001; Briz-Redon and Serrano-Aroca, 2020) is used for mapping vertices in the surveyed area by the intersection method. The `rgdal` package is used to perform projection and coordinate transformation operations by means of the `PROJ` library (Bivand et al., 2021), in order to transform coordinate vertices into spatial points with attributes and map them. The `sp` class objects are exported to KML and ESRI Shapefile formats and mapped in R via the `maptools` package (Bivand et al., 2020a) and in Google Earth.

10.7.1   Installing R packages

The `install.packages` function is used to install the R packages `circular`, `LearnGeom`, `rgdal` and `maptools` in the R console.

```
## install.packages("circular")
## install.packages("LearnGeom")
## install.packages("rgdal")
## install.packages("maptools")
```

10.7.2   Enabling R packages

The `library` function is used to enable the R packages `circular`, `LearnGeom`, `rgdal` and `maptools` in the R console.

```
library(circular)
library(LearnGeom)
library(rgdal)
library(maptools)
```

10.7.3   Calculate the coordinates X_P, Y_P

The following UTM coordinates and azimuth observations are made: $X_A = 503142.10$ m E; $Y_A = 7654216.99$ m E; $X_B = 503211.00$ m E; $Y_B = 7654195.00$ m N; $Az_{AP} = 198°22'26.04"$; $Az_{BP} = 198°41'36.24"$.

An illustration of the area of interest is made to verify the tower in the Serra da Bocaina to the South, south of Lavras city, as well as vertices A and B, to the North of the region (Figure 10.4).

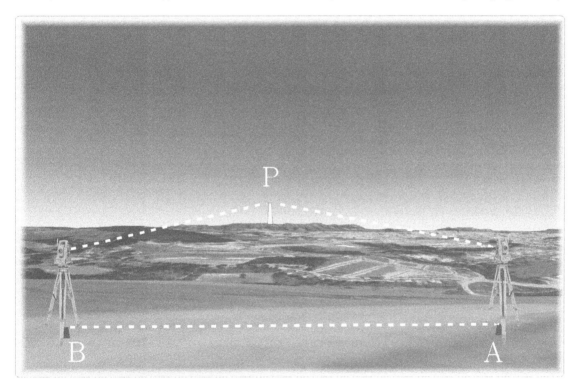

FIGURE 10.4: Illustration of intersection survey performed from vertices A and B to determine the UTM coordinates of the vertex P.

A database is created with information of plane coordinates of longitude and latitude of vertices A and B, in UTM projection, UTM zone 23S.

```
# Create a database with information on vertices A and B
verticesAB <- c('A', 'B')
XAB <- c(503142.1000, 503211.0000)
YAB <- c(7654216.9900, 7654195.0000)
intAB<-data.frame(verticesAB, XAB, YAB)
intAB
```

```
##    verticesAB      XAB     YAB
## 1           A 503142.1 7654217
## 2           B 503211.0 7654195
```

The difference between the longitude and latitude coordinates of vertices A and B on the X and Y axes is determined and stored in the same table created earlier.

```
# Calculate delta XAB and delta YAB
dx1 <- intAB$X[2] - intAB$X[1]
dy1 <- intAB$Y[2] - intAB$Y[1]
intAB<-cbind(intAB, dx1, dy1)
intAB
```

```
##    verticesAB      XAB     YAB dx1    dy1
## 1           A 503142.1 7654217 68.9 -21.99
## 2           B 503211.0 7654195 68.9 -21.99
```

The horizontal distance between vertices A and B is determined using the Pythagorean theorem, according to the difference between the longitude and latitude coordinates of A and B.

```
# Determine the distance between vertices A and B
HAB <- sqrt((intAB$dx1[1]^2)+(intAB$dy1[1]^2))
HAB
```

```
## [1] 72.32406
```

The azimuth between vertices A and B is calculated by the arc tangent function of the differences between the longitude and latitude coordinates of A and B. Since the values of dy are negative, the value of 180° is added to the end of the azimuth determination equation.

```
# Determine AB azimuth
AZAB <- deg(atan(intAB$dx1[1]/intAB$dy1[1]))+180
AZAB
```

```
## [1] 107.7009
```

The internal angle at A is determined as a function of the difference between the azimuth AB (Az_{AB}) and the azimuth AP (Az_{AP}). The interior angle at B is determined by extending the azimuth AB at vertex B, adding 180° and subtracting the azimuth BP. The internal angle at P is determined by considering that the sum of the internal angles of a triangle equals 180°, that is, the angle at P is determined by the difference between 180° and the sum of the internal angles A and B.

```
# Determine the interior angles of triangle A, B, and P
AZAP <- (198+22/60+26.04/3600)
AZAP
```

```
## [1] 198.3739
```

```
AZBP <- (198+41/60+36.24/3600)
AZBP
```

```
## [1] 198.6934
```

```
A <- AZAP - AZAB
A
```

```
## [1] 90.67299
```

```
B <- (AZAB + 180) - (AZBP)
B
```

```
## [1] 89.00751
```

```
P <- 180 - (A + B)
P
```

```
## [1] 0.3195
```

Having the internal angles at B, P and the horizontal distance between AB, the distance AP is determined by the law of sines.

```
# Determine the distance between A and P by the law of sines
HAP <- HAB*sin(rad(B))/sin(rad(P))
HAP
```

```
## [1] 12967.96
```

Then the X and Y coordinates of vertex P are determined, using the coordinate transport from A to P, through the absolute coordinate determination of P.

```
# Register X and Y coordinates of A and B
XA <- 503142.10
YA <- 7654216.99
XB <- 503211.00
YB <- 7654195.00
# Determine X and Y coordinates of P
XP <- XA + sin(rad(AZAP))*HAP
XP
```

```
## [1] 499054.4
```

```
YP <- YA + cos(rad(AZAP))*HAP
YP
```

```
## [1] 7641910
```

```
XP_1<-format(XP, digits=4, nsmall=4)
XP_1
```

```
## [1] "499054.3815"
```

```
YP_1<-format(YP, digits=4, nsmall=4)
YP_1
```

```
## [1] "7641910.1401"
```

The X and Y coordinate results of P are checked by performing the coordinate transport from B to P and the equation for determining the absolute coordinates of P.

```
# Determine the distance between B and P
HBP <- HAB*sin(rad(A))/sin(rad(P))
HBP
```

```
## [1] 12969.01
```

```
# Determine X and Y of P for checking
XP1 <- XB + sin(rad(AZBP))*HBP
XP1
```

```
## [1] 499054.4
```

```
YP1 <- YB + cos(rad(AZBP))*HBP
YP1
```

```
## [1] 7641910
```

```
# Check the results based on segment BP
XP1<-format(XP1, digits=4, nsmall=4)
XP1
```

```
## [1] "499054.3815"
```

```
YP1<-format(YP1, digits=4, nsmall=4)
YP1
```

```
## [1] "7641910.1401"
```

```
# check ok!
```

The results with X and Y coordinates of P are merged into a single database, along with the vertices A and B.

```
# Merge the results into one database
vertices <- c('A', 'B', 'P')
X <- c(503142.1000, 503211.0000, 499054.3815)
Y <- c(7654216.9900, 7654195.0000, 7641910.1401)
points<-data.frame(vertices, X, Y)
points
```

```
##    vertices        X        Y
## 1         A 503142.1 7654217
## 2         B 503211.0 7654195
## 3         P 499054.4 7641910
```

The dataset is exported for later use as .txt file.

```
write.table(points, file = "E:/Aulas/Topografia/Aula9/pontos.txt", sep = " ",
            row.names = TRUE, col.names = TRUE)
```

10.7.4   Mapping vertices by the R package LearnGeom

After obtaining the UTM coordinates of vertices A, B and P, the dimensions of X, Y axis are defined to map the surveyed vertices. Then the polygon A, B, P is drawn, in a gray color (Figure 10.5).

```
# Set dimensions of x,y axis
x_min <- 498950.00
x_max <- 503400.00
y_min <- 7641500.00
y_max <- 7654500.00
# Draw the x,y axis
CoordinatePlane(x_min, x_max, y_min, y_max)
A <- c(points$X[1],points$Y[1])
B <- c(points$X[2],points$Y[2])
P <- c(points$X[3],points$Y[3])
# Draw the polygon
Draw(CreatePolygon(A,B,P), "gray", label=T)
```

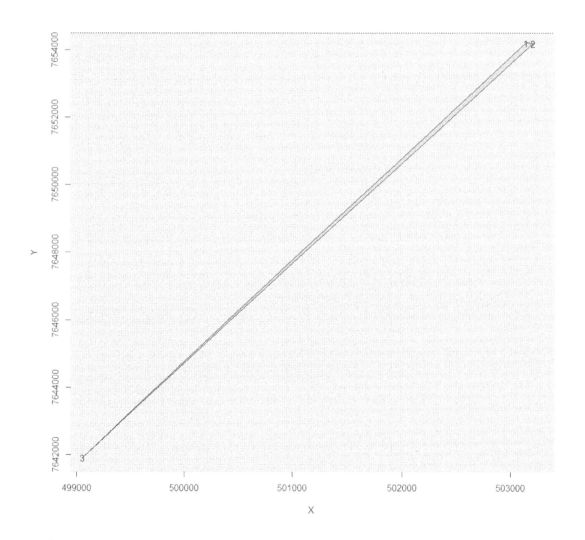

FIGURE 10.5: Intersection coordinate mapping using the LearnGeom package.

10.7.5   Converting vertices to spatial points with attributes and KML

Before mapping the cadastral vertices and `.kml` file, it is necessary to define the data columns with coordinates in the dataset and apply the UTM map projection system to the vertices. The coordinates of the vertices are re-projected to the WGS-84 geographic projection with the `spTransform` function for the subsequent mapping.

```r
# Define columns with coordinates
pointsGeo<-points
coordinates(pointsGeo) <- c("X", "Y")
# Define and apply UTM map projection system
prj_mg <- CRS("+init=epsg:32723") #South of MG
proj4string(pointsGeo) <- prj_mg
# Reproject to WGS84 geographic projection
prj_wgs84 <- CRS("+proj=longlat +ellps=WGS84 +datum=WGS84")
points_wgs84 <- spTransform(pointsGeo, CRS= prj_wgs84) #Transform to WGS84
```

The vertices with the geographic coordinates are exported as two distinct file types: ESRI Shapefile and KML.

```r
## # Export as SpatialPointsDataFrame
## writeOGR(obj=pontosGeo,
##          dsn="E:/Aulas/Topografia/Aula9/pontosGeo.shp",
##          layer="pontosGeo", driver="ESRI Shapefile")
## # Export as KML
## writeOGR(pontos_wgs84,
##          dsn="E:/Aulas/Topografia/Aula9/pontos_wgs84.kml",
##          layer= "pontos_wgs84", driver ="KML",
## dataset_options=c("NameField=name"))
```

10.7.6   Mapping vertices by the R package `maptools`

The vertices are also mapped by the `maptools` package and the labels are used to identify the vertices on the map (Figure 10.6).

```r
plot(pointsGeo, axes=T)
pointLabel(coordinates(pointsGeo),labels=pointsGeo$vertices)
```

10.7.7   Mapping the vertices with Google Earth

The file `points_wgs84.kml` has been loaded in the source directory, requiring the Google Earth application installed on the computer to visualize it (Figure 10.7).

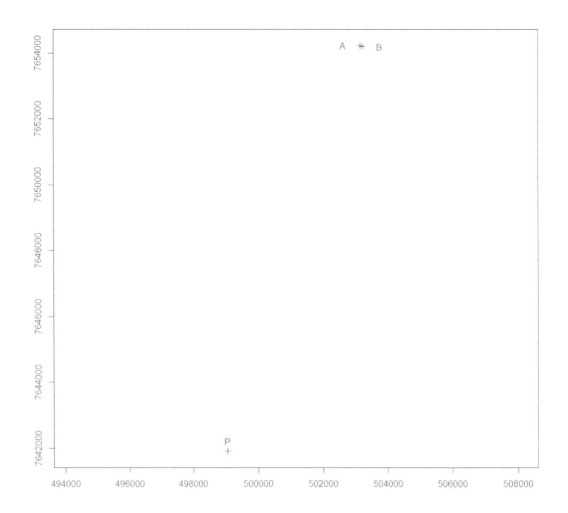

FIGURE 10.6: Mapping coordinates defined by the intersection method in `maptools`.

10.8   Solved Exercises

10.8.1   A trigonometric intersection survey is performed between landmarks A, B and C, installed on a mountain. Determine the level differences between A and B, A and C, and B and C, based on the following data:

A: - With the instrument on A: $\alpha = 87°10'$; $\beta = 83°20'$; $\phi_A = +5°\ 12'$; $AiA = 1.50$ m; $H_{AB} = 40.00$ m. - With the instrument on B: $\alpha = 87°10'$; $\beta = 83°20'$; $\phi_B = +4°\ 55'$; $AiB = 1.48$ m; $H_{AB} = 40.00$ m.

A: The differences in level between A and B, A and C, and B and C are 23.4068, 22.3026, and 1.1042 m, respectively.

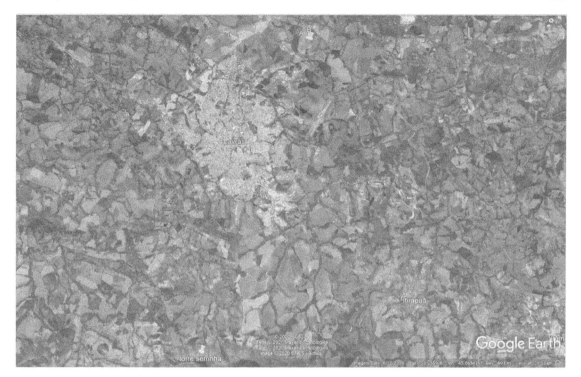

FIGURE 10.7: Mapping coordinates defined by the intersection method in Google Earth.

$$d_{AC} = \frac{H_{AB}sen\beta}{sen[180° - (\alpha + \beta)]}tg\Phi_A + Ai = +23.4068 \text{ m} \qquad (10.18)$$

$$d_{BC} = \frac{H_{AB}sen\alpha}{sen[180° - (\alpha + \beta)]}tg\Phi_B + Ai = +22.3026 \text{ m} \qquad (10.19)$$

$$d_{AB} = d_{AC} - d_{BC} = +1.1042 \text{ m} \qquad (10.20)$$

```
# Calculations performed in R
dAC<-((40*(sin(rad(83+20/60))))/
    (sin(rad(180-(87+10/60+83+20/60)))))*tan(rad(5+12/60))+(1.5)
dAC
```

```
## [1] 23.40684
```

```
dBC<-((40*(sin(rad(87+10/60))))/
    (sin(rad(180-(87+10/60+83+20/60)))))*tan(rad(4+55/60))+(1.48)
dBC
```

```
## [1] 22.30263
```

```
dAB<-dAC-dBC
dAB
```

```
## [1] 1.10421
```

10.9   Homework

Choose one exercise presented by the teacher and solve the question with different input values. Compare the results obtained. Perform a practice with the intersection method in the field. Determine X, Y coordinates of the inaccessible point and check the results.

10.10   Resources on the Internet

As a study guide, slides and illustrative videos are presented about the subject covered in the chapter in Table 10.1.

TABLE 10.1: Slide shows and video presentations on intersection coordinate geometry.

Guide	Address for Access
1	Slides on polygonation and coordinate geometry in intersection surveying[1]
2	Intersection computations[2]
3	Surveying orientation and intersection[3]

10.11   Research Suggestion

The development of scientific research on geomatics is stimulated by the activity proposals that can be used or adapted by the student to assess the applicability of the subject matter covered in the chapter (Table 10.2).

---

[1] http://www.sergeo.deg.ufla.br/geomatica/book/c10/presentation.html#/
[2] https://youtu.be/AC3unwXFMuI
[3] https://youtu.be/u7XLrQg39hM

TABLE 10.2: Practical and research activities used or adapted by students using intersection coordinate geometry.

Activity	Description
1	Apply the intersection method to define the coordinates of an inaccessible point and present the advantages and disadvantages of the method
2	Apply the intersection survey method in the field using topographic equipment. Evaluate the difficulties obtained according to the options available for surveying
3	Conduct a survey of how the intersection method can be used in leveling situations. Perform a data simulation according to the example provided by the teacher in the chapter's theoretical approach

10.12 Learning Outcome Assessment Strategy

Perform a summary of the chapter, "Coordinate Geometry of Intersection Surveying with Geomatics and R", on a single A4 page in order to show the student's abilities to summarize a subject presenting key points considered of greater importance today.

11

Traverse Area Evaluation and Surveying Memorial

11.1 Learning Questions

The emergent learning questions answered through reading the chapter are as follows:

- How to determine the area of closed-path polygon from topographic survey data.
- How to draw up a descriptive report and inverse calculation from topographic survey data.
- What is the difference between a textual descriptive report and a spreadsheet?
- How to determine area from polygon and geospatial polygon survey data automatically in R and R packages.

11.2 Learning Outcomes

The learning outcomes expected from reading the chapter are as follows:

- Evaluate area of closed-path polygon from topographic survey data.
- Prepare a descriptive report and inverse calculation from topographical survey data.
- Prepare a descriptive report in text and spreadsheet format.
- Compare calculated areas of polygons by the functions `gArea`, from package `st_area`, and `st_area`, from package `sf`.
- Perform a computational practice of area calculation using the R packages `circular`, `sf`, `rgdal`, `rgeos` and `maptools`.

11.3 Introduction

Determining the area of polygons is important for describing property or details of interest in neighboring areas, as well as for providing support for volume calculations. Based on this measurement, we can determine how to manage rural and urban areas or use the area as a criterion to define property values and tax payments. Field and map measurements can be used to determine the area of a location.

Various geometric, analytical, and mechanical methods can be used to determine areas, such as (Alves and Silva, 2016):

- Division of land into simple figures of triangles, rectangles, and trapezoids;

- Removal from a straight line;
- Calculating coordinates;
- Double meridian distances;
- Counting pixels of an image referring to the map;
- Digitizing map coordinates;
- Applying a planimeter on map.

It should be noted that for legal and administrative purposes, the area of a plot of land is calculated as a function of horizontal projections of the boundary lines that delimit the land on the horizontal plane of reference, i.e., it is different from the surface area obtained from inclined distances (Silva and Segantine, 2015).

An area calculation method is considered analytic when instead of using known equations to calculate areas of elementary geometric figures, data such as measurements of directions, distances, or known coordinates of polygon vertices are used to calculate area (Silva and Segantine, 2015). These methods are discussed below to exemplify how to calculate the area obtained by surveying.

11.4   Area Assessment by Radial Traverse Surveying

The calculation of area by irradiation can be done based on the angle and distance measurements used when making a radial polygon (Silva and Segantine, 2015) (Figure 11.1):

$$A = \frac{1}{2} \sum [H_{i+1} H sen(Az_{i+1} - Az_i)] \tag{11.1}$$

where H is the horizontal distance from the anterior vertex, H_{i+1}, the horizontal distance from the posterior vertex, Az_i, the observed direction at the anterior vertex and, Az_{i+1}, the observed direction at the posterior vertex.

The calculation of the area of a polygon obtained by the irradiation method can also be performed in a practical way, by arranging the coordinates of the points in two columns X and Y, repeating the coordinates of the first point at the last point as performed in the irradiation Chapter 8, by the method of matrix determinants of order two (Alves and Silva, 2016).

11.5   Area Assessment by Closed-Path Traverse Surveying

The analytical method of area calculation by coordinates is one of the most used. This method can be applied for computational calculation of areas of geometric figures where coordinates are digitized. Only one equation can be applied for all geometric configurations of closed polygon shapes.

The area calculation can be performed in a practical way by arranging the coordinates of the points in two columns X and Y, repeating the coordinates of the first point at the last point as performed in the previous chapters by matrix determinants of order two.

Another method used in area calculation can be accomplished by the double meridian distance method (Figure 11.2). In this method, an expression is used to calculate the double area, based

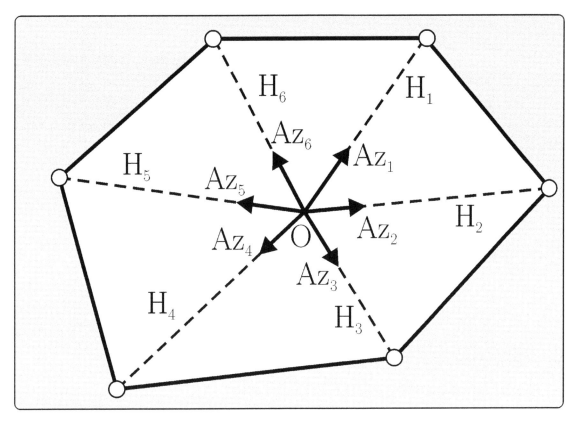

FIGURE 11.1: Radial triangles used in area calculation in the irradiation method.

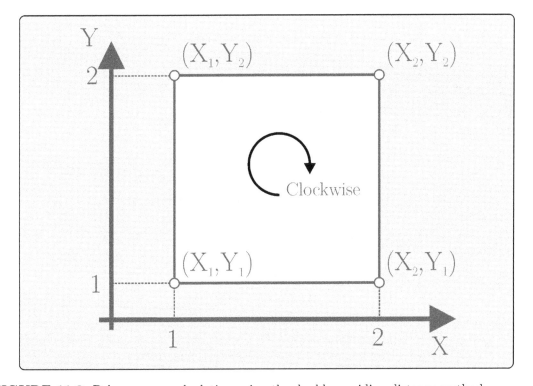

FIGURE 11.2: Polygon area calculation using the double meridian distance method.

on the binary sums of X, multiplied by the binary differences of Y (Table 11.1) (Alves and Silva, 2016).

TABLE 11.1: Area calculation using the double meridian distance method with clockwise registration of coordinates in the spreadsheet from the first to the last vertex.

Station	Absolute Abscissa	Absolute Ordinate	Binary Sum		Binary Difference		Double Area	
N	X	Y	X	Y	X	Y	$X.Y$	$Y.X$
1	1	1	2	3	$+0$	$+1$	$+2$	$+0$
2	1	2	3	4	$+1$	$+0$	$+0$	$+4$
3	2	2	4	3	$+0$	-1	-4	$+0$
4	2	1	3	2	-1	$+0$	$+0$	-2
–	–	–	–		–		–	
					sum(+)		$+2$	$+4$
					sum(-)		-4	-2
					double area		-2	$+2$
					area $=$		$+2/2 = 1$ m^2	

11.6 Trapezoidal Polygon Area Assessment

Trapezoidal polygon area assessment can be made by the trapezoids or Bezout's method. In this case, a succession of trapezoids measured along the polygon is assumed with a perpendicular offset at fixed distances of 10 x 10 m or 20 x 20 m. The practical approximation lies in assuming that the points A, B, C, D, ..., n are connected by straight lines, which is not strictly accurate. This equation is easy to apply and widely used, especially when using graph paper (Borges, 2013) (Figure 11.3).

The area calculated by the Bezout equation (A_B) is (Garcia and Piedade, 1987; Borges, 2013) (Figura 6.21):

$$A_B = \frac{y_1 + y_2}{2} H + \frac{y_2 + y_3}{2} H + ... + \frac{y_n + y_{n+1}}{2} H \tag{11.2}$$

Putting $H/2$ in evidence,

$$A_B = \frac{H}{2}(y_1 + 2y_2 + 2y_3 + ... + 2y_{n+1} + y_n) \tag{11.3}$$

where the total area is equal to $H/2$ times the sum of the y, with the extreme y values (E) being summed once and the middle y (M) being summed twice:

$$A_B = \frac{H}{2}(E + 2M) \tag{11.4}$$

The calculation of the depth of marine profiles has been analyzed by engineers and researchers over decades due to its relevance in maritime works such as beach sustainability or coastal defense projects. Quantification of transverse sediment transport was performed from profile surveys in a coastal region of Spain. The differences between various criteria based on the change in sea bottom elevation were studied in order to assess the relevance for engineering works (Aragonés et al., 2019).

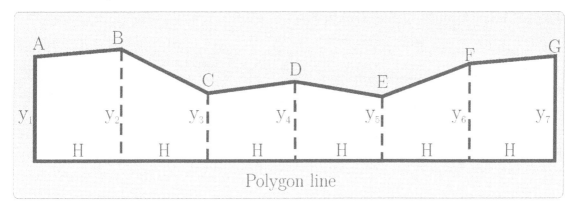

FIGURE 11.3: Succession of trapezoids along a polygon line with the same distance H.

11.7   Area Computation

Most area calculations of polygons can be performed by computer programs to generate consistent results with simple operations. However, it may be necessary to convert the geometry from point-type attribute vectors to polygon-type attribute vectors before performing area calculation. If the polygon on which we wished to calculate the area is already available in digital form, depending on the program used, we can calculate the area of the polygon with a simple press of the mouse button, or by scrolling the edges of the polygon with the mouse pointer. Another option in some situations, is to send the file to cloud programs on the Internet in specific extensions where it is possible to calculate the area, as an example of the KML or shapefile formats.

11.8   Inversing Calculation and Surveying Memorial

In the polygonal adjustment, corrections are applied to longitudes and latitudes to obtain adjusted values. These values are used to calculate X and Y coordinates of the polygon stations. As the longitude and latitude line values are modified in the adjustment process, the length and azimuth values are also modified. This made it necessary to calculate the final or adjusted lengths and directions.

The vertices used in the area calculation can be defined based on the map of points surveyed in the area of interest. Next, a descriptive memorial is prepared, including the boundary line, longitude, latitude, horizontal distance, azimuth, boundary type and confrontation between the alignment of each vertex with the subsequent vertex (Table 11.2) (Alves and Silva, 2016).

TABLE 11.2: Example of table used to elaborate a descriptive memorial.

Boundary Line	Longitude (m)	Latitude (m)	Azimuth (°)	Distance (m)	Boundary Type	Confrontation

If the longitude (ΔX) and latitude (ΔY) of an alignment i are known, the length (H), bearing (B) and azimuth (Az) can be obtained by the inverse calculation:

$$H_i = \sqrt{\Delta X_i^2 + \Delta Y_i^2} \tag{11.5}$$

$$B_i = tg^{-1}(\frac{|\Delta X_i|}{|\Delta Y_i|}) \tag{11.6}$$

where it is necessary to observe the sign of the partial longitude and latitude to determine the quadrant of the bearing as Northeast (NE), if ΔX and $\Delta Y > 0$, Southeast (SE), if $\Delta Y < 0$ and $\Delta X > 0$, Southwest (SW), if $\Delta X < 0$ and $\Delta Y < 0$ and, Northwest (NW), if $\Delta X < 0$ and $\Delta Y > 0$.

$$Az_{AB} = tg^{-1}(\frac{\Delta X_i}{\Delta Y_i}) + C \tag{11.7}$$

where C is 0°, if ΔX and $\Delta Y > 0$; 180°, if $\Delta Y < 0$; and 360°, if $\Delta X < 0$ and $\Delta Y > 0$.

The longitude and latitude equations can be expressed in terms of rectangular plane coordinate differences:

$$\Delta X_i = X_{i+1} - X_i \tag{11.8}$$

$$\Delta Y_i = Y_{i+1} - Y_i \tag{11.9}$$

where $i + 1$ and i are the back and front vertices that defined the alignment, respectively.

Azimuth calculation has been used to determine the direction of buried pipelines through synthetic array of emitters, single survey line and scattering matrix formalism. These results are relevant in civil engineering applications where accurate azimuth is required and it is not possible to acquire data following 2D grids due to obstacles on the ground surface (Bullo et al., 2016).

11.9   Computation

As a computation practice, the objective is to calculate the area of a soccer field with the measurements obtained by irradiation survey and, of a closed-path polygon traverse obtained walking survey with irradiations of a building in the campus of the Federal University of Mato Grosso, Brazil. In the case of the soccer field survey, the vertices are converted to polygon type features and the area is calculated by the `gArea` function of the `rgeos` package. In the case of the area demarcated by irradiations in a walking survey, the double meridian distance method is used for area calculation from the longitude and latitude plane coordinates of vertices. The inverse calculation of azimuths and clockwise distances is performed, and a descriptive data table and an example in textual format are prepared. The vertices are converted into polygon type features to check the area calculation by the `gArea` function.

The R package `circular` (Lund et al., 2017a) is used for angle conversion to perform the inverse calculation. The package `rgdal` is used to perform coordinate projection operations via the library PROJ (Bivand et al., 2021), in order to transform coordinate vertices into vector attribute features. Polygon area calculation is performed using the R packages `rgeos` (Bivand et al., 2018) and `sf` (Pebesma et al., 2021). The polygons are mapped by the R package `maptools` (Bivand et al., 2020a).

11.9.1 Installing R packages

The `install.packages` function is used to install the `circular`, `sf`, `rgdal`, `rgeos` and `maptools` packages in the R console.

```
## install.packages("circular")
## install.packages("sf")
## install.packages("rgdal")
## install.packages("rgeos")
## install.packages("maptools")
```

11.9.2 Enabling R packages

The `library` function is used to enable the `circular`, `sf`, `rgdal`, `rgeos` and `maptools` packages in the R console.

```
library(circular)
library(sf)
library(rgdal)
library(rgeos)
library(maptools)
```

11.9.3 Convert vertices to polygons and calculate the area using the `rgeos` package

The survey data with irradiations are obtained from the computing practice performed in Chapter 6 about stadia technique, where the edges of a polygon are surveyed by vertices irradiated from a base installed in the center of a soccer field. The `read.table` function is used to import the file to perform the area calculation based on the length and direction measurements.

```
field<-read.table("files/irr1.txt", header = TRUE, sep = " ", dec = ".")
```

The UTM map projection system, UTM zone 21S, is applied to the vectors that defined the X and Y coordinate vertices of the survey database.

```
# Define columns with coordinates
irrField <-field
coordinates(irrField) <- c("X", "Y")
# Define and apply UTM map projection system
prj_mt <- CRS("+init=epsg:32721")##MT
proj4string(irrField) <- prj_mt
```

The vertices are converted into a spatial polygon (simple feature) by the `SpatialPolygons-DataFrame` function. A map projection is associated with the polygon, as well as a database for cadastral use.

```
# Convert points to polygon
pField <-Polygon(irrField)
psField <- Polygons(list(pField),1)
spsField <- SpatialPolygons(list(psField))
# Associate a map projection to the polygon
proj4string(spsField) <- CRS("+init=epsg:32721")
# Associate a database
data <- data.frame(f=99.9)
# Join the database to the polygon
spdfField <- SpatialPolygonsDataFrame(spsField,data)
```

The area defined by the polygon is determined by the `gArea` function from the R package `rgeos`.

```
gArea(spdfField)
```

```
## [1] 9779.432
```

The polygon is mapped with identifying information for the vertices that defined the geometry of the polygon, using the `plot` and `pointLabel` functions from the R package `maptools` (Figure 11.4).

```
plot(spdfField, axes=T, col="gray")
pointLabel(coordinates(spdfField),labels=spdfField$vertices)
```

11.9.4   Calculate the area from vertices of irradiations defined by closed-path polygon traverse

Another area is calculated from the vertices of irradiations using the double meridian distance method. The irradiation data are obtained from the computing practice performed in Chapter 9, in which the extremities of a polygon are surveyed by vertices irradiated from a counterclockwise walking survey by closed-path polygon traverse. Four bases installed around a building in the Federal University of Mato Grosso, Brazil are used in the surveying. The `read.table` function is used to import the file to perform area calculation based on length and direction measurements.

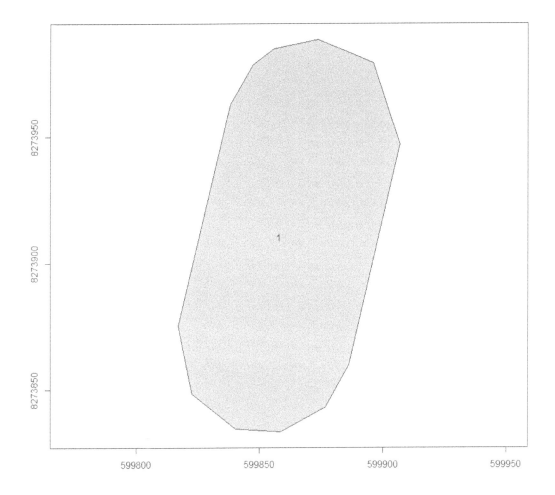

FIGURE 11.4: Mapping a soccer field and polygon identifier code with the `sp` and `maptools` packages.

```
irr<-read.table("files/irr2.txt", header = TRUE, sep = " ", dec = ".")
```

The double meridian distance method is used to calculate area from plane longitude and latitude coordinates of the vertices of the building's radiations. Binary sum and difference are determined for X and Y and, the double area XY, YX. The summation of the positive and negative results of the XY and YX columns is performed and then the area determination with the result check.

```
# Determine the binary sum for Xi and Yi
sx6_1<-irr$Xi[6]+irr$Xi[1]
sx5_6<-irr$Xi[5]+irr$Xi[6]
sx4_5<-irr$Xi[4]+irr$Xi[5]
sx3_4<-irr$Xi[3]+irr$Xi[4]
sx2_3<-irr$Xi[2]+irr$Xi[3]
```

```
sx1_2<-irr$Xi[1]+irr$Xi[2]
sx<-c(sx6_1, sx5_6, sx4_5, sx3_4, sx2_3, sx1_2)
sy6_1<-irr$Yi[6]+irr$Yi[1]
sy5_6<-irr$Yi[5]+irr$Yi[6]
sy4_5<-irr$Yi[4]+irr$Yi[5]
sy3_4<-irr$Yi[3]+irr$Yi[4]
sy2_3<-irr$Yi[2]+irr$Yi[3]
sy1_2<-irr$Yi[1]+irr$Yi[2]
sy<-c(sy6_1, sy5_6, sy4_5, sy3_4, sy2_3, sy1_2)
# Determine the binary difference for Xi and Yi
dx6_1<-irr$Xi[6]-irr$Xi[1]
dx5_6<-irr$Xi[5]-irr$Xi[6]
dx4_5<-irr$Xi[4]-irr$Xi[5]
dx3_4<-irr$Xi[3]-irr$Xi[4]
dx2_3<-irr$Xi[2]-irr$Xi[3]
dx1_2<-irr$Xi[1]-irr$Xi[2]
dx<-c(dx6_1, dx5_6, dx4_5, dx3_4, dx2_3, dx1_2)
dy6_1<-irr$Yi[6]-irr$Yi[1]
dy5_6<-irr$Yi[5]-irr$Yi[6]
dy4_5<-irr$Yi[4]-irr$Yi[5]
dy3_4<-irr$Yi[3]-irr$Yi[4]
dy2_3<-irr$Yi[2]-irr$Yi[3]
dy1_2<-irr$Yi[1]-irr$Yi[2]
dy<-c(dy6_1, dy5_6, dy4_5, dy3_4, dy2_3, dy1_2)
# Determine the double area Xi*Yi and Yi*Xi
# Xi*Yi
adXY1<-sx6_1*dy6_1
adXY1
```

```
## [1] -9178547
```

```
adXY2<-sx5_6*dy5_6
adXY2
```

```
## [1] -4020481
```

```
adXY3<-sx4_5*dy4_5
adXY3
```

```
## [1] -11335801
```

```
adXY4<-sx3_4*dy3_4
adXY4
```

```
## [1] -5740400
```

```
adXY5<-sx2_3*dy2_3
adXY5
```

```
## [1] 20100264
```

```
adXY6<-sx1_2*dy1_2
adXY6
```

```
## [1] 10174495
```

```
adXY <- c(adXY1, adXY2, adXY3, adXY4, adXY5, adXY6)
# Sum of the positive and negative results of columns XY and YX
sadXYplus <- sum(adXY[which(adXY>0)])
sadXYminus <- sum(adXY[which(adXY<0)])
# Yi*Xi
adYX1<-sy6_1*dx6_1
adYX1
```

```
## [1] 84637648
```

```
adYX2<-sy5_6*dx5_6
adYX2
```

```
## [1] -87441673
```

```
adYX3<-sy4_5*dx4_5
adYX3
```

```
## [1] 105082388
```

```
adYX4<-sy3_4*dx3_4
adYX4
```

```
## [1] -124701851
```

```
adYX5<-sy2_3*dx2_3
adYX5
```

```
## [1] -182520957
```

```
adYX6<-sy1_2*dx1_2
adYX6
```

```
## [1] 204944915
```

```
adYX <- c(adYX1, adYX2, adYX3, adYX4, adYX5, adYX6)
# Sum of the positive and negative results of columns XY and YX
sadYXplus <- sum(adYX[which(adXY>0)])
sadYXminus <- sum(adYX[which(adXY<0)])
# Determining the area
area<-(sadXYplus + sadXYminus)/2
area
```

```
## [1] -235.323
```

```
# Checking
area<-(sadYXplus + sadYXminus)/2
area
```

```
## [1] 235.323
```

The area obtained is 235.323 m^2.

The results of data columns used to determine binary sum, binary difference and double area are joined to the original irradiation measurement data by means of the cbind function.

```
irrArea<-cbind(irr,sx, sy, dx, dy, adXY, adYX)
```

The write.table function is used to organize the database into file in .txt extension.

```
## write.table(irrArea,file= "E:/Aulas/Topografia/Aula10/irrArea.txt",
##              sep = " ",   row.names = TRUE, col.names = TRUE)
```

11.9.5   Perform the inverse calculation and prepare the descriptive memorial

The inverse calculation of azimuths and distances is performed from the first vertex, clockwise, in order to draw up a descriptive memorial referring to the polygon surveyed in the field. For this, a database with the measurements and irradiation adjustment calculation memory of a building is imported through the read.table function.

```
irrArea<-read.table("files/irrArea.txt", header = TRUE,
                    sep = " ", dec = ".")
```

The azimuth and distance calculations for alignments made around the building are determined from inverse calculation equations to determine longitude and latitude angular and linear measurements obtained from the absolute X and Y coordinates of vertices.

```
# Azimuth calculation
Az1_6<-deg(atan(irrArea$dx[1]/irrArea$dy[1]))+180
Az1_6
```

```
## [1] 146.2243
```

```
Az6_5<-deg(atan(irrArea$dx[2]/irrArea$dy[2]))+180
Az6_5
```

```
## [1] 237.6286
```

```
Az5_4<-deg(atan(irrArea$dx[3]/irrArea$dy[3]))+180
Az5_4
```

```
## [1] 146.0846
```

```
Az4_3<-deg(atan(irrArea$dx[4]/irrArea$dy[4]))+180
Az4_3
```

```
## [1] 237.5982
```

```
Az3_2<-deg(atan(irrArea$dx[5]/irrArea$dy[5]))+360
Az3_2
```

```
## [1] 326.6306
```

```r
Az2_1<-deg(atan(irrArea$dx[6]/irrArea$dy[6]))+0
Az2_1
```

```
## [1] 55.60936
```

```r
Az<-c(Az1_6, Az6_5, Az5_4, Az4_3, Az3_2, Az2_1)
# Calculating distances
H1_6 <- sqrt((irrArea$dx[1]^2)+(irrArea$dy[1]^2))
H1_6
```

```
## [1] 9.200217
```

```r
H6_5 <- sqrt((irrArea$dx[2]^2)+(irrArea$dy[2]^2))
H6_5
```

```
## [1] 6.256556
```

```r
H5_4 <- sqrt((irrArea$dx[3]^2)+(irrArea$dy[3]^2))
H5_4
```

```
## [1] 11.38115
```

```r
H4_3 <- sqrt((irrArea$dx[4]^2)+(irrArea$dy[4]^2))
H4_3
```

```
## [1] 8.925581
```

```r
H3_2 <- sqrt((irrArea$dx[5]^2)+(irrArea$dy[5]^2))
H3_2
```

```
## [1] 20.05342
```

```r
H2_1 <- sqrt((irrArea$dx[6]^2)+(irrArea$dy[6]^2))
H2_1
```

```
## [1] 15.00859
```

```
H<-c(H1_6, H6_5, H5_4, H4_3, H3_2, H2_1)
P<-sum(H)
P
```

```
## [1] 70.82552
```

```
# Display X and Y coordinates with 4 decimal places
X1<-format(irr$Xi[1], digits=4, nsmall=4)
X1
```

```
## [1] "600105.9978"
```

```
X6<-format(irr$Xi[6], digits=4, nsmall=4)
X6
```

```
## [1] "600111.1126"
```

```
X5<-format(irr$Xi[5], digits=4, nsmall=4)
X5
```

```
## [1] "600105.8283"
```

```
X4<-format(irr$Xi[4], digits=4, nsmall=4)
X4
```

```
## [1] "600112.1787"
```

```
X3<-format(irr$Xi[3], digits=4, nsmall=4)
X3
```

```
## [1] "600104.6427"
```

```
X2<-format(irr$Xi[2], digits=4, nsmall=4)
X2
```

```
## [1] "600093.6126"
```

```
X<-c(X1, X6, X5, X4, X3, X2)
Y1<-format(irr$Yi[1], digits=4, nsmall=4)
Y1
```

```
## [1] "8273801.4628"
```

```
Y6<-format(irr$Yi[6], digits=4, nsmall=4)
Y6
```

```
## [1] "8273793.8154"
```

```
Y5<-format(irr$Yi[5], digits=4, nsmall=4)
Y5
```

```
## [1] "8273790.4656"
```

```
Y4<-format(irr$Yi[4], digits=4, nsmall=4)
Y4
```

```
## [1] "8273781.0208"
```

```
Y3<-format(irr$Yi[3], digits=4, nsmall=4)
Y3
```

```
## [1] "8273776.2380"
```

```
Y2<-format(irr$Yi[2], digits=4, nsmall=4)
Y2
```

```
## [1] "8273792.9854"
```

```
Y<-c(Y1, Y6, Y5, Y4, Y3, Y2)
```

With the results of longitude, latitude, azimuth and horizontal distance, information about boundary type and the confrontation for each alignment are added to the descriptive memorial. Then, a database is organized with rows and columns of these variables.

```
# Descriptive memorial in table
Vertice <- c("i1", "i6", "i5", "i4", "i3", "i2")
Boundary <-c("masonry", "masonry", "masonry", "masonry", "masonry",
             "masonry")
Confrontation <- c("UFMT campus", "UFMT campus", "UFMT campus",
                   "UFMT campus", "UFMT campus", "UFMT campus")
memorial <-cbind(Vertice, X, Y, Az, H, Boundary, Confrontation)
```

The descriptive memorial database is exported by the `write.table` function in `.txt` file format.

```
## write.table(memorial,file= "E:/Aulas/Topografia/Aula10/memorial.txt",
##             sep = " ", row.names = TRUE, col.names = TRUE)
```

11.9.6 Descriptive textual memorial

The descriptive textual memorial is made with information from the database previously used in the inverse calculation, as follows:

The perimeter description is started by means of longitude (X) and latitude (Y) coordinates from vertex 1 ($X = 600105.9978$ m; $Y = 8273801.4628$ m) going to vertex 6 ($X = 600111.1126$ m; $Y = 8273793.8154$ m) with azimuth of $146.224277173901°$ and distance of 9.20021694736941 m, masonry border and confronting the university campus; from vertex 6, followed to vertex 5 ($X = 600105.8283$ m; $Y = 8273790.4656$ m), with azimuth of $237.62857327939°$ and distance of 6.25655590885889 m, masonry boundary and confronting the university campus; from vertex 5, proceeding to vertex 4 ($X = 600112.1787$ m; $Y = 8273781.0208$ m), with azimuth of $146.084587371097°$ and distance of 11.3811485699373 m, masonry boundary and confronting the university; from vertex 4, proceeding to vertex 3 ($X = 600104.6427$ m; $Y = 8273776.2380$ m), with azimuth of $237.598199434916°$ and distance of 8.92558141699616 m, masonry boundary and confronting the University campus; from vertex 3, followed to vertex 2 ($X = 600093.6126$ m; $Y = 8273792.9854$ m), with azimuth of $326.63061226206°$ and distance of 20.0534227306883 m, masonry boundary and confronting the university campus; from vertex 2, followed to vertex 1, with azimuth of $55.6093648494757°$ and distance of 15.0085933710381 m, masonry boundary and confronting the university campus. The perimeter is 70.82552 m and the area, 235.323 m^2. The coordinates are described in the 21S UTM plane coordinate system. The map of the surveyed property and the data spreadsheet are presented below.

11.9.7 Descriptive memorial in spreadsheet

The descriptive memorial can also be described in table format as an annex to the descriptive textual memorial. To do this, you must import the table to view the summary using the `read.table` function.

```
memorial<-read.table("files/memorial.txt", header = TRUE, sep = " ",
                     dec = ".")
memorial # Viewing table in R
```

```
##    vertice      X        Y         Az         H     divisa    confrontante
```

```
## 1      i1 600106.0 8273801 146.22428   9.200217 alvenaria campus da UFMT
## 2      i6 600111.1 8273794 237.62857   6.256556 alvenaria campus da UFMT
## 3      i5 600105.8 8273790 146.08459 11.381149 alvenaria campus da UFMT
## 4      i4 600112.2 8273781 237.59820   8.925581 alvenaria campus da UFMT
## 5      i3 600104.6 8273776 326.63061 20.053423 alvenaria campus da UFMT
## 6      i2 600093.6 8273793   55.60936 15.008593 alvenaria campus da UFMT
```

11.9.8   Mapping the polygon

The polygon mapping described in the textual descriptive memorial and in the table is used to characterize the vertices used in the area computation. To map the polygon, irradiations are previously converted from point to polygon type features.

```
# Import shapefile of irradiation vertices
irrGeo <- readOGR(dsn="files/irrGeo.shp", "irrGeo")
# Convert points into polygon and assign UTM coordinate system
pIrr<-Polygon(irrGeo)
psIrr<- Polygons(list(pIrr),1)
spsIrr <- SpatialPolygons(list(psIrr))
proj4string(spsIrr) <- CRS("+init=epsg:32721")
# Create database for the polygon
data = data.frame(f=99.9)
# Convert to attribute polygon
spdfIrr <- SpatialPolygonsDataFrame(spsIrr,data)
```

Next, we map the polygon and the vertices' information (Figure 11.5).

```
plot(spdfIrr, axes=T, col="gray")
pointLabel(coordinates(spdfIrr),labels=spdfIrr$vertices)
```

11.9.9   Calculate polygon area using **gArea** function

For checking purposes, the polygon area that characterized the irradiated building is checked by using the gArea function of the rgeos package.

```
gArea(spdfIrr)
```

```
## [1] 235.323
```

11.9.10   Calculate polygon area using **st_area** function

The st_area function from the R package sf is applied to calculate the polygon area allowing to obtain the same building area value by different methods and functions.

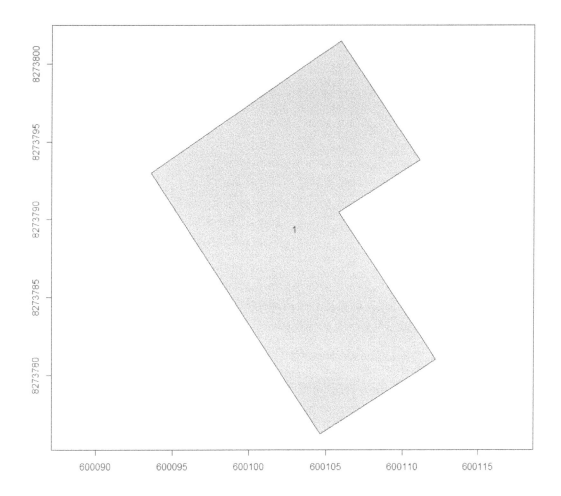

FIGURE 11.5: Mapping polygon geometry and label information of a building using the maptools R package.

```
# Convert to sf feature
Irrsf<-st_as_sf(spdfIrr)
```

```
## Warning in CPL_crs_from_input(x): GDAL Message 1: +init=epsg:XXXX syntax is
## deprecated. It might return a CRS with a non-EPSG compliant axis order.
```

```
# Calculate the polygon area
st_area(Irrsf)
```

```
## 235.323 [m^2]
```

```
# Check Ok!
```

11.9.11   Export the polygon

Finally, the `SpatialPolygonsDataFrame` of the irradiated university campus building is exported as ESRI Shapefile with the `writeOGR` function, for further use.

```
## writeOGR(spdfIrr,dsn="E:/Aulas/Topografia/Aula10/spdfIrr.shp",
##          layer="spdfIrr", driver="ESRI Shapefile")
```

11.10   Solved Exercises

11.10.1   Calculate the area of field survey data by the radiation method. Check the results by another method.

Field survey data used is based on the following data (Table 11.3):

TABLE 11.3: Field survey data by the radiation method.

Vertex	Horizontal Distance (m)	Horizontal Angle
1	882.371	0°0′0″
2	808.679	129°14′01″
3	1157.491	203°56′42″
4	825.571	255°03′37″

A: The data is organized into rows and columns to perform area calculation by different methods using the function `data.frame`. The angle values in degrees, minutes and seconds are converted to decimal degrees to perform the calculations.

```
pto<-c(1, 2, 3, 4)
H<-c(882,371, 808,679, 1157,491, 825,571)
AH<-c(0, 129+14/60+1/3600, 203+56/60+42/3600, 255+3/60+37/3600)
data<-cbind(pto, H, AH)
data<-data.frame(data)
```

The area of 1451645.6153 m^2 is obtained using the radial triangles method for area calculation.

```
# Perform area calculation by radial triangles
A1<-data$H[2]*data$H[1]*sin(rad(data$AH[2]-data$AH[1]))
A2<-data$H[3]*data$H[2]*sin(rad(data$AH[3]-data$AH[2]))
A3<-data$H[4]*data$H[3]*sin(rad(data$AH[4]-data$AH[3]))
```

```
A4<-data$H[1]*data$H[4]*sin(rad(data$AH[1]-data$AH[4]))
# Determine area
A<-(sum(A1,A2,A3,A4))/2
A
```

```
## [1] 774156.1
```

```
# Determine area with 4 decimal places
A<-format(A, digits=4, nsmall=4)
A
```

```
## [1] "774156.0529"
```

Then, the polygon area is determined by the matrix determinants of order two method to evaluate the results. The longitude and latitude values are calculated.

```
x<- data$H*(sin(rad(data$AH)))
x
```

```
## [1]     0.0000   287.3668 -327.9345 -656.0482     0.0000   380.3156 -334.8341
## [8] -551.6988
```

```
y<- data$H*(cos(rad(data$AH)))
y
```

```
## [1]   882.0000 -234.6515 -738.4599 -175.0480 1157.0000 -310.5495 -753.9968
## [8] -147.2053
```

In the calculation of absolute coordinates, the value 2000 is added to X and Y in order to obtain positive absolute coordinates for each vertex.

```
X<- 2000+x
X
```

```
## [1] 2000.000 2287.367 1672.066 1343.952 2000.000 2380.316 1665.166 1448.301
```

```
Y<- 2000+y
Y
```

```
## [1] 2882.000 1765.349 1261.540 1824.952 3157.000 1689.450 1246.003 1852.795
```

The calculated absolute coordinates are added to the original dataset.

```
data<-cbind(data, X, Y)
```

The area obtained by the matrix determinants of order two is 1451645.6153 m^2.

```
# Determine sum of the ascending products
X2_Y1 <-data$X[2]*-data$Y[1]
X3_Y2 <-data$X[3]*-data$Y[2]
X4_Y3 <-data$X[4]*-data$Y[3]
X1_Y4 <-data$X[1]*-data$Y[4]
sumAsc<-sum(X2_Y1, X3_Y2, X4_Y3, X1_Y4)
# Determine sum of the descendant products
X1_Y2 <-data$X[1]*-data$Y[2]
X2_Y3 <-data$X[2]*-data$Y[3]
X3_Y4 <-data$X[3]*-data$Y[4]
X4_Y1 <-data$X[4]*-data$Y[1]
sumDesc<-sum(X1_Y2, X2_Y3, X3_Y4, X4_Y1)
# Determine area
area<-(sumAsc-sumDesc)/2
# Determine area with 4 decimal places
area<-format(area, digits=4, nsmall=4)
area
```

```
## [1] "-774156.0529"
```

The same result is obtained by different area calculation methods based on checking the results.

11.10.2 A polygon traversing was defined by a succession of trapezoids. Determine the polygon area by Bezout's method, based on the following data:

- The values obtained are $H = 20$ m, $y1 = 1.8$ m, $y2 = 3.5$ m, $y3 = 4.7$ m, $y4 = 5.5$ m, $y5 = 5.8$ m, $y6 = 5.4$ m, $y7 = 3.8$ m.

A: The area is 554 m^2.

```
# Calculations performed in R
A<-(20/2)*((1.8+3.8)+(2*(3.5+4.7+5.5+5.8+5.4)))
A
```

```
## [1] 554
```

11.11 Homework

Choose one exercise presented by the teacher and solve the question with different input values. Compare the results obtained. Perform a practice with a closed-path polygon traversing method in the field. Determine X, Y coordinates of the vertices, perform area calculation, descriptive memorial and mapping.

11.12 Resources on the Internet

As a study guide, it is requested to view the slides and illustrative videos about the subject covered in the chapter in Table 11.4.

TABLE 11.4: Slide shows and video presentations on area calculation in surveying.

Guide	Address for Access
1	Slides on area calculation in surveying[1]
2	Area computation in surveying using coordinates[2]
3	Computation of areas[3]

11.13 Research Suggestion

The development of scientific research on geomatics is stimulated by the activity proposals that can be used or adapted by the student to assess the applicability of the subject matter covered in the chapter (Table 11.5).

TABLE 11.5: Practical and research activities used or adapted by students using area calculation in surveying.

Activity	Description
1	Check different methodologies for area calculation and evaluate which is the best method, providing advantages and disadvantages
2	Determine the area of a closed polygon based on a topographical survey conducted in the field using plane coordinate information
3	Determine the area of a closed polygon based on a topographical survey carried out in the field using an automatic calculation method based on an R package. Compare the results obtained by another methodology

---

[1] http://www.sergeo.deg.ufla.br/geomatica/book/c11/presentation.html#/
[2] https://www.youtube.com/watch?v=jGhqu84oyxM
[3] https://www.youtube.com/watch?v=Kg_CNxx3Jqc

Activity	Description
4	Perform the inverse calculation of coordinates and prepare the text surveying memorial of a topographic survey

11.14 Learning Outcome Assessment Strategy

Perform a summary of the chapter, "Traverse Area Evaluation and Surveying Memorial with Geomatics and R", on a single A4 page in order to show the student's abilities to summarize a subject presenting key points considered of greater importance today.

12

Coordinate Reference Systems for Geodetic

12.1 Learning Questions

The emergent learning questions answered through reading the chapter are as follows:

- How to define coordinate reference systems.
- How to check and assign a coordinate reference system in geographic data with R.
- How to determine Euclidean and geodesic distance between vertices with R.
- How to determine the geodetic area of closed-path polygon traverses with R.
- What is the applicability of EPSG geodetic parameter dataset and the function `proj4string` in topographic surveys?
- Whether there is a difference between area calculated on SIRGAS-2000, SAD-69, Chua, and Córrego Alegre 1961 ellipsoids.

12.2 Learning Outcomes

The learning outcomes expected from reading the chapter are as follows:

- Know concepts involved in coordinate reference systems (CRSs) applied to vector and raster type data with R.
- Check and assign a coordinate reference system in geographic data.
- Determine Euclidean and geodetic distance between vertices.
- Calculate geodetic area of polygons.
- Evaluate the applicability of EPSG geodetic parameter dataset and the function `proj4string` in topographic surveys with R.
- Compare area calculated on SIRGAS-2000, SAD-69, Chua, and Córrego Alegre 1961 ellipsoids.

12.3 Introduction

The "datum" is a term that meant geometric reference in geomatics and has been used to define a referential for position of geometric elements in space or in a topographic mapping. The set of information defining shapes, size, origin and orientation of coordinate systems established positioning of points on the Earth's surface with geodetic reference by a geodetic datum. After determining and deploying a geodetic datum for a region or a country, a network of points with

DOI: 10.1201/9781003184263-12

coordinates referenced to that datum must be deployed, called the "geodetic coordinate reference system" (Silva and Segantine, 2015).

Geodetic surveys have been used to establish horizontal and vertical accuracy of reference monuments and have been used as a basis for generating or checking survey projects such as topographic and hydrographic mapping, boundary demarcation, route definition, construction planning, among others. The established references have been essential as support for the use of geographic and Earth information systems (Ghilani and Wolf, 1989; Alves and Silva, 2016).

General types of geodetic datum are: horizontal and vertical. With the horizontal datum, planimetric positions of points on the Earth's surface are established through latitude, geodetic longitude, direction and parameters that defined a reference ellipsoid. With the vertical datum, references are established to determine orthometric altitudes of points over large areas. Plane coordinates, such as those of the Universal Transverse Mercator (UTM) system can be calculated based on values defined in geodetic surveys (Silva and Segantine, 2015; Alves and Silva, 2016).

Different methods can be used to define a network of geodetic control points in order to extend the understanding about the Earth's shape and mass. Extending the network of geodetic control points can be accomplished by leveling, accurate polygonals, very long base interferometry, satellite laser mapping, Global Navigation Satellite System (GNSS) observations, and gravity surveys. The accumulation of these data can be used to update and extend existing control data, as well as to accurately define the geometric shape of the Earth (Kavanagh and Slattery, 2015). GNSS has been used for control surveys over extensive areas, nationally and continentally. GNSS surveys have been rapidly replacing other existing control surveying methods with advantages related to ease of use, speed, and extremely high accuracy over large distances. GNSS surveying can also be used to establish vertical control, but with limitations due to the need to obtain an accurate geoidal model as a reference. The horizontal control survey with GNSS has been used to determine geodetic latitudes and longitudes of points associated with geoid and ellipsoid (Alves and Silva, 2016).

Control grids can be used as a reference for the point structure used in (Schofield et al., 2007):

- Topographic mapping and large-scale production planning;
- Dimension control in construction;
- Surveying deformation of structures;
- Augmentation and densification of existing control networks.

To establish a geodetic reference system, it is necessary to include definitions about (Silva and Segantine, 2015):

- The topographic or physical surface, with relief irregularities, mountain ranges, valleys, fields, oceanic hollows and marshes;
- The geoidal surface, determined as function of the equipotential surface of the Earth's gravitational field, is considered the real Earth's shape;
- The geometric surface, determined by a sphere flattened at the poles, is called the "ellipsoid of revolution".

12.4   Geoid and Ellipsoid

The horizontal control survey can be used to determine geodetic latitudes and longitudes of vertices. Geoid and ellipsoid definition is necessary for geodetic latitude and longitude determination.

The geoid is an equipotential gravitational surface at any location, perpendicular to the direction of gravity. The geoid is irregularly shaped due to variations in the Earth's mass distribution and rotation (Figure 12.1) (Ghilani and Wolf, 1989; Schofield et al., 2007; Alves and Silva, 2016).

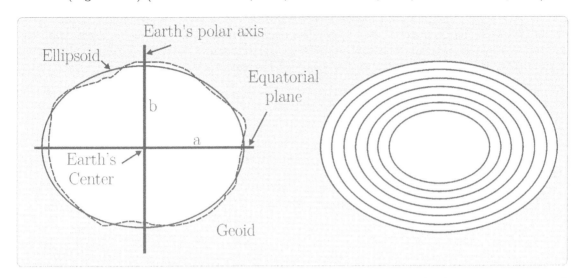

FIGURE 12.1: Illustration of setting ellipsoid and geoid on the Earth (left) and an equipotential surface (right).

The magnitude of gravity force is greater in relation to the Earth's center of mass than to the latitude. Gravity is also affected by the Earth's mass distribution and density variations. The shape of the Earth is defined by gravity. If the Earth is a molten material with homogeneous density, unaffected by the gravitational field of external bodies, and does not rotate on its own axis, then by gravitational attraction, the Earth's surface can be described as a sphere. The biggest source of error in this ideal model occurs because of the Earth's rotation. The centrifugal force at the Equator acts in the opposite direction to the gravitational force, so that the Earth is best described by an ellipse of rotation around its minor axis, called by the synonymous terms "ellipsoid" or "spheroid" (Schofield et al., 2007; Alves and Silva, 2016).

By imagining the oceans, which make ~70% of the Earth's surface, as a flow of channels interconnected by land masses, the equipotential surface would be formed approximately at mean sea level, disregarding frictional effects, winds and tides, among others. On this surface, the gravitational potential would be the same at any point. Thus, the level lines never cross. Therefore, the geoid is a surface of physical reality where the surface can be measured. Although the gravitational potential is the same at any location, the value of gravity is not. The magnitude of the gravity vector at any point is corresponding to the rate of change of the potential gravity at that point. The geoid surface is smoother than the Earth's physical surface, but still contains many irregularities that make it undesirable for planimetric mathematical position location. These irregularities are caused by mass anomalies along the Earth (Schofield et al., 2007; Alves and Silva, 2016).

Since the direction of the gravity vector, called "vertical", is normal to the geoid, the surveyor's plumb line is defined by this vector. Therefore, any instrument leveled by plumb line will be referenced by the equipotential surface that passes through the instrument. The elevation defined by the equipotential surface passing through mean sea level is called "orthometric altitude". Orthometric altitude is the linear distance measured along the gravity vector of a point on the surface in relation to the equipotential surface used as a reference datum. As the geoid characterized an equipotential surface adjusted to mean sea level, the measured heights refer to altitudes above or below mean sea level. Thus, it is observed that orthometric altitudes are datum dependent. Similarly, mean sea level varies on the geoid by about 3 m at the poles and Equator, mainly as a function of temperature variation. Therefore, elevations referring to a specific datum cannot be the same as those established by another country's datum (Schofield et al., 2007; Alves and Silva, 2016).

The ellipsoid is defined as a mathematical surface obtained by the revolution of an ellipse on the Earth's polar axis. The dimensions of the ellipsoid are selected to determine a good fit of the ellipsoid on the geoid over large areas, based on surveys conducted in the area. The ellipsoid approximated the geoid and can be mathematically defined to calculate positions of points separated by large distances in control and geodetic surveys (Alves and Silva, 2016).

The first accurate terrestrial ellipsoid was determined by German astronomer Friedrich Wilhelm Bessel (1784-1846) in 1841. In Brazil, three ellipsoids are considered (Silva and Segantine, 2015):

- International Ellipsoid of 1924 (Hayford), on which the Córrego Alegre datum is based;
- Reference Ellipsoid of 1967, on which the Chuá datum is based, referenced to the South American Datum of 1969 (SAD-69);
- Ellipsoide Geodetic Reference System of 1980 (GRS-80), on which is based the Geocentric Reference System for the Americas (SIRGAS-2000).

Currently, the Geodetic Reference System of 1980 (GRS-80) and the Geodetic Global Reference System of 1984 (WGS-84) have been used as reference in several countries because they showed satisfactory fit to the geoid and ease of use in GNSS surveys (Alves and Silva, 2016). The size and shape of ellipsoids can be defined by two parameters called semi-axis a, b and flattening f, where $f = 0$ describes a sphere geometry (Table 12.1) (Silva and Segantine, 2015; Alves and Silva, 2016).

TABLE 12.1: Definition of ellipsoid parameters used in geomatics.

Ellipsoid	Semi-axis a (m)	Semi-axis b (m)	Flattening f	Use
GRS-80	6378137.0	6356752.314140356	1/298.25	Global (ITRS), Brazil, United States
WGS-84	6378137.0	6356752.314245179	1/298.25	Global (GPS)
Hayford-1924	6378388.0	6356911.946127947	1/297.00	United States / Europe
Córrego Alegre	6378388.0	6356911.946127947	1/297.00	Brazil
GRS-1967	6378160.0	6356774.516000000	1/298.24	Australia
SAD-1969	6378160.0	6356774.719000000	1/298.25	South America

The flattening of the ellipsoid can be described by (Ghilani and Wolf, 1989; Hager et al., 1989; Schofield et al., 2007; Alves and Silva, 2016):

$$f = 1 - \frac{b}{a} = \frac{a - b}{a} \tag{12.1}$$

where a is half of the major axis and, b, half of the minor axis of the Earth's ellipsoid of revolution around the PP' axis (Figure 12.2) (Schofield et al., 2007; Alves and Silva, 2016).

Other variables used in ellipsoid calculation are the first eccentricity (e), the first quadratic eccentricity (e^2), the second eccentricity (e') and the second quadratic eccentricity (e'^2) (Ghilani and Wolf, 1989; Hager et al., 1989; Schofield et al., 2007; Alves and Silva, 2016):

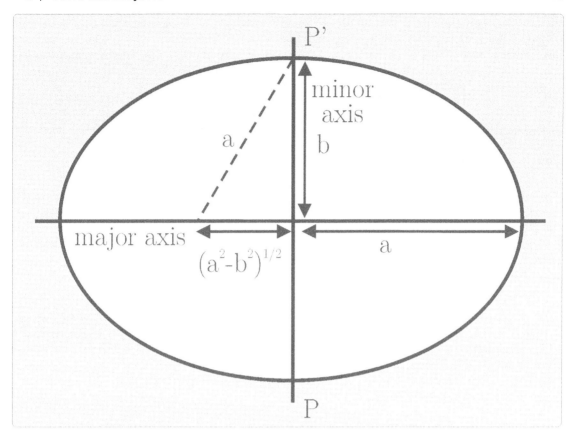

FIGURE 12.2: Parameters a and b of an ellipse used to characterize the Earth's shape.

$$e = \frac{\sqrt{a^2 - b^2}}{a} = \frac{a^2 - b^2}{a^2} = \sqrt{2f - f^2} = \sqrt{f(2 - f)} \tag{12.2}$$

$$e^2 = \frac{a^2 - b^2}{a^2} = 2f - f^2 \tag{12.3}$$

$$e' = \frac{\sqrt{a^2 - b^2}}{b} = \frac{e}{\sqrt{1 - e^2}} = \frac{e^2}{1 - e^2} = \frac{f(2 - f)}{(1 - f)^2} \tag{12.4}$$

$$e'^2 = \frac{a^2 - b^2}{b^2} = \frac{e^2}{1 - e^2} \tag{12.5}$$

where the polar semi-axis (b) is approximately 21 km smaller than the equatorial semi-axis (a) for each ellipsoid. This means that the ellipsoid approaches the shape of a sphere. This assumption can be used in some situations with length calculations on the order of 50 km away.

By assuming the ellipsoid as a sphere, the radius and volume of the reference ellipsoid are equated to those of the sphere (Ghilani and Wolf, 1989; Alves and Silva, 2016):

$$r = \sqrt[3]{a^2 b} \tag{12.6}$$

The spherical distance between two points on the Earth's surface that are at the same distance from the center of the spheroid representing the Earth is measured over the arc generated by the spheroid containing the two points. In practice, reductions from topographic to ellipsoidal distances are performed by considering the ellipsoid as a spheroid of radius equal to the local mean radius of the Earth at the latitude of the point. It may be irrelevant whether the Earth is a spheroid or an ellipsoid for conventional topographical calculations (Silva and Segantine, 2015).

12.5   Conventional Terrestrial Reference System

The ellipsoid is defined on the basis of the size of an ellipse rotated around the Earth's polar axis. Since the Earth's principal axis of inertia does not coincide with the Earth's axis of rotation, the polar axis does not show a fixed position at a given time. This motion can be subdivided into two categories: precession and nutation. Precession is the largest of the movements that ran along the polar axis for a long period of time. By international convention, the Earth's mean axis of rotation is defined between the years 1900 and 1905. This position is called the "Earth's Conventional Terrestrial Pole" (CTP) (Ghilani and Wolf, 1989; Alves and Silva, 2016).

CTP defines the Z axis of a 3D Cartesian coordinate globe with the positive north portion. The positive X axis is located in the mid-equatorial plane and started at the Earth's center of mass, passing through the Greenwich meridian. The Y axis is located in the mid-equatorial plane, generating a Cartesian coordinate system to the right. This coordinate system has come to be known as the "Conventional Terrestrial Reference System" (CTS) (Figure 12.3) (Ghilani and Wolf, 1989; Alves and Silva, 2016).

12.6   Geodetic Position and Ellipsoidal Radii of Curvature

The position of a vertex P on the surface can be defined by means of the latitude, longitude and geodetic altitude values, ϕ_P, λ_P, h_P, respectively. The vertex P located on the Earth's surface is represented by P' on the ellipsoid along the normal (Figure 12.4).

Meridians are large circles on the circumference of the ellipsoid that pass through the north and south poles. Any plane containing a meridian and the polar axis is a meridian plane. The angle in the equatorial plane from the Greenwich meridian passes through P defining the geodetic longitude (λ_P) of the vertex. The plane defined by the vertical circle passing through the vertex P, perpendicular to the meridian plane in the ellipsoid is called the "main vertical" or "normal section". The radius of the main vertical at point P is called the "normal radius" (R_N) because it is perpendicular to a plane tangent to the ellipsoid at P. The geodetic latitude (ϕ_P) is the angle in the meridian plane containing P between the equatorial plane and the normal at P. The geodetic altitude (h_P) should be included to define the location of point P on the Earth's surface. The geodetic altitude is the distance measured along the length of the normal from P' on the ellipsoid to P on the Earth's surface. The geodetic altitude is not equal to the elevation determined by differential leveling. The great circle that defined the prime vertical at P has a radius of the normal section (R_N) that differs from the radius at the meridian (R_M) at P. The radius of the great circle (R_α) at any azimuth α in relation to the meridian is different from R_N or R_M (Ghilani and Wolf, 1989; Hager et al., 1989; Alves and Silva, 2016):

$$R_N = \frac{a}{\sqrt{1 - e^2 sen^2\phi}} = R_M(1 + e'^2 cos^2\phi) \tag{12.7}$$

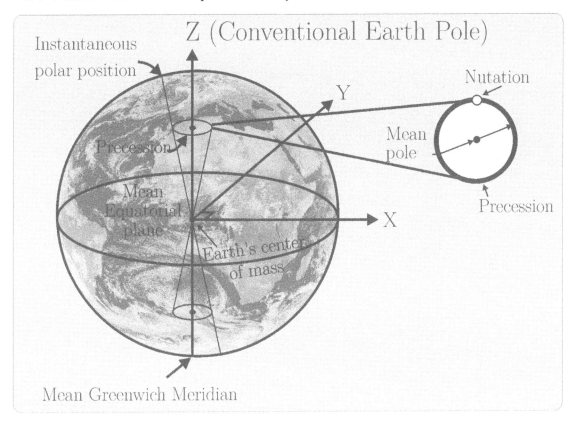

FIGURE 12.3: Conventional Earth system with nutation and precession movements of the Earth's polar axis.

$$R_M = \frac{a(1 - e^2)}{(1 - e^2 \, sen^2 \phi)^{3/2}} \tag{12.8}$$

$$R_\alpha = \frac{R_N R_M}{R_N cos^2 \alpha + R_M sen^2 \alpha} \tag{12.9}$$

where a and e are parameters of the ellipsoid and, ϕ, the geodetic latitude of the station at which the radius is calculated.

We observed that R_N is equal to R_M at the poles under $\phi = 90°$. Considering that $(1 - e^2)$ is less than one, the radius of the prime vertical R_N is greater than the meridian radius R_M at all other locations under $\phi \neq 90°$.

The local mean radius of the Earth's curvature can be understood as the radius of a sphere tangent to the reference ellipsoid at the point of interest. The mean radius is calculated by geometric averaging the radius at the meridian and the radius of the normal section (Silva and Segantine, 2015):

$$R_0 = \sqrt{R_M R_N} \tag{12.10}$$

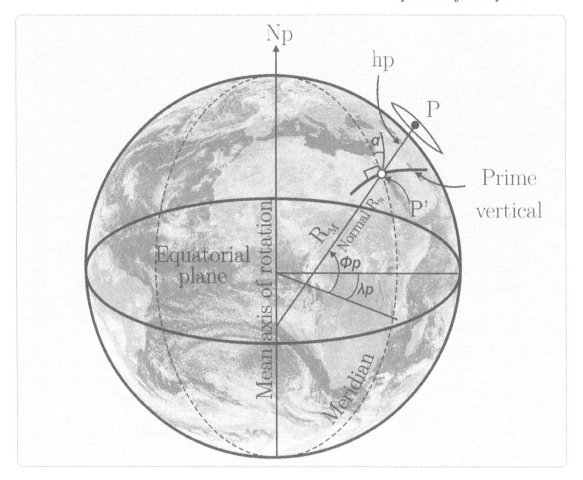

FIGURE 12.4: Definition of different radii and a vertex P on the ellipsoid.

12.7   Geoid Undulation and Deflection of the Vertical

Measurements with topographic equipment can be made relative to the ellipsoid or the geoid, yielding different results. Therefore, it is important to know the conceptual differences between height, altitude, normal line and vertical line of a place (Silva and Segantine, 2015):

- Height: Dimension of a body from the base to the top;
- Altitude: Value of the elevation of a point relative to a vertical datum;
- Normal line or Normal of the point: Line that intercepts the topographic surface, perpendicular to the ellipsoidal surface;
- Vertical line of the point: Tangent line to the line of gravitational force that intercepted the topographic surface, perpendicular to the geoidal surface and that can be materialized by a plumb line.

The geoid is an equipotential surface defined by gravity. If the Earth is represented by a perfect ellipsoid with no internal density variations, the geoid would correspond perfectly to the ellipsoid. However, the geoid may be 100 m or more away from some ellipsoids, depending on the locality. Traditional survey instruments have been oriented in reference to gravity and therefore the observations obtained will be relative to the geoid. The separation between geoid and ellipsoid determines the difference between the height of the point above the ellipsoid (geoidal altitude) and above the geoid (orthometric altitude), known as elevation. This difference, called "geoidal altitude" or "geoidal separation"'geoidal separation", can be observed when comparing the geodetic altitude of a point obtained by GNSS survey with the elevation determined by leveling. The relationship between the orthometric altitude (H) and the geodetic altitude (h) at any point is (Figure 12.5) (Ghilani and Wolf, 1989; Silva and Segantine, 2015; Alves and Silva, 2016):

$$h = N + H \qquad (12.11)$$

where N is the geoidal altitude.

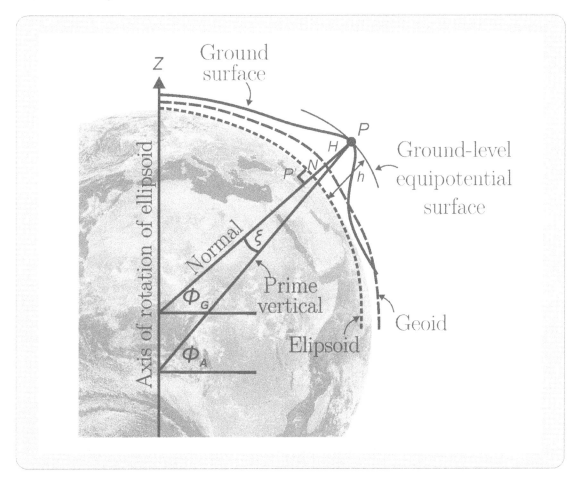

FIGURE 12.5: Relationship between the ellipsoid and the geoid.

The altitude on the geoid surface is established from a permanent landmark, natural or artificial, called "vertical datum". The vertical datum is determined from measurements of tidal variations over a period of time. In Brazil, the Brazilian Institute of Geography and Statistics (IBGE) used information collected by a tide gauge station in Imbituba, Santa Catarina, to define the vertical datum of the Brazilian Geodetic System. Since the geoid and the ellipsoid rarely coincide, it is necessary to consider the height difference (N) between them for geodetic and topographic calculations. The height difference between the geoid and the ellipsoid is named "geoidal undulation" or "geoidal height" (Silva and Segantine, 2015).

In Brazil, the MAPGEO geoidal undulation model is used as a computational system to provide the geoidal undulation (N) used in the conversion of geometric altitudes obtained by geospatial positioning techniques by positioning satellite systems (GNSS) into altitudes consistent with the High Precision Altimetric Network (RAAP) of the SGB, made available by the IBGE. The geoidal undulation can be obtained for one or more points quickly and practically, requiring fundamental knowledge associated with the surfaces (physical, geoidal and ellipsoidal) and references adopted in Geodesy as the Geocentric Reference System for the Americas (SIRGAS-2000) and the South American Datum 1969 (SAD-69), enabling the user to analyze more consistently the results obtained (Chuerubim, 2013).

Another geometric effect arising from the geometric inconsistency between the geoid and the ellipsoid is the vertical deflection. The deflection of the vertical at any point P is the angle between the vertical (vertical deflection) and the normal in relation to the ellipsoid. This angle is generally determined by two components: the orthogonal projections onto the meridian and the normal planes. The zenith of the equipotential surface at field level is called the "astronomical zenith" Z_A, because it corresponds to the direction of gravity (zenith) of a leveled instrument during astronomical observations. The Z_G is the normal at point P. The projected components of the total vertical deviation in the meridian and normal planes are denoted by ξ and η, along the NS and EW directions, respectively. The meridian (NS) component (ξ) is positive when located north of the normal to the ellipsoid. The prime vertical (EW) component (η) is positive when located east of the normal to the ellipsoid (Figure 12.6) (Ghilani and Wolf, 1989; Silva and Segantine, 2015; Alves and Silva, 2016).

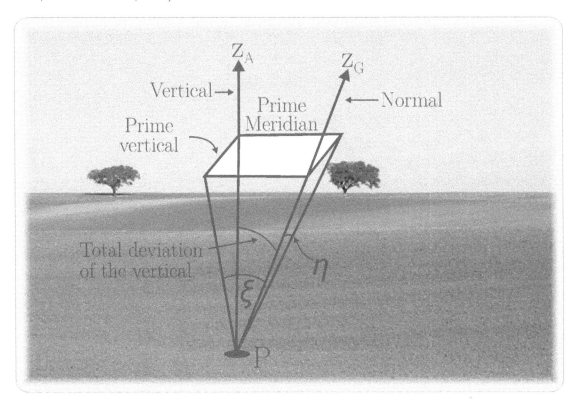

FIGURE 12.6: Components of the vertical deflection.

With the leveling of topographic equipment relative to the vertical of the site, there is an angular difference between the tangent of the vertical of the site and the line normal to the ellipsoid, called the "vertical offset". The vertical offset can vary from approximately 5" in flat regions to 30" in mountainous regions (Silva and Segantine, 2015).

The components, ξ, on the meridian line and, η, perpendicular to the meridian section, can be calculated by (Ghilani and Wolf, 1989; Silva and Segantine, 2015; Alves and Silva, 2016):

$$\xi = \phi_A - \phi_G \tag{12.12}$$

$$\eta = (\lambda_A - \lambda_G)cos\phi = (Az_A - Az_G)cot\phi \tag{12.13}$$

Laplace's equation can be derived from this equation by:

$$Az_G = Az_A - (\lambda_A - \lambda_G)sen\phi = Az_A - \eta tan\phi \tag{12.14}$$

where ϕ_A and ϕ_G are the astronomical and geodetic latitudes, λ_A and λ_G, the astronomical and geodetic longitudes, and Az_A and Az_G, the astronomical and geodetic azimuths, respectively.

12.8 Coordinate Reference Systems

The use of a coordinate system made it possible to (Silva and Segantine, 2015):

- Determine the position of a topographic point through coordinates;
- Standardize calculation methods to define topographic points;
- Unify individual systems into a single overall system to simplify the identification and management of topographic points in a single project.

In geomatics, the following coordinate systems have been described (Silva and Segantine, 2015):

- Cartesian plane coordinate system or plane-rectangular system;
- Plane polar coordinate system;
- Spatial Cartesian coordinate system;
- Geodetic geographic coordinate system.

12.8.1 Cartesian plane coordinate system

The Cartesian plane coordinate system or plane-rectangular system is one of the best known in geomatics. This system consisted of two geometric axes in the same plane and perpendicular to each other forming quadrants. The primary, horizontal axis is the abscissa (X) and the vertical axis is the ordinate (Y). The vertical axis is adopted as the origin axis for the directions and the clockwise angular direction as positive. The direction is indicated by the azimuth angle formed by the vertical axis and the alignment considered (Silva and Segantine, 2015).

12.8.2 Plane polar coordinate system

The plane polar coordinate system is defined by a fixed point of origin or pole, a direction (polar angle) relative to a reference axis, and by a distance (radius vector) between the origin and the point at which the coordinates are to be determined. This system has been used in observations of horizontal distances with field-sighted surveying instruments (Silva and Segantine, 2015).

12.8.3   Spatial Cartesian coordinate system

The spatial positioning of a point can be determined in a Cartesian system with the addition of a third Z axis to the Cartesian plane coordinate system or the polar plane coordinate system. The Z axis is added perpendicular to the plane established by the X, Y axes. The geocentric spatial Cartesian system is used to determine the position of points in space by positioning satellites and GNSS signal receiving antennas. In this case, the system origin is defined at the center of the Earth, the X, Y axes at the Equator and the Z axis, coincident with the Earth's medium rotation axis and the X axis, intersecting a meridian adopted as reference. In this system, the Z coordinate is perpendicular to the plane of the Equator and the ellipsoidal height, h, is normal to the reference surface. Since there is a difference between the Z coordinate and the ellipsoidal height h, an increase in h will produce an equal increase in Z only at the poles (Silva and Segantine, 2015).

12.8.4   Geodetic geographic coordinate system

In the geodetic geographic coordinate system, the coordinates have been determined on a surface of a reference ellipsoid rather than on a spherical surface, as in the case of the geographic coordinate system. The geodetic coordinate system is based on the axis of rotation of the reference ellipsoid and the plane of the Equator. The meridian lines pass through the poles determining elliptical surfaces. Parallels occur perpendicular to meridians, with a maximum circle in the equatorial plane. For geodetic latitudes and longitudes, there is a specific meridian and the Equator plane as the origin for determining arcs over the reference surface. The geodetic latitude (ϕ) of a point on the reference surface is defined by the angular value of the arc formed by the straight line normal to that surface between the point and the Equator plane. Geodetic latitudes are referenced from the Equator from 0 to 90° in the Northern Hemisphere (N), and from 0 to -90° in the Southern Hemisphere (S). The geodetic longitude (λ) of a point on the reference surface is the value of the dihedral angle formed between the meridian plane passing through the point, with the plane passing through the meridian plane of origin. Geodetic longitudes are referenced to the Greenwich meridian and ranged from 0 to 360° for East (E) and, from 0 to -180° for West (W). Since the reference ellipsoid is flattened, the normal line that defined the geodetic latitude of a point does not pass through the center of the ellipsoid (Silva and Segantine, 2015).

12.9   Coordinate Reference System Transformation

In coordinate transformation, a mathematical operation is used to relate two different coordinate systems in order to characterize the position of a point in different systems. The main coordinate transformations between plane coordinate systems are the transformations of coordinates from Cartesian system to polar system, between two Cartesian coordinate systems, and the transformation of spatial Cartesian coordinates to geodetic geographic coordinates and vice versa. The transformation from rectangular to polar coordinates and vice versa is widely used in geomatics when measurements are performed in the field in a polar coordinate system and the topographic map representation is on rectangular coordinate system. With the advent of GNSS, there is a need to transform spatial Cartesian coordinates into geodetic coordinates and vice versa. The positions determined by GNSS are initially obtained in spatial coordinates (X, Y, Z) which are later transformed into plane coordinates (E, N, H) for use in engineering projects (Silva and Segantine, 2015).

12.10   Geodetic Calculation Algorithms

In geodetic position calculations, there are direct and inverse problems. In direct problems, the latitude and longitude of a station can be calculated as a function of the latitude and longitude of another station, the geodetic length and azimuth of the alignment between the two stations. For lines determined over large extents, it is necessary to consider the shape of the ellipsoid to obtain satisfactory accuracy. In inverse geodetic problems, the latitude and longitude values of an alignment are used to determine the geodetic length and azimuth between two points (Ghilani and Wolf, 1989; Alves and Silva, 2016). Algorithms for computing geodetic position on an ellipsoid of revolution with accurate results, robust and fast solutions to direct and inverse geodetic problems are obtained in Karney (2013).

Algorithms for solving geodesic problems using calculators are summarized in the work of Vincenty (1975), based on earlier work such as that of Helmert (1964). The algorithms used to calculate geodetic distance, azimuth and area are adapted from these works for use on modern computers. Some improvements over previous methods are (Karney, 2013):

- Precision is increased to match the standard precision of most computers;
- An inverse problem solution is used to converge all pairs of points overcoming shortcomings observed by Vincenty's method;
- Differential and integral properties of geodesics are calculated in order to enable the calculation of scales of geodesic projections without resorting to numerical differentiation;
- Differential properties and integrals are used in solving the inverse problem, with a method to obtain the area of a geodesic polygon, extending the work of Danielsen (1989).

A number of functions have been developed in `sf` with ellipsoidal geometries via the `lwgeom` R-package (Pebesma and Dunnington, 2020):

- Calculation of area of polygons with the function `st_geod_area`;
- Calculation of line length with the function `st_geod_length`;
- Calculation of distance between features with the function `st_geod_distance`;
- Segmentation of lines along great circles with the function `st_geod_segmentize`.

Geocomputation binary predicates, geometry operators and neighbor functions can be used as if in the equivalent equirectangular projection or the WGS-84 datum, for example (Pebesma and Dunnington, 2020):

- Binary predicates: (`intersects, touches, covers, contains, equals, equals_exact, relate`);
- Geometry generation operators: (`centroid, intersection, union, difference, sym_difference`);
- Neighbor functions: (`nearest_point, nearest_feature`);
- Other functions: (`st_filter, st_join, agreggate`).

In traditional cartography, based on map projections, functions are used to map points on the Earth's surface onto a plane map. Map projections created distortions because the shape of the Earth is not plane. For example, the well-known Mercator projection is discontinuous along the 180° meridian, has large-scale distortions at high latitudes, and cannot represent the north and south poles. With the R `s2` package, this problem is addressed using exclusively spherical projections. As the name implies, spherical projections are used to map points on the Earth's surface onto a mathematically perfect sphere. In these mappings there is still some distortion, since the Earth is not entirely spherical, but is much closer to a sphere than a plane. With spherical projections, it is possible to approximate the entire surface of the Earth with maximum distortion of 0.56%, preserving the correct topology of the Earth. Although an ellipsoid represents the Earth better than a sphere, ellipsoidal operations are slower than the corresponding operations on a sphere. Furthermore, robust geometric algorithms require the implementation of exact geometric predicates that are not subject to numerical errors.

The geometry library **S2**[1] can be used for geospatial operations in R (Dunnington et al., 2021). This library is written by Google for use in Google Earth, Google Maps, Google Earth Engine, and Google BigQuery GIS. In the s2 geometry, the straight lines between points on the globe are not formed by straight lines in the equirectangular projection, but by large circles, according to the shortest path on the sphere (Pebesma and Dunnington, 2020).

In the R package sf up to version 0.9-x, in geographic coordinate data, one degree of longitude is equal to one degree of latitude, regardless of its global location, as in this example (Figure 12.7) (Pebesma and Dunnington, 2020):

```
library(sf)
library(rnaturalearth)
```

```
Earth <- countries110 %>%
  st_as_sf() %>%
  st_geometry()
```

```
plot(Earth, axes = TRUE)
```

This means working in a projection similar to the equirectangular projectionequirectangular projection, also called equidistant cylindrical projection (in French, la carte parallélogrammatique), plate carrée or geographic projection (Snyder, 1993), as in this example (Figure 12.8):

```
EarthP<-st_transform(Earth, "+proj=eqc")
```

```
plot(EarthP, axes = TRUE)
```

Some advantages of the s2 geometry library are:

- Flexible support for spatial indexing;
- Fast in-memory spatial indexing of collections of points, polylines, and polygons;
- Robust constructive operations as intersection, union, simplification and Boolean predicates;
- Efficient query operations for finding nearby objects, measuring distances, calculating centroids;
- Flexible and robust implementation of instantaneous rounding;
- Collection of efficient and exact mathematical predicates for testing relationships between geometric primitives;

---

[1]https://s2geometry.io

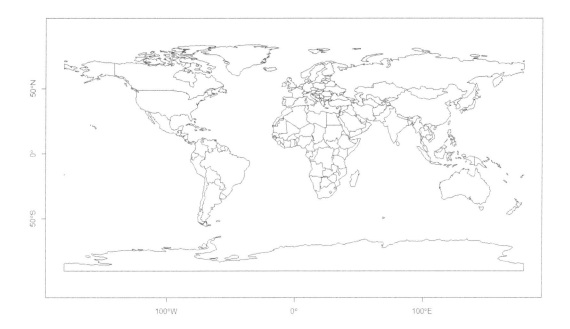

FIGURE 12.7: Mapping the Earth in geographic projection.

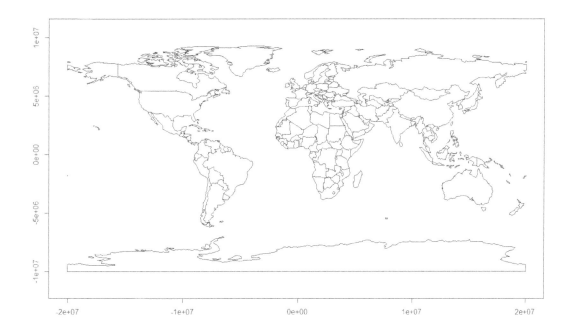

FIGURE 12.8: Mapping the Earth in equirectangular projection.

- Extensive testing on Google's vast collection of geographic data.

The `install.packages` function is used to install the `s2` package.

```
install.packages("s2")
```

The `library` function is used to enable the `s2` package.

```
library(s2)
```

The polygon of Brazil is obtained from the world map, from the R package `spData` and used to calculate the geodesic and spherical area of Brazil in a comparative way, considering the WGS-84 ellipsoid. The map of Brazil with the limits is mapped by the `plot` function (Figure 12.9).

```
library(spData)
```

```
# Perform a subset of Brazil
BrazilWGS84 = world[world$name_long == "Brazil",]
```

```
# Mapping
plot(BrazilWGS84[1], axes=T, col="grey92")
```

The `st_area` and `s2_area` functions are used to perform the geodetic and spherical area calculation in Brazil, obtaining area values of 8508557 and 8540954 km^2 for geodetic and spherical areas, respectively.

```
a1<-units::set_units(st_area(BrazilWGS84),km^2)
a1
```

```
## 8540954 [km^2]
```

```
a2<-units::set_units(s2_area(BrazilWGS84)/1000000,km^2)
a2
```

```
## 8540954 [km^2]
```

In the `s2` package, the Earth's radius is taken as 6371.01 km (Yoder, 1995) as the default value in functions that return a distance or accept a distance as input. Based on this, a difference of 0.37% is observed between area calculations considering the ellipsoid or the sphere.

FIGURE 12.9: Mapping Brazil from a subset of all countries.

```
ellipsoid<-8508557
sphere<-8540954
ellipsoid/sphere
```

```
## [1] 0.9962069
```

```
# Difference in percent
d<-((1-(ellipsoid/sphere))*100)
d
```

```
## [1] 0.3793136
```

A package with algorithms for spherical trigonometry calculations for geographic applications has been developed to calculate distances and related measurements to angular locations of longitude and latitude. There are a number of functions to calculate distance and direction (bearing, azimuth) along great circles (shortest distance on a sphere) and along rhumb lines (lines of constant direction). There are also functions that calculate distances on a spheroid. Other functions include calculating the location of an object according to direction and distance, area, perimeter and centroid of a spherical polygon. Geographic locations must be specified in degrees of longitude and latitude (Hijmans et al., 2019).

There are four distinct functions for calculating the distance between two points, in order of increasing complexity from the spherical cosine law algorithm, Haversine (Sinnott, 1984), Vincenty's sphere, and Vincenty's ellipsoid (Vincenty, 1975). The first three assumed the Earth as a sphere, while in Vincenty's ellipsoid, as an ellipsoid. The results of the first three functions are identical for practical purposes. In Haversine's formula, with the computational precision of the time it is developed, the accuracy is lower than today, of 15 decimal places. With today's accuracy, with the spherical cosine law formula, equal or better results are obtained for very small distances (Hijmans et al., 2020).

Another option for geodetic, spherical and 3D Cartesian distance and direction measurements is observed in the R package `nvctr` (Spinielli and EUROCONTROL, 2020). In this package non-singular position calculation functions are implemented with functions for geographic position calculation for ellipsoidal and spherical models (Gade, 2010) in which the normal vector to the Earth's ellipsoid (called n vector) is used. This non-singular position representation is convenient for practical position calculations, simplicity of use, and exact answers for global positions and distance, in the ellipsoidal, spherical, and Cartesian 3D Earth coordinate models (Spinielli and EUROCONTROL, 2020).

The n-vector method (Gade, 2010) presented simple and non-singular solutions for problems of geographical position calculations as (Spinielli, 2020):

- Given two positions A and B, find the exact vector from A to B in meters north, east and down, and find the direction (azimuth/bearing) to B, relative to north using WGS-84 ellipsoid;
- Given positions A and B, find the surface distance (i.e., great circle distance) and the Euclidean distance;
- Given three positions A, B, and C, find the mean position (center/midpoint);
- Given position A, an azimuth/bearing and a great circle distance, find the destination point B.

Area calculations can also be applied for large geodetic polygons on the ellipsoid. Test calculations of geodetic area are performed on administrative units in Poland. For small distances between points, accurate calculation results are obtained; however, for larger areas, it is recommended to use equal area map projections to project the ellipsoid into the plane (Pedzich and Kuźma, 2012).

12.11   Geodetic Control Network

A geodetic reference network is formed by a set of geodesic points that characterized the Earth's topographic surface through landmarks implanted and materialized on the terrain with the following purposes (Silva and Segantine, 2015):

- Service to international scientific projects;
- Tightening and control of geodesic and cartographic works;
- Support for topographic surveys where accuracy criteria are used on Earth simplifications.

In a geodetic reference network, geodetic points are determined according to operational procedures associated with a geodetic coordinate system, calculated by precision geodetic models, compatible with the purpose and with a reference ellipsoid that defined the Earth's geometry. A geodetic reference system is considered worldwide when the Cartesian variables X, Y, Z are geocentric and without regional characteristics, with model fit to the global geoid. This system is considered national when the scope is a specific region or country in which the adopted geometric model best fitted the geoid in the considered region. The system is considered multinational when it serves as a basis for geodetic work in several countries (Silva and Segantine, 2015).

The horizontal and vertical datum consists of a network of monuments and control landmarks whose horizontal positions or elevations are determined by accurate geodetic control surveys. These monuments are used as reference points to generate new surveys of all types, and are referred to as "reference networks". The process of estimating the coordinates of the physical points used to define a given benchmark is accompanied by the calculation of a network related to the surveyed points. The result established by adjustment of observations is a set of coordinate values for the stations that constituted the materialization of the geodesic system. Usually, it is common to adopt a single denomination for the definition and materialization of the system by an abbreviation acronym. In this way, several adjustments of geodetic networks can be performed in the same referential defined with different injunctions or the same data can be adjusted with several definitions (Ghilani and Wolf, 1989; Alves and Silva, 2016).

Since 2005 the Brazilian Institute of Geography and Statistics (IBGE), through the Geodesy Coordination, incorporated in its activities the weekly processing of data from all stations of the Brazilian Network of Continuous Monitoring of GNSS Systems (RBMC), in order to evaluate the quality of GNSS observations and the maintenance of the new Geocentric Reference System of the Brazilian Geodetic System, SIRGAS-2000, officially in use in Brazil from February 2005 (Alves and Silva, 2016).

The historical goal of geodesy is to obtain a coordinate reference network for common use. However, many countries or regions developed independent reference networks. With this, it became necessary to transform coordinates obtained in GNSS surveys to those used in a given reference network. To perform datum transformations, it is necessary to know the geodetic coordinates of both reference networks. If a sufficient number of common stations are known, a 3D coordinate transformation system can be used to convert the coordinates of stations from one reference network to another (Schofield et al., 2007).

It is worth noting that although the GRS-80 ellipsoid is adopted in SIRGAS-2000, there is also compatibility between the GRS-80 and WGS-84 ellipsoids at the centimeter level, i.e., the difference between the coordinates calculated in both systems is of the order of 1.0 cm (Silva and Segantine, 2015).

12.12 Accuracy Standards for Control Surveys

The standard of accuracy for a control survey initially depended on the purpose of the survey. Some of the factors that affected accuracy are the type and condition of the equipment used, the field processes adopted, and the experience and capability of the users. Accuracy standards and geodetic survey specifications are used to:

- Standardize a set of minimum acceptance specifications for control surveys for different purposes;
- Establish specifications for instrumentation, field procedures and error checking in order to ensure that the intended level of accuracy is achieved.

In Brazil, geodetic points have been classified by IBGE according to the surveying quality as (Silva and Segantine, 2015):

- High precision geodetic surveys (national);
- Precision geodetic surveys (regional);
- Geodetic surveys for topographic purposes (local features).

The established horizontal accuracy standards for GNSS surveying can vary in accuracy according to the relative accuracy between the surveyed vertices (Table 12.2):

TABLE 12.2: Horizontal accuracy of control surveying.

Order	Precision
Superior	1:100000000
Average	1:100000
Inferior	1:5000

The established vertical accuracy standards vary according to the order of accuracy, from highest to lowest, where K is the distance between the landmarks in kilometers (Table 12.3) (Ghilani and Wolf, 1989; Alves and Silva, 2016):

TABLE 12.3: Vertical accuracy of control surveying.

Order	Precision	Tolerance (m)
First	> 1:10000000	0.5 - 0.7 $\sqrt{K}$
Second	1:10000000	1.0 - 1.3 $\sqrt{K}$
Third	< 1:10000000	2.0 $\sqrt{K}$

The success of a mapping or engineering project is dependent on the control survey used. The higher the order of accuracy required, the greater the time and cost. Therefore, it is necessary to select the appropriate order of accuracy for a given project and carefully follow the specifications. It can be seen that regardless of how the control survey is conducted, there will be errors in the calculated positions, but a higher order of accuracy assumed that smaller errors would occur.

Surveying equipment must undergo periodic testing followed by the issuance of calibration certificates. These tests have been performed using comparators and specialized measuring stations. An instrument is developed to evaluate the calibration of geodetic leveling systems. The instrument is characterized by a practical interferometric measurement system to comparatively evaluate digital geodetic leveling equipment from manufacturers such as Leica, Trimble, Topcon, Sokkia and Zeiss (Kuchmister et al., 2020).

The management of a large and complex research infrastructure requires high accuracy through geodetic surveying for instrument positioning and rearrangement. For this, a precision geodetic network based on GNSS and electronic total station measurements has been established to obtain the positioning and alignment of different elements of the 3 km long Advanced Virgo interferometer used to detect gravitational waves. Monitoring activity has been carried out over the years by means of periodic high-precision leveling, compared with differential interferometry results based on data from on-board synthetic aperture radar (Marsella et al., 2020).

12.13 Control Point Description

Control vertices should be materialized in favorable locations for later use. Vertices should be monumented and described appropriately for future users. Reference monuments can be marked with bronze discs attached to concrete or rock structures. Procedures for establishing the monuments vary according to the type of soil or rock, climatic conditions, and the use application of the monument. In cases where the ground could be excavated, the monuments are set in concrete of greater depth. Another option used is a stainless steel rod as a landmark. A complete description of the control stations should be given, including the installation location in relation to neighborhoods, the precise location of the monument in relation to nearby objects, datum used, geodetic latitude and longitude, plane coordinates, convergence angle, scale factor, UTM coordinates, approximate elevation and geoidal elevation (m). It is recommended to place the landmarks 10 cm above the ground (**Figure** 12.10)(Alves and Silva, 2016).

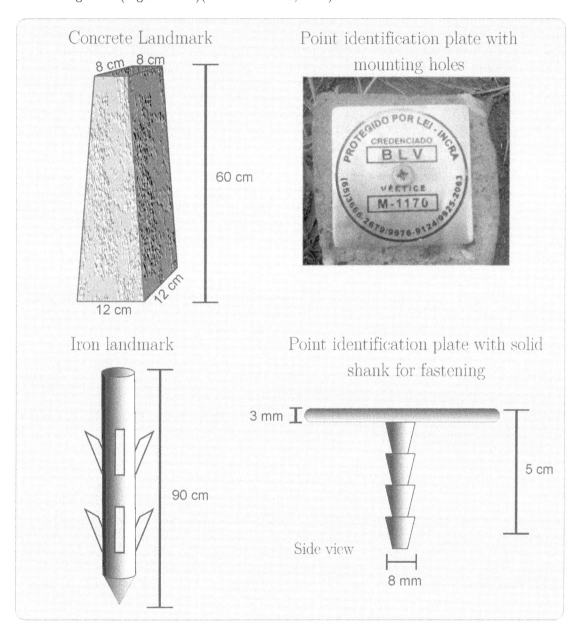

FIGURE 12.10: Metallic discs used to describe horizontal and vertical control stations in surveying.

12.14   Computation

As a computation practice, coordinate reference systems (CRSs) are applied to vector attribute data, referring to the vertices surveyed point-by-point, and to raster data, referring to georeferenced orthoimagery from orbital surveys. CRS enable to define the spatial relationship of objects to the Earth's surface. CRSs can be geographic or projected and used at any location on Earth, described by longitude and latitude values. Longitude is the location in the East-West direction measured relative to angular distance from the plane of the prime meridian. Latitude is the angular distance in the North or South direction relative to the equatorial plane. Therefore, geographical distances in the CRS are not measured in meters. The Earth's surface in geographic coordinate systems is represented by spherical or ellipsoidal surface. In spherical models, the surface is represented by a perfect sphere conforming to a specific radius. These models are simple but inaccurate, since the shape of the Earth is not a sphere. Ellipsoidal models are defined by the equatorial and polar radii. The equatorial radius is approximately 11.5 km larger than the polar radius. The ellipsoid parameters are part of the *ellps* component of the CRS. The precise relationship between Cartesian coordinates and location on the Earth's surface are stored in the `towgs84` argument of the `proj4string`. With this, local variations on the Earth's surface, for example of mountains, are accounted for in a local CRS. There are two types of datum: local and geocentric. In a local datum, such as SIRGAS-2000, the ellipsoidal surface is shifted to align with the surface at a specific location. In a geocentric datum such as WGS-84, the center of gravity of the Earth is the center of the coordinate system projection (Lovelace et al., 2019a).

The CRS can be described in R by an EPSG code or a definition of `proj4string`. Both approaches have advantages and disadvantages. An EPSG code is shorter and simpler to execute when compared to `proj4string`. The code also referred to only one well-defined coordinate reference system. On the other hand, with a definition of `proj4string` there is more flexibility in specifying different parameters, such as the type of projection, the datum and the ellipsoid. This made it possible to specify different projections and modify existing ones. The R packages include support for a wide variety of CRS and use of the PROJ library. EPSG codes can be obtained from the Internet or by functions available in R packages, such as in `rgdal` (Lovelace et al., 2019a).

The EPSG dataset was originally created by European Petroleum Survey Group (EPSG) in 1985 and was made public in 1993. EPSG geodetic parameter dataset is a public registry of geodetic datums, spatial reference systems, Earth ellipsoids, coordinate transformations and related units of measurement. Each entity is assigned an EPSG code and the EPSG dataset is actively maintained by the International Association of Oil and Gas Producers (IOGP) Geomatics Committee (IOGP, 2009, 2012, 2021). For example, the EPSG corresponding to the WGS-84 datum is 4326.

The syntax of PROJ4 consisted of a list of parameters, each prefixed with the + character. For example, one might cite a UTM projection (+ proj = utm) for UTM zone 23S (+ zone = 23) and on a WGS-84 datum (+ datum = WGS84). Other bits of information that could be gleaned from the projection sequence are the units (meters) and the underlying ellipsoid (WGS-84). Some of the PROJ parameters used in defining a coordinate system are described below in Table 12.4 (Lovelace et al., 2019a):

TABLE 12.4: Parameters and description of `PROJ4` to define coordinate systems.

Parameter	Description
+a	Radius of semimajor axis of ellipsoid
+b	Radius of semiminor axis of ellipsoid
+datum	Datum name
+ellps	Elipsoid name
+lat_0	Latitude of origin
+lat_1	Latitude of first standard parallel
+lat_2	Latitude of second standard parallel
+lat_ts	True scale latitude
+lon_0	Central meridian
+over	Longitude beyond -180 to 180, turning off distortion
+proj	Projection name
+south	UTM zone in Southern Hemisphere
+units	meters, etc.
+x_0	False east
+y_0	False north
+zone	UTM zone

As a computation practice, the objective is to observe some datum definitions available in the `sf` package, observe EPSG codes interactively in R, check and assign a coordinate system to vertex data of vector attribute type, determine the Euclidean distance and compare with the geodetic distance between vertices, calculate the geodetic area of a polygon and of Brazil in SIRGAS-2000, SAD-69, Chua and Córrego Alegre 1961 ellipsoids. Finally, a Shuttle Radar Topography Mission (SRTM) digital elevation model is used as an example of transforming a raster in WGS-84 geographic projection to SIRGAS-2000.

The package `rgdal` is used to perform coordinate projection operations through the library `PROJ` (Bivand et al., 2021), in order to transform coordinate vertices into spatial points with attributes. Polygon area calculation is performed using the `sf` package (Pebesma et al., 2021). The calculations of Euclidean distance, geodetic distance and geodetic area are performed with the `lwgeom` package (Pebesma et al., 2020b). The algorithms for geodesic calculations are described in Karney (2013). Spherical area and distance calculations considering the Earth as a sphere are also performed with the `s2` package (Dunnington et al., 2021). Polygons are mapped by the R package `tmap` (Tennekes, 2018; Tennekes et al., 2020). The `spData` package is used as a source of polygon data with Brazilian borders (Bivand et al., 2020b). The `units` package is used in area unit transformation (Pebesma et al., 2020a). The `raster` package is used to obtain an SRTM digital elevation model (NASA, 2013) of an area and to transform the geodetic coordinates of the raster from WGS-84 to SIRGAS-2000 (Hijmans et al., 2020).

12.14.1 Installing R packages

The `install.packages` function is used to install the `sf`, `s2`, `rgdal`, `units`, `lwgeom`, `spData`, `tmap`, and `raster` packages in the R console.

```
## install.packages("sf")
## install.packages("s2")
## install.packages("rgdal")
## install.packages("units")
```

```
## install.packages("lwgeom")
## install.packages("spData")
## install.packages("tmap")
## install.packages("raster")
```

12.14.2 Enabling R packages

The `library` function is used to enable the `sf`, `s2`, `rgdal`, `units`, `lwgeom`, `spData`, `tmap`, `raster` packages in the R console.

```
library(sf)
library(s2)
library(rgdal)
library(units)
library(lwgeom)
library(spData)
library(tmap)
library(raster)
```

12.14.3 Observing available datum definitions in sf

The function `sf_proj_info` is used to enable the package `sf`.

```
sf_proj_info(type = "ellps")
```

12.14.4 Viewing EPSG codes interactively

The EPSG codes available in the `rgdal` package are obtained using the `make_EPSG` function. Then the results are visualized using the `View` function. Another way to obtain the EPSG code for a given geographic or plane coordinate system can be through searches on Internet websites such as Google.

```
crs_data = rgdal::make_EPSG()
View(crs_data)
```

12.14.5 Checking and assigning a coordinate system to point vector attribute data

Vector attribute point data used in surveying with the calculation of coordinates by the intersection method (Chapter 10) are used to check and assign a coordinate system to vertices and are imported from the Internet.

```
## download.file(url=
##   "http://www.sergeo.deg.ufla.br/geomatica/downloads/pontosGeo.zip",
##   destfile = "pontosGeo.zip")
```

The files are unzipped by the `unzip` function.

```
## unzip(zipfile = "p.zip")
```

The directory into which data is imported can be observed with the `getwd` function.

```
## getwd()
```

The topographic survey data are imported with the `st_read` function and then we observed information about the file, such as the name of the data columns associated with the vector file.

```
# Import simple feature point data
pointsGeo <- st_read("files/pontosGeo.shp")
```

```
## Reading layer `pontosGeo' from data source
##   `C:\bookdown\surveying-with-geomatics-and-r_R1_03102021\files\pontosGeo.shp'
##   using driver `ESRI Shapefile'
## Simple feature collection with 3 features and 1 field
## Geometry type: POINT
## Dimension:     XY
## Bounding box:  xmin: 499054.4 ymin: 7641910 xmax: 503211 ymax: 7654217
## Projected CRS: WGS 84 / UTM zone 23S
```

```
pointsGeo
```

```
## Simple feature collection with 3 features and 1 field
## Geometry type: POINT
## Dimension:     XY
## Bounding box:  xmin: 499054.4 ymin: 7641910 xmax: 503211 ymax: 7654217
## Projected CRS: WGS 84 / UTM zone 23S
##   vertices                 geometry
## 1        A POINT (503142.1 7654217)
## 2        B   POINT (503211 7654195)
## 3        P POINT (499054.4 7641910)
```

```
# Look up data column names
names(pointsGeo)
```

```
## [1] "vertices" "geometry"
```

The coordinate reference system (CRS) in the imported file is evaluated with regard to the projection type and parameters, as well as EPSG, by the `st_crs` function . This verified the EPSG: 32723 of the data, referring to World Geodetic System datum, 1984 (WGS-84) and UTM zone 23 South (23S).

```
# Check the reference coordinate system on vector data
st_crs(pointsGeo) # get CRS
```

```
## Coordinate Reference System:
##   User input: WGS 84 / UTM zone 23S
##   wkt:
## PROJCRS["WGS 84 / UTM zone 23S",
##     BASEGEOGCRS["WGS 84",
##         DATUM["World Geodetic System 1984",
##             ELLIPSOID["WGS 84",6378137,298.257223563,
##                 LENGTHUNIT["metre",1]]],
##         PRIMEM["Greenwich",0,
##             ANGLEUNIT["degree",0.0174532925199433]],
##         ID["EPSG",4326]],
##     CONVERSION["UTM zone 23S",
##         METHOD["Transverse Mercator",
##             ID["EPSG",9807]],
##         PARAMETER["Latitude of natural origin",0,
##             ANGLEUNIT["Degree",0.0174532925199433],
##             ID["EPSG",8801]],
##         PARAMETER["Longitude of natural origin",-45,
##             ANGLEUNIT["Degree",0.0174532925199433],
##             ID["EPSG",8802]],
##         PARAMETER["Scale factor at natural origin",0.9996,
##             SCALEUNIT["unity",1],
##             ID["EPSG",8805]],
##         PARAMETER["False easting",500000,
##             LENGTHUNIT["metre",1],
##             ID["EPSG",8806]],
##         PARAMETER["False northing",10000000,
##             LENGTHUNIT["metre",1],
##             ID["EPSG",8807]]],
##     CS[Cartesian,2],
##         AXIS["(E)",east,
##             ORDER[1],
##             LENGTHUNIT["metre",1]],
##         AXIS["(N)",north,
##             ORDER[2],
##             LENGTHUNIT["metre",1]],
##     ID["EPSG",32723]]
```

If CRS is not configured in the file of interest, we can configure CRSs with the `st_set_crs`.

```
## pointsGeo <- st_set_crs(pointsGeo, 32723) # atribuir CRS
```

Another way to assign the CRS can be to describe the coordinate system in the +proj parameter

```
## pointsGeo <- st_set_crs(pointsGeo, "+proj=utm +zone=23 +south +
##                         datum=WGS84 +units=m +no_defs")
```

Or configure the coordinate system with the proj4string function, according to the EPSG code for the coordinate system for spatial data of sp class.

```
## proj4string(pointsGeo) <- CRS("+init=epsg:32723")# For sp data
```

12.14.6 Determining Euclidean and geodesic distance between vertices

First of all, the data are organized in rows and columns to calculate the longitude and latitude between vertices in order to determine the Euclidean and geodesic distance between vertices A and P, defined by intersection survey from UTM coordinates. Then the Pythagoras theorem is used to determine the horizontal distance between vertices AP of 12967.96 m.

```
# Organize data in rows and columns
vertices <- c('A', 'B', 'P')
X <- c(503142.10, 503211.00, 499054.3815)
Y <- c(7654216.99, 7654195.00, 7641910.1401)
int<-data.frame(vertices, X, Y)
dx <- int$X[3] - int$X[1]
dy <- int$Y[3] - int$Y[1]
# Determine the distance between vertices A and P
HAP <- sqrt((dx^2)+(dy^2))
HAP
```

```
## [1] 12967.96
```

The distance calculation performed between vertices is checked with the st_distance function from the lwgeom package, obtaining the same value of 12967.96 m.

```
# Determine Euclidean distance with the lwgeom package
st_distance(pointsGeo$geometry[1], pointsGeo$geometry[3])
```

```
## Units: [m]
##            [,1]
## [1,] 12967.96
```

The intersection surveying data in plane coordinate system are transformed to geographic coordinate system, in WGS-84 datum, through the `st_transform` function.

```
pointsWGS84 <- st_transform(pointsGeo,
        crs= CRS("+proj=longlat +ellps=WGS84 +datum=WGS84"))
```

The geodesic distance between vertices AP is calculated, considering the terrestrial curvature characterized by the WGS-84 ellipsoid, in Lavras, Minas Gerais, Brazil. The `st_geod_distance` function is used to calculate the distance, obtaining the value of 12973.15 m. We observed that the geodetic distance is greater than the Euclidean distance by 5.19 m.

```
# Calculate the geodesic distance between vertices A and P
st_geod_distance(pointsWGS84$geometry[1],
      pointsWGS84$geometry[3], tolerance = 0, sparse = FALSE)
```

```
## Units: [m]
##            [,1]
## [1,] 12973.15
```

```
s2_distance(pointsWGS84$geometry[1], pointsWGS84$geometry[3],
            radius=s2_earth_radius_meters())
```

```
## [1] 13021.24
```

The vertices, in geographic and plane coordinate systems, are mapped comparatively by means of the `qtm` and `tmap_arrange` functions of the R package `tmap` (Figure 12.11).

```
# Map the points in geographic and plane coordinates
w1 <- qtm(pointsGeo, symbols.size = 0.1, symbols.col = "red",
        title ="Plane Projection")+ tm_grid(col = "gray70") +
  tm_xlab("Longitude") +    tm_ylab("Latitude")
w2 <- qtm(pointsWGS84,symbols.size = 0.1, symbols.col = "red",
        title ="Geographic Projection")+tm_grid(col = "gray70") +
  tm_xlab("Longitude") +    tm_ylab("Latitude")
current.mode <- tmap_mode("plot")
tmap_arrange(w1, w2, widths = c(1, 1))
```

12.14.7   Calculating geodesic area of polygon

The vector data of irradiation vertices used in a topographical survey, with the calculation of coordinates by the walking method of a closed-path polygon of a building at the Federal University of Mato Grosso, are imported from the Internet.

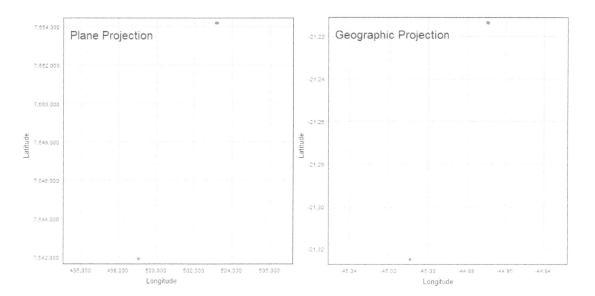

FIGURE 12.11: Comparative mapping of attribute points defined by the intersection method in the plane and geographic projections.

```
## # download training areas
## download.file(url=
## "http://www.sergeo.deg.ufla.br/geomatica/downloads/spdfIrr.zip",
##      destfile = "spdfIrr.zip")
```

The files have been unzipped by the unzip function.

```
## unzip(zipfile = "spdfIrr.zip")
```

Data are imported into the directory by the getwd function.

```
## getwd()
```

The file is imported into R with the st_read function from the sf package and named "areaPol".

```
areaPol <- st_read("files/spdfIrr.shp")
```

```
## Reading layer `spdfIrr' from data source
##    `C:\bookdown\surveying-with-geomatics-and-r_R1_03102021\files\spdfIrr.shp'
##    using driver `ESRI Shapefile'
## Simple feature collection with 1 feature and 1 field
## Geometry type: POLYGON
## Dimension:     XY
## Bounding box:  xmin: 600093.6 ymin: 8273776 xmax: 600112.2 ymax: 8273801
## Projected CRS: WGS 84 / UTM zone 21S
```

The polygon area calculation based on Euclidean distances that defined the surveyed perimeter is determined using the `st_area` function. The polygon area is 235.323 m².

```
st_area(areaPol)
```

```
## 235.323 [m^2]
```

The polygon, initially in plane UTM 21S projection, is transformed to WGS-84 geographic projection to calculate the polygon geodetic area.

```
areaGeo <- st_transform(areaPol, 4326) #Reproject to WGS-84
```

The polygon geodetic area is calculated using the function `st_geod_area`. There is a difference between the calculated geodetic area and the area based on Euclidean distance, of the order of 0.1300 m².

```
st_geod_area(areaGeo) # Calculate geodetic area
```

```
## 235.4532 [m^2]
```

The spherical area of the polygon of the building calculated by the `s2_area` function presented a difference between the geodetic area and the area based on Euclidean distance being 0.9581 m².

```
s2_area(areaGeo)
```

```
## [1] 236.2811
```

12.14.8   Calculate the geodesic area of Brazil

The polygon of Brazil is obtained from the world map, from the R package `spData` to calculate the geodetic area of Brazil comparatively, in the ellipsoids SIRGAS-2000, SAD-69, Chua and Córrego Alegre 1961. The world map with the country boundaries in each continent is mapped with the `plot` function (Figure 12.12).

```
plot(world["continent"], key.pos=1)
```

A subset of South America is conducted, to assess Brazil and the context of its neighborhood.

```
world_america <- world[world$continent == "South America",]
```

Maps are made of all data columns available in the archive and of the South American countries in isolation, with the `plot` function (Figure 12.13).

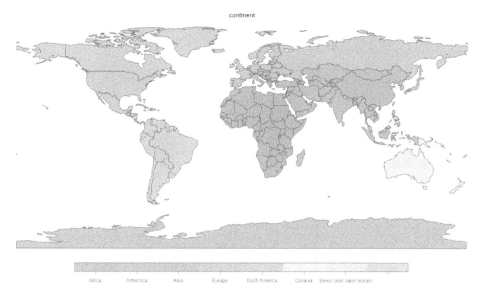

FIGURE 12.12: Mapping continents from cadastral map of countries in the world.

```
plot(world_america, max.plot = 10) # Mapping attributes
plot(world_america["name_long"], key.pos=1, axes=T) # Mapping countries
```

Then a subset is made for Brazil with the [] operator.

```
BrazilWGS84 <- world[world$name_long == "Brazil",]
```

The Brazil polygon has been exported with the st_write function for later use.

```
## st_write(BrazilWGS84, dsn=
##          "E:/Aulas/Topografia/Aula11/BrazilWGS84.shp",
##          layer = "BrazilWGS84.shp", driver = "ESRI Shapefile")
```

The subset of Brazil is mapped against all available data in the archive and also of Brazil with other South American countries, using the plot function (Figure 12.14).

```
plot(BrazilWGS84, max.plot = 10) # Mapping Brazil's attributes
plot(st_geometry(BrazilWGS84), expandBB = c(0.2, 0.2, 0.1, 1),
     col = "gray", lwd = 3, axes=T) # Mapping Brazil's geometry
plot(world_america[0], add = TRUE) # Mapping all countries
```

The Brazil polygon is transformed into different Brazilian geographic projection reference systems in order to perform geodetic area determination comparisons. The st_transform function is used for the different coordinate EPSG codes.

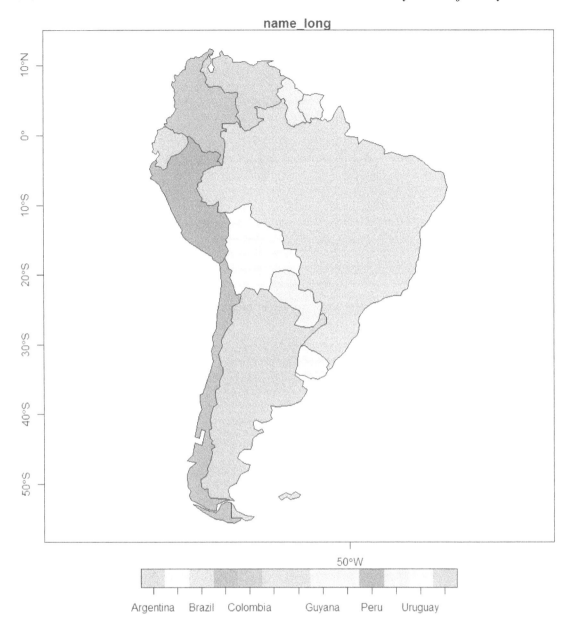

FIGURE 12.13: Attribute mapping of South American countries.

```
BrazilSIRGAS<-st_transform(BrazilWGS84, 4674) # WGS-84 in SIRGAS-2000
BrazilSAD69<-st_transform(BrazilWGS84, 4618) # WGS-84 in SAD-69
BrazilChua<-st_transform(BrazilWGS84, 4224) # WGS-84 in Chua
BrazilCA<-st_transform(BrazilWGS84, 5524) # WGS-84 in Corrego Alegre-1961
```

The geodetic area of Brazil is calculated in square meters and square kilometers using the set_units function from the units package. The geodetic areas of Brazil in the SIRGAS-2000, SAD-69, Chuá, and Córrego Alegre-1961 geographic projections are 8508557, 8508525, 8509011, and 8509009 km², respectively. The largest area of Brazil is obtained in the Chuá projection, followed by Córrego Alegre-1961, SIRGAS-2000 and SAD-69.

FIGURE 12.14: Mapping Brazil in relation to its South American neighbors.

```
# Calculate geodetic area for SIRGAS-2000
st_geod_area(BrazilSIRGAS)
```

```
## 8.508557e+12 [m^2]
```

```
st_geod_area(BrazilSIRGAS)/1000000
```

```
## 8508557 [m^2]
```

```
units::set_units(st_area(BrazilSIRGAS), km^2)
```

```
## 8540954 [km^2]
```

```
# Calculate geodetic area for SAD-69
st_geod_area(BrazilSAD69)
```

```
## 8.508525e+12 [m^2]
```

```
st_geod_area(BrazilSAD69)/1000000
```

```
## 8508525 [m^2]
```

```
units::set_units(st_area(BrazilSAD69), km^2)
```

```
## 8540861 [km^2]
```

```
# Calculate geodetic area for Chuá
st_geod_area(BrazilChua)
```

```
## 8.509011e+12 [m^2]
```

```
st_geod_area(BrazilChua)/1000000
```

```
## 8509011 [m^2]
```

```
units::set_units(st_area(BrazilChua), km^2)
```

```
## 8540956 [km^2]
```

```
# Calculate geodetic area for Córrego Alegre-1961
st_geod_area(BrazilCA)
```

```
## 8.509009e+12 [m^2]
```

```
st_geod_area(BrazilCA)/1000000
```

```
## 8509009 [m^2]
```

```
units::set_units(st_area(BrazilCA), km^2)
```

```
## 8540954 [km^2]
```

The subset of Brazil, on the different transformed ellipsoids, is mapped by the `qtm` and `tmap_arrange` functions from the R package `tmap` (Figure 12.15).

```
## p1 <- qtm(BrazilSIRGAS, title="SIRGAS")+ tm_grid(col = "gray70") +
##   tm_xlab("Longitude") +  tm_ylab("Latitude") # SIRGAS-2000
## p2 <- qtm(BrazilSAD69, title="SAD69") + tm_grid(col = "gray70") +
##   tm_xlab("Longitude")+  tm_ylab("Latitude") # SAD-69
## p3 <- qtm(BrazilChua, title="Chuá")+ tm_grid(col = "gray70") +
##   tm_xlab("Longitude")+  tm_ylab("Latitude") # Chua
## p4 <- qtm(BrazilCA, title="Corr.Aleg.")+ tm_grid(col = "gray70") +
##   tm_xlab("Longitude")+  tm_ylab("Latitude") # Corrego Alegre-1961
## current.mode <- tmap_mode("plot")
## tmap_arrange(p1, p2, p3, p4, widths = c(1, 1))
```

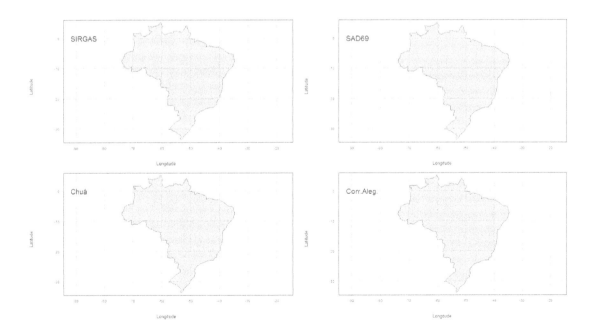

FIGURE 12.15: Mapping Brazil in the SIRGAS-2000, SAD-69, Chuá, and Córrego Alegre coordinate systems.

12.14.9   Transforming the geographic projection of raster data

In the raster data projection transformation, data are obtained from a SRTM topographic radar mission used to determine the altitude of Lavras, state of Minas Gerais, Brazil. The `getData` function is used to obtain the vectors of Brazilian municipalities, to subsequently make a subset in the region of interest.

```
brazilRegions3 <- getData('GADM', country='BRA', level=3)
```

The Brazilian districts are mapped in order to visualize the spatial distribution of vectors in the region of interest (Figure 12.16).

```
plot(brazilRegions3, axes=T, asp=1)
```

FIGURE 12.16: Mapping political boundaries of Brazilian municipalities obtained with the `getData` function.

A subset of the vectors at the boundary boundaries defining the city of Lavras is performed in order to generate a subset of the raster data in that region.

```
lavras<-subset(brazilRegions3, NAME_2=="Lavras")
```

Then, Lavras city is mapped with the `plot` function (Figure 12.17).

```
plot(lavras, axes=T, col="grey92")
```

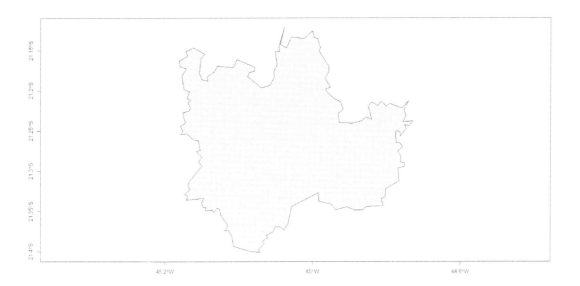

FIGURE 12.17: Mapping of Lavras city, Minas Gerais, Brazil, obtained from a subset of Brazilian municipalities.

The vectors of Lavras city are converted into `sf` class files to export the file defined as the region of interest in Brazil using `st_as_sf` function.

```
##
## lavras_sf<-st_as_sf(lavras)
```

The vectors for the district of Lavras are exported in ESRI Shapefile format with the `st_write` function.

```
## st_write(lavras_sf, dsn = "E:/Aulas/Topografia/Aula11/lavras_sf.shp",
##          layer = "lavras_sf.shp", driver = "ESRI Shapefile")
```

The `getData` function is used to obtain the digital elevation model of the area. Due to the geographical area position, it is necessary to obtain files referring to longitudes -45 and -46 and latitude -21.

```
## demLavras1 <- getData('SRTM', lon=-46, lat=-21, download=TRUE)
## demLavras2 <- getData('SRTM', lon=-45, lat=-21, download=TRUE)
```

Image mosaicking is performed using pixel averaging in regions with overlapping image data.

```
## demLavras<-mosaic(demLavras1, demLavras2, fun=mean)
## demLavras
```

We cropped and masked the image mosaic from the Lavras vectors with the `crop` and `mask` functions, respectively.

```
## demLavras_clip <- crop(demLavras, lavras) # Crop
## demLavras_mask=mask(demLavras_clip, lavras) # Mask
```

The digital elevation model mask is mapped within the Lavras boundaries with the `plot` function (Figure 12.18).

```
## plot(demLavras_mask, col=grey((0:48)/48))
```

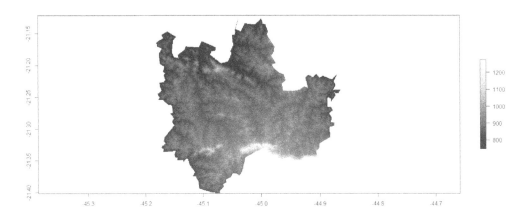

FIGURE 12.18: SRTM altitude mapping in Lavras, Minas Gerais, Brazil.

The map projection of the digital elevation model at Lavras is evaluated with the `crs` function.

```
## crs(demLavras_mask)
```

After, the `projectRaster` function is used to perform the transformation of the geographic projection of the digital elevation model from WGS-84 to SIRGAS-2000.

```
## demLavrasSirgas <- projectRaster(demLavras_mask, crs=
## "+proj=longlat +ellps=GRS80 +towgs84=0,0,0,0,0,0,0 +no_defs")
```

We can use parameters of the `+proj` function to set a value of `crs` to assign a map projection to the raster when it does not exist.

```
## crs(demLavrasSirgas) <-
##    "+proj=longlat +ellps=GRS80 +towgs84=0,0,0,0,0,0,0 +no_defs"
```

Finally, the digital elevation model is exported for further use by means of the `writeRaster` function.

```
## writeRaster(demLavras_mask,
##      filename="E:/Aulas/Topografia/Aula11/demLavras_mask.tif",
##             overwrite=TRUE)
```

12.15   Solved Exercises

12.15.1   Determine the first eccentricity of the GRS-80 ellipsoid.

A: The parameters given in Table 12.1 are used to determine the first eccentricity of the GRS-80 ellipsoid. The first eccentricity of the GRS-80 ellipsoid is 0.08181919.

```
# Data
f<- 1/298.257222101
# Calculate the first eccentricity
e<-sqrt((2*f)-(f^2))
e
```

```
## [1] 0.08181919
```

12.15.2   Considering the reference ellipsoid in the SIRGAS-2000 geodetic system, determine the minor semi-axis to 6 decimal places and the second eccentricity.

A: The minor semi-axis is 6356752.314140 m. The second eccentricity is 0.08209444.

```
# Data
f<- 1/298.257222101
a<-6378137.0
# Calculate minor semi-axis
b<-(1-f)*a
# Calculate minor semi-axis with 6 decimal places
b6<-format(b, nsmall=6)
b6
```

```
## [1] "6356752.314140"
```

```
# Calculate the second eccentricity
ex<-sqrt(a^2-b^2)/b
```

12.15.3 Determine the radius of the meridian and prime vertical of a point of latitude 41°18′15.0132″ N, using parameters from the GRS-80 ellipsoid. What is the radius of the great circle located at an azimuth of 142°14′36″ from the point? Present the results to 6 decimal places.

A: The radius of the meridian is 6363257.346179 m; the radius of the prime vertical is 6387458.536282 m; the radius of the great circle is 6372309.400720 m.

```
library(circular) # Enable circular Package
```

```
# Calculate e^2 for GRS-80
e2<-e^2
# Data
a<-6378137.0
phi<-41+18/60+15.0132/3600
alpha<-142+14/60+36/3600
# Calculate RM
RM<-(a*(1-e2))/(1-(e2*(sin(rad(phi)))^2))^(3/2)
# Present RM with 6 decimal places
RM6<- format(RM, nsmall=6)
RM6
```

```
## [1] "6363257.346179"
```

```
# Calculate RN
RN<-a/sqrt(1-(e2*(sin(rad(phi)))^2))
# Present RN with 6 decimal places
RN6<- format(RN, nsmall=6)
RN6
```

```
## [1] "6387458.536282"
```

```
# Calculate Ra
Ra<-(RN*RM)/((RN*(cos(rad(alpha)))^2)+(RM*(sin(rad(alpha)))^2))
Ra6<-format(Ra, nsmall=6)
Ra6
```

```
## [1] "6372309.400720"
```

12.15.4 Determine the radius of the meridian, the prime vertical, and the mean radius of a point with latitude $-22°00'17.8160"$ in the **SIRGAS-2000** geodetic system. Present results with 6 decimal places.

A: The radius of the meridian is 6344159.215385 m, the radius of the prime vertical is 6381061.877162 m, the average radius is 6362583.792135 m.

```
# Calculate e^2 for GRS-80
e2<-e^2
# Data
a<-6378137.0
phi<--22+17/60+0.8160/3600
# Calculate RM
RM<-(a*(1-e2))/(1-(e2*(sin(rad(phi)))^2))^(3/2)
# Present RM with 6 decimal places
RM6<- format(RM, nsmall=6)
RM6
```

```
## [1] "6344159.215385"
```

```
# Calculate RN
RN<-a/sqrt(1-(e2*(sin(rad(phi)))^2))
# Present RN with 6 decimal places
RN6<- format(RN, nsmall=6)
RN6
```

```
## [1] "6381061.877162"
```

```
# Calculate R0
R0<-sqrt(RN*RM)
R06<-format(R0, nsmall=6)
R06
```

```
## [1] "6362583.792135"
```

12.15.5 Determine the Earth's radius for the ellipsoid GRS-80 with 20 decimal places.

A: The Earth's radius is 6371000.78997413441538810730 m.

```
# Enter parameters of the ellipsoid
a<-6378137.000000000
b<-6356752.314140356
# Calculate the radius
r<-((a^2)*b)^(1/3)
r
```

```
## [1] 6371001
```

```
# Present the result with 20 decimal places
r1<-format(r, nsmall=20)
r1
```

```
## [1] "6371000.78997413441538810730"
```

12.15.6 Determine the tolerance of a second-order standard survey for measuring elevations of two geodetic landmarks 25 km apart.

A: The elevation tolerance is ± 6.5 mm.

```
L<-25
T<-1.3*(L^0.5)
T
```

```
## [1] 6.5
```

12.16 Homework

Choose one exercise presented by the teacher and solve the question with different input values. Compare the results obtained. Perform plane, geodetic and spherical area calculation of vector attribute data of interest. Compare the results obtained.

12.17 Resources on the Internet

As a study guide, slides and illustrative videos are presented about the subject covered in the chapter in Table 12.5.

TABLE 12.5: Slide shows and video presentations on geodesy and coordinate reference systems.

Guide	Address for Access
1	Slides on coordinate reference systems in geodetic surveys[2]
2	How to define a geodetic datum[3]
3	Lecture on geodetic surveying[4]
4	Accurate geodetic datum development plans[5]
5	Determining geodetic landmarks[6]

12.18 Research Suggestion

The development of scientific research on geomatics is stimulated by the activity proposals that can be used or adapted by the student to assess the applicability of the subject matter covered in the chapter (Table 12.6).

TABLE 12.6: Practical and research activitie used or adapted by students using geodesy and coordinate reference systems.

Activity	Description
1	In the content on coordinate reference systems, interest may arise in doing the work based on examples of geodetic calculations presented
2	Determine the Euclidean and geodetic distance from the intersection survey example performed by the teacher in the previous chapter. Compare the results obtained
3	Determine the geodetic area of a polygon surveyed in the field. Compare the results with the area obtained from plane measurements

12.19 Learning Outcome Assessment Strategy

Perform a summary of the chapter, "Coordinate Reference Systems for Geodetic Surveying with Geomatics and R", on a single A4 page in order to show the student's abilities to summarize a subject presenting key points considered of greater importance today.

---

[2]http://www.sergeo.deg.ufla.br/geomatica/book/c12/presentation.html#/
[3]https://youtu.be/kXTHaMY3cVk
[4]https://youtu.be/VeBRfIu5jZ8
[5]https://youtu.be/w69xc_U1Rao
[6]https://youtu.be/-oUFqg1Lw1U

13

Cartographic Coordinate Projection Systems

13.1 Learning Questions

The emergent learning questions answered through reading the chapter are as follows:

- What does cartographic representation mean?
- How to classify cartographic projections.
- What is the difference between spherical, plane and Universal Transverse Mercator cartographic projections?
- How to obtain geospatial data with the R packages `rnaturalearth` and `spData`.
- How to assign, transform and map cartographic projections in R.

13.2 Learning Outcomes

The learning outcomes expected from reading the chapter are as follows:

- Understand the concepts of cartographic representation, classification of cartographic projections and definition of spherical, plane and Universal Transverse Mercator projections.
- Obtain geospatial data with the `rnaturalearth` and `spData` R packages.
- Assign, transform and map cartographic projections with `rgdal`, `units`, `tmap` and `raster` R packages.

13.3 Introduction

Cartography is defined as the science, technique, and art of representing the Earth's surface. In Brazil, the concept of cartography was introduced in 1839 by Visconde de Santarém, Manuel Francisco de Barros and Sousa de Mesquita de Macedo Leitão, with the etymological significance of mapping (IBGE, 1999).

The cartographic representation of information regarding the phenomena distributed on the Earth's surface is carried out in two phases (IBGE, 1999):

- In the first phase (Geodesy), points on the Earth's surface is projected onto a previously selected reference sphere or ellipsoid;

DOI: 10.1201/9781003184263-13

- In the second phase (Mathematical Cartography), the reference sphere or ellipsoid is projected onto a flat surface.

The plane representation of the Earth's surface is a legacy attributed to Eratosthenes (280 BC to 194 BC), Hipparchus (194 BC to 126 BC) and Marino (1st century BC). More details on the history of mathematical cartography can be found at Casaca et al. (2007).

Eratosthenes was the first to calculate the dimensions of the Earth 200 years BC. Eratosthenes concluded that the Egyptian cities of Alexandria and Syene are located approximately on the same meridian. He also observed that on the summer solstice, the sun is directly over Syene. Thus, at that time, the sun, Syene and Alexandria are on the same meridian plane. Measuring the length of the arc between the two cities and the subtended angle at the center of the Earth, it would be possible to calculate the Earth's circumference. The angle between the two cities and the center of the Earth was determined by the length of the shadow of a vertical rod of known length in Alexandria (Figure 13.1) (Ghilani and Wolf, 1989; Alves and Silva, 2016).

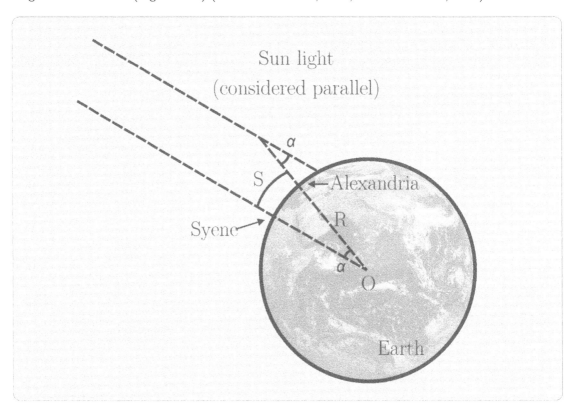

FIGURE 13.1: Procedure used by Eratosthenes to determine the Earth's circumference.

Through this procedure, Eratosthenes applied the following equation to determine the circumference of a circle (C) (Ghilani and Wolf, 1989; Slocum et al., 2014; Alves and Silva, 2016):

$$C = 360° \frac{S}{\alpha} \tag{13.1}$$

where S is the distance between two locations on the Earth's surface and, α, the angle that separated the two locations that defined latitude.

The α angle obtained by measuring a shadow in Alexandria was 7° 12'. The distance determination was based on the number of days traveling by camel between the two cities in ~810 km. Replacing these values in the equation, the Earth's estimated circumference was 40500 km.

$$C = 360° \frac{810}{7°12'} = 40500 \text{ km} \tag{13.2}$$

This calculation performed in R is as follows:

```
C<-360*(810/(7+12/60))
C
```

```
## [1] 40500
```

The Earth's circumference result is similar to the current measurements of 40075 km (Alves and Silva, 2016).

Fundamental contributions to modern mathematical cartography were made by Johann Heinrich Lambert, Carl Friedrich Gauss and Auguste Tissot. Lambert in 1772 presented and described seven cartographic projections including Mercator cross-sectional projection, conformal conic projection and azimuthal equal-area projection. Gauss (1827) contributed with differential geometry to analyze cartographic projections. Tissot (1860, 1881) developed the theory of cartographic deformation (Casaca et al., 2007).

Most surveys in small areas adopted the hypothesis of flat terrain surface. However, when the survey area is large, it becomes necessary to consider the Earth's curvature. The calculations used to determine geodetic positions from survey observations with distance and azimuth data can be laborious, so practical surveyors may not be familiar with these procedures. A system for specifying positions of geodetic stations using plane coordinates has been desirable, provided that calculations are performed using simple coordinate geometry equations (Ghilani and Wolf, 1989; Alves and Silva, 2016).

In a plane coordinate system, a common reference datum is used for horizontal control over large areas, just as in the geoid, with single reference point for vertical control. This procedure eliminates individual surveys based on different reference coordinates not related to other coordinate systems. Plane coordinate systems are available for all control points in a national reference space system and for other existing control points. These systems are widely used as reference points to initiate surveys of different types, such as road construction projects, border demarcation and photogrammetric mapping. The use of plane coordinates enables to initiate projects for high-speed highways starting at a control station close to another station with the same used coordinate system. In boundary surveys, if the ends of a reference plot are in a plane coordinate system, the locations of the points can be easily found. The positions of monuments, iron poles and other types of marks may disappear, but the original locations could be recovered based on surveys nearby landmarks using a plane coordinate system. For this reason, the plane coordinate system is recommended when making new territorial subdivisions or as a reference system for data entry in geographic information systems registration and spatial analysis (Ghilani and Wolf, 1989; Alves and Silva, 2016).

13.4 Cartographic Representation

The cartographic representation can be represented by trace or image. The trace representation is divided into globe, map, chart and plant (IBGE, 1999). Globe refers to a cartographic representation of natural and artificial aspects in a planetary figure, for cultural and illustrative purposes, on spherical surfaces and small scale. Map is defined as the representation of the geographical, natural, cultural and artificial aspects of an area obtained from the surface of a planetary figure, delimited by physical, political and administrative elements, intended for thematic, cultural and illustrative use, on the plane, usually with small scale. Chart is defined as the representation of the artificial and natural aspects of an area in a planetary surface, subdivided into sheets, delimited by conventional, parallel and meridian lines, with scale compatible precision, on a medium or large scale. Plan is a particular case of a chart that represents an area of restricted extension, with constant scale and in which the Earth's curvature is not considered.

The image representation is divided into mosaic, photochart, orthophotochart, orthophotomap, photo index and image chart. The mosaic refers to a set of photos from a certain area, organized in such a way that the whole set comprised a single photograph. Photochart refers to a controlled mosaic on which a cartographic treatment was carried out. Orthophotochart refers to a photograph resulting from the transformation of an original photo as a terrain central perspective, in orthogonal projection on a plane, georeferenced, complemented by symbols and planimetric information. Orthophotomap refers to a set of several adjacent orthophotocharts in a given region. Photo index refers to a set of overlapped photographs, usually on a small scale, used for quality control of aerial surveys and to produce charts using photogrammetric methods. In this case, the scale is normally reduced 3 to 4 times in relation to the flight scale. Image chart refers to a referenced image based on identifiable points with known coordinates, projection grids, symbology and toponymy.

13.5   Cartographic Projection Classification

The basic problem with cartographic projections has been to represent a curved surface in a plane. The ideal would be to build a chart with all the properties of representing the surface strictly similar to the Earth. This chart should (IBGE, 1999):

- Maintain the true shape of the areas (conformal);
- Not change the areas (equivalente);
- Maintain constant relationships between the distances from the represented points (equidistant).

As the Earth's surface is not plane, it was not possible to build a chart with all the desired conditions for all situations.

Cartographic representations comprise the following steps (IBGE, 1999):

- Adopt a simplified mathematical model of the Earth's surface, in general the sphere or ellipsoid of revolution;
- Design all the elements of the Earth's surface on the chosen model;
- Relate the points of the mathematical model with the plane representation, with a scale and coordinate system.

In mathematical cartography, the applications of the sphere and the ellipsoid in a plane and the resulting deformations are studied and classified. The relative positioning of an object on the surface (sphere or reference ellipsoid) and on the image surface (cartographic plane) is defined by a point preferably located in the center of the region to be represented, designated by a central point. The meridian and the parallel of the central point are called "meridian" and "central parallel". The cartographic projection must be bijective, that is, the images of the continuous lines on the object surface must be equally continuous in the cartographic plane. The cartographic abscissas (X) are designated by distances to the meridian, symbolized by M and counted on the perpendicular. The cartographic ordinates (Y) are called "perpendicular distances", symbolized by P and counted over the meridian. In Anglo-Saxon countries, the designations corresponding to the abscissa and ordinate are Easting and Northing, symbolized by E and N, respectively (Casaca et al., 2007).

When the origin of the coordinates of the cartographic projection is subjected into a translation, in order to place the region in the first quadrant to avoid negative coordinates or to minimize differences to an older coordinate system, the new coordinates should not be symbolized by M and P, avoiding confusion with the distances to the meridian and the perpendicular. This same situation, in the Anglo-Saxon countries, determined modification of the new coordinates E, N, by the terms of False-Easting and False-Northing (Casaca et al., 2007).

The following parameters must be specified for a cartographic projection to be operational (Casaca et al., 2007):

- Geodesic datum;
- Central point;
- Scale factor;
- A new origin for cartographic coordinates.

The set formed by the projection and the parameters is called the "cartographic projection system" (Casaca et al., 2007).

Cartographic projections are classified by different methodologies in order to determine a better fit of the represented surface (IBGE, 1999; Fitz, 2008).

13.5.1 Classification by methods

Cartographic projections are classified according to the following methods:

- Geometric: The projection is based on a plane according to a point of view, based on the intersection on the projection surface of the straight beam that passes through points of the reference surface starting from a perspective center;
- Analytical: Mathematical formulations are used to calculate the projections.

13.5.2 Classification by deformations

Considering the impossibility of developing a spherical or ellipsoid surface on a plane without deformations, in practice, projections are used to reduce or eliminate part of deformations according to different applications, as the following properties (Fitz, 2008):

- Conformal or similar projections: The true shape of the represented areas is maintained by not deforming the angles on the map;
- Equidistant projections: There is constant variation in the represented distance, without linear deformations;
- Equivalent projections: There is constant variation in the relative dimensions of the represented areas;
- Azimuthal projections: The directions and azimuths of all lines coming from the central point of the projection were the same as the corresponded lines in the terrestrial sphere;
- Arbitrary projections: Areas, angles, distances or azimuths are not preserved in the projections.

13.5.3   Classification by point of view

The cartographic projections are classified in terms of the following points of view (Alves and Silva, 2016):

- Gnomic or central: The point of view is located in the center of the ellipsoid;
- Stereographic: The point of view is located at the opposite end of the projection surface;
- Orthographic: The point of view is infinite.

13.5.4   Classification by type of projection surface

Cartographic projections are classified according to the following types of projection surfaces (Alves and Silva, 2016):

- Plane: The projection surface is a plane;
- Conical: The projection surface is a cone;
- Cylindrical: The projection surface is a cylinder;
- Polyhedral: Several projection planes are brought together to form a polyhedron.

13.5.5   Classification by position of the projection surface

Cartographic projections are classified according to the following positions of the projection surfaces (Alves and Silva, 2016):

- Equatorial: The center of the projection surface is located on the Earth's Equator;
- Polar: The center of the projection plane is a pole;
- Transverse: The axis of the projection surface is perpendicular to the axis of the Earth's rotation;
- Oblique: Part of the Earth's surface is projected onto a plane tangential to it between the poles and the Equator.

13.5.6   Classification by situation of projection surface

Cartographic projections are classified according to the following situations of the projection surfaces (Alves and Silva, 2016):

- Tangent: The projection surface touches the ellipsoid at a point (plane) or in a line (cylinder or cone);
- Secant: The projection surface cuts the ellipsoid in two points (plane) or in two lines (cylinder or cone) of secancy.

13.6   Spherical Cartographic Projections

Spherical cartographic projections are used on small-scale geographic maps. In this case, a reference sphere is applied to the cartographic plane. The charts and larger topographic plans are based on ellipsoidal projections, that is, a reference ellipsoid is applied in the cartographic plane (Casaca et al., 2007).

The sphere is an adequate representation of the Earth for many applications that do not require the greatest precision. The first approximation made to the shape and size of the Earth was a sphere with a radius of 6371 km. The 3D spherical coordinates of longitude (λ), latitude (ϕ) and altitude (h) are defined in the sphere. The shortest route between two points on the sphere is called the "great circle". The great circle is defined by the intersection plane that passed through both points and the center of the sphere (Figure 13.2) (Iliffe and Lott, 2008).

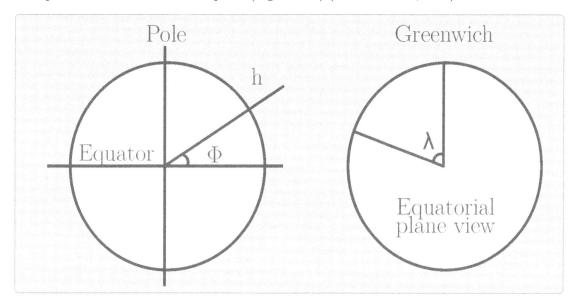

FIGURE 13.2: 3D spherical coordinates of longitude, latitude and altitude.

Considering the coordinates of two points A and B, $(\phi_A - \lambda_A)$ and $(\phi_B - \lambda_B)$, respectively, the distance L_{AB} between two points is (Iliffe and Lott, 2008):

$$cosL_{AB} = sin\phi_A sin\phi_B + cos\phi_A cos\phi_B cos\Delta\lambda \tag{13.3}$$

where $\Delta\lambda$ means the difference in longitude between the two points (Iliffe and Lott, 2008):

$$\Delta\lambda = \lambda_B - \lambda_A \tag{13.4}$$

With that, the response in angular unit can be converted into distance using the angle in radians and multiplying by an appropriate value of the Earth's spherical radius (Iliffe and Lott, 2008):

$$L_{(km)} = 6371\frac{\pi}{180}L_{(degrees)} \tag{13.5}$$

The azimuth AB is defined by the hour angle between the meridian at A and the large circle up to B (Iliffe and Lott, 2008):

$$cot_{AB} = \frac{cos\phi_A tan\phi_B - sin\phi_A cos\Delta\lambda}{sin\Delta\lambda} \tag{13.6}$$

In conjunction with the coordinate system λ, ϕ and h, the following terms are defined (Iliffe and Lott, 2008):

- Parallels of latitude: Lines of equal latitude on the sphere surface;
- Meridians: Lines of equal length on the sphere surface.

13.7   Plane Cartographic Projections

The conversion of positions on the Earth's surface to plane rectangular coordinates could be accomplished by projecting the points mathematically from the ellipsoid to a developed imaginary surface. A surface can theoretically be developed without distortion of shape or size. A rectangular grid can be superimposed on the developed plane surface so that the positions of points on the plane were specified in reference to the grid's X and Y axes. A plane grid can be developed using a mathematical process called "map projection".

All map projections must contain at least two mathematical equations: one defining the X value and the other, the Y value. A simple projection involving the two equations is (Slocum et al., 2014; Alves and Silva, 2016):

$$X = R(\lambda - \lambda_0) \tag{13.7}$$

$$Y = R\phi \tag{13.8}$$

where λ is the longitude value, ϕ, the latitude, λ_0, the value of the central meridian and, R, the radius of the reference globe.

To calculate the X and Y values based on these equations, the longitude value of the central meridian must be obtained. If the longitude value is 0°, the central meridian value will coincide with the first meridian. All latitude and longitude values must be converted to radians. This conversion is necessary to perform geocomputation (Alves and Silva, 2016).

Through the `sf`, `rnaturalearth` and `tmap` packages it is possible to determine and map different projections of country boundaries on Earth.

```
library(sf)
library(rnaturalearth)
library(tmap)
```

The Earth's spatial polygons with attributes is obtained on a small scale from the `rnaturalearth` package, in WGS-84 geographic projection, and converted to `sf` for later calculation of different plane cartographic projections.

```
Earth <- countries110 %>%
  st_as_sf() %>%
  st_geometry()
```

The Plate Carrée projection is one of the oldest projections developed by the Greeks. We observed that in the longitude and latitude lines of the Plate Carrée projection there is spacing at equal intervals. The longitude lines presented equal sizes, but with no convergence at the poles, as it should be under conditions of an ideal projection (Slocum et al., 2014; Alves and Silva, 2016). As an example, the Earth is re-projected and mapped in the Plate Carrée (Equidistant Cylindrical) projection in R using `qtm` function and `+proj=eqc` (Figure 13.3).

```
qtm(Earth, projection = "+proj=eqc")+tm_grid()
```

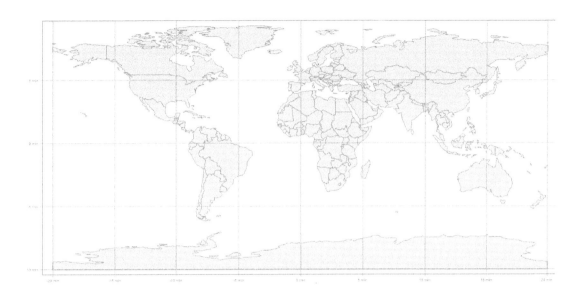

FIGURE 13.3: Mapping Earth's continents in the Plate Carrée projection.

Just by adding trigonometric functions, it is possible to expand the map based on simple equations. For example, with the addition of cosine in longitude, the equation can be modified by (Slocum et al., 2014; Alves and Silva, 2016):

$$X = R(\lambda - \lambda_0)cos\phi \qquad (13.9)$$

$$Y = R\phi \qquad (13.10)$$

Lambert's cylindrical projection of equal area, developed in 1772 by Johann H. Lambert, is determined with this mathematical transformation. In this projection, the meridians are equally spaced in a similar way to the Plate Carrée projection, but the spacing between the parallels decreased with increasing distance from the Equator. The Equal Area Cylindrical projection can be mapped with `qtm` function and `+proj=cea` (Figure 13.4).

```
qtm(Earth, projection = "+proj=cea")+tm_grid()
```

An alternative can be realized by introducing a sine function in simple equations, changing the way of determining latitude (Slocum et al., 2014; Alves and Silva, 2016):

$$X = R(\lambda - \lambda_0) \qquad (13.11)$$

$$Y = Rsin\phi \qquad (13.12)$$

The change in latitude results in the sinusoidal pseudocylindrical projection. In this case, the lines

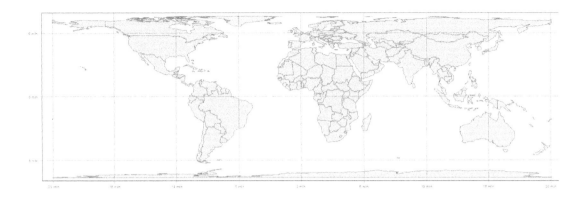

FIGURE 13.4: Mapping Earth's continents in the Equal Area Cylindrical projection.

of latitude are parallel and equally spaced. However, the meridians are curved converging towards the north and south poles, but the spacing between the parallels decreased with increasing distance from the Equator. The sinusoidal projection can be mapped with `qtm` function and `+proj=sinu` (Figure 13.5).

```
qtm(Earth, projection = "+proj=sinu")+tm_grid()
```

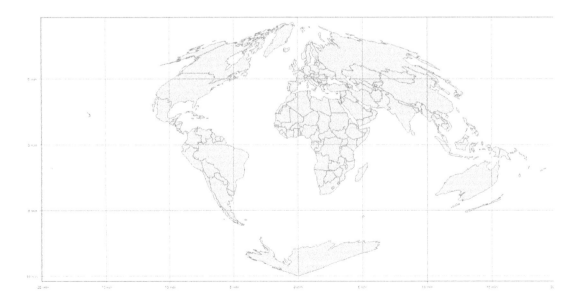

FIGURE 13.5: Mapping Earth's continents in the sinusoidal projection.

With the combination of sine and cosine functions in simple equations, the polycylindrical or Mollweide projection is obtained (Slocum et al., 2014; Alves and Silva, 2016):

$$X = R(\lambda - \lambda_0)cos\phi \tag{13.13}$$

$$Y = Rsin\phi \tag{13.14}$$

The meridians are curved as in the sinusoidal projection, and the parallels are similar to the other two cylindrical projections presented. The Mollweide projection can be mapped with `qtm` function and `+proj=moll` (Figure 13.6).

```
qtm(Earth, projection = "+proj=moll")+tm_grid()
```

FIGURE 13.6: Mapping Earth's continents in the Mollweide projection.

Projection parameters can be modified using CRS definitions (Lovelace et al., 2019b). For example, modifying the center of the Azimuth Lambert projection from an area equal to longitude and latitude 0, 0 to longitude -50, -17, the map is centered in Lavras, Minas Gerais, Brazil (Figure 13.7).

```
qtm(Earth,projection="+proj=laea +x_0=0 +y_0=0 +lon_0=-44.9998 +lat_0=-21.2457")
+tm_grid()
```

The projection concepts described previously applied equally to data with raster geometry. However, important differences in the re-projection of vectors and rasters are (Lovelace et al., 2019b):

- In the geometric transformation of a vector object, the coordinates of each vertex have changed;
- Rasters are composed of rectangular cells of the same sizes and it is not possible to transform pixel coordinates separately;

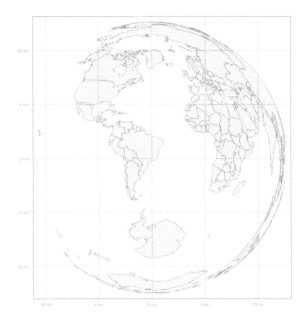

FIGURE 13.7: Modification of the Azimuth Lambert projection center for longitude and latitude coordinates in Lavras, Minas Gerais, Brazil.

- In raster re-projection, a new raster object is created, usually with a number of columns and rows different from the original;
- In the raster re-projection, the attributes must be subsequently re-estimated, so that new pixels can be filled with other values.

In the re-projection of raster data, two separate spatial operations are considered (Lovelace et al., 2019b):

- Vectorial re-projection of centroid pixels to another CRS;
- Calculation of new pixel values by re-sampling.

13.8   State Plane Coordinate System

The United States has adopted its own specialized coordinate systems for applications such as surveying that require very high accuracy (Longley et al., 2001). The State Plane Coordinate System (SPCS) is a system of large-scale conformal map projections originally created in the 1930s to support surveying, engineering, and mapping activities throughout the United States and its territories. The system was revised in 1983 to accommodate the shift to new North American Datum (NAD-83) with the GRS-80 ellipsoid parameters (Longley et al., 2001; NOAA, 2020b). The R packages USAboundaries and USAboundariesData include contemporary state, county, and Congressional district boundaries, as well as zip code tabulation area centroids for use in R. The packages also include historical boundaries from 1629 to 2000 for states and counties from the Newberry Library's Atlas of Historical County Boundaries and historical city population data from Erik Steiner's United States Historical City Populations, 1790-2010. It is also possible to acess helper data, including a table of state names, abbreviations, FIPS codes, functions and data to get SPCS projections as `EPSG` codes or `proj4string` (Mullen and Bratt, 2018). Geographical vector boundaries of the United States can also be obtained with the `maps` R package (Brownrigg, 2018) and re-projected from WGS-84 datum to the US National Atlas Equal Area projection (`+proj=laea +lat_0=45 +lon_0=-100 +x_0=0 +y_0=0 +a=6370997 +b=6370997 +units=m +no_defs`) and displayed using `ggplot2`. In this same plot, we included California state vector boundaries transformed from WGS-84 datum to the NAD-83/California zone 3 SPCS projection with the `USAboundaries` and `sf` packages (Figure 13.8).

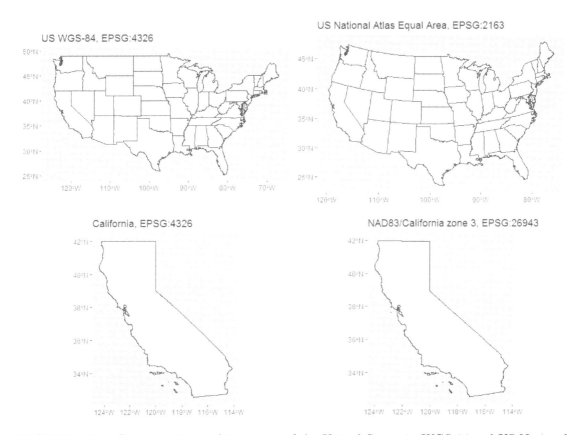

FIGURE 13.8: Boundary lines of the states of the United States in WGS-84 and US National Atlas Equal Area projection and the California state in WGS-84 and State Plane Coordinate System projection.

Some large states decided that distortions are still too great and designed their SPCS with internal zones. For example, Texas has five zones based on the Lambert conformal conic projection. Hawaii has five zones based on the Transverse Mercator Projection. Many other countries have adopted their own coordinate systems. The UK uses a single projection and coordinate system known as the "National Grid", based on Transverse Mercator Projection. Canada uses a uniform coordinate system based on the Lambert Conformal Conic Projection (Longley et al., 2001).

13.9 Universal Transverse Mercator Projection

The Universal Transverse Mercator (UTM) projection is another Earth projection system used. The UTM system became prominent in surveys after the inclusion of UTM coordinates as metric unit projection for different existing ellipsoids (datum). The UTM system was initially developed by the US Department of Defense (DoD) for military use, with global coverage from 80° S latitude to 84° N latitude. Each zone has longitudinal width of 6°. With this, 60 zones are needed to represent the entire Earth (Figure 13.9) (Ghilani and Wolf, 1989; Alves and Silva, 2016).

Other characteristics of the UTM projection are (Silva and Segantine, 2015):

- Latitude of origin: 0° (Equator);
- Longitude of origin: The longitude of the central meridian of the fuse;
- False-Northing (North translation): 10000000 m for the Southern Hemisphere;
- False-Easting (East translation): 500000 m;
- Scale factor at the central meridian: $k_0 = 0.9996$;
- The Equator and the central meridian line of each fuse were represented by straight lines on the projection and the other meridians and parallels by concave lines with respect to the central meridian and nearest pole, respectively.

The central meridian of the UTM projection can be determined 6° by 6°, with the first central meridian at longitude 177° and the last at longitude 3°. The relationship between the fuse and central meridian can be obtained by (Silva and Segantine, 2015):

$$Fuse = \frac{183° - MC}{6} \tag{13.15}$$

$$MC = 183° - (6Fuse) \tag{13.16}$$

where MC is the central meridian of the fuse.

The UTM fuses, excluding the polar ice caps, are projected on cylinders tangent to their central meridian. The 60 UTM cartographic planes, with the images of fuses, are discontinuous at the borders, so that the projection used is polysurface or interrupted. The UTM zones are numbered from 1 to 60, starting at longitude 180° W. The central meridian of each zone assigned a Fals-Easting offset value E_0 of 500000 m (Casaca et al., 2007). A False-Northing offset value of zero is applied for each zone in the Northern Hemisphere and 10000000 m for the Southern Hemisphere, in order to avoid negative Y coordinates. We must specify the zone number as well as the Northing and Easting offset to specify any point in the UTM system (Ghilani and Wolf, 1989; Alves and Silva, 2016) (Figure 13.10). In the UTM system, there is overlap of adjacent zones of 0°30′ creating an overlap area of 1° width. The overlap area made field work easier in certain activities. Evaluating the scale deformation in a UTM fuse with a tangent, we verified a scale factor equal to one in the central meridian and approximately equal to 1.0015 or (1/666), in the extremes of the fuse. Assigning the k_0 scale factor of 0.9996 to the central meridian of the UTM system transformed the tangent cylinder into a secant cylinder, determining a more favorable scale deformation pattern along the fuse. The scale error is limited to 1/2500, at the central meridian and, 1/1030 at the fuse extremities (IBGE, 1999).

13.9.1 Transformation of geodetic coordinates into UTM coordinates

The transformation of geodetic coordinates into UTM coordinates is exemplified by means of rigorous equations (IBGE, 1995; Silva and Segantine, 2015):

$$N' = (I) + (II)p^2 + (III)p^4 + (A_6)p^6 \tag{13.17}$$

$$E' = (IV)p + (V)p^3 + (B_5)p^5 \tag{13.18}$$

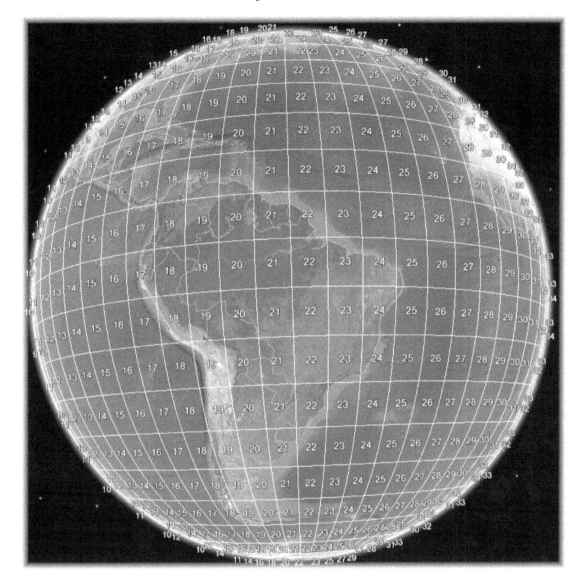

FIGURE 13.9: Fuse zones of the Universal Transverse Mercator system in South America, in a spherical representation of the Earth in the Google Earth software.

where the ordinate N = N' for the Northern Hemisphere, the ordinate N = 10000000 - N' for the Southern Hemisphere, the abscissa E = 500000 + E' for points east of the central meridian, E = 500000 - E' for points west of the central meridian.

$$(I) = k_0 S \tag{13.19}$$

$$S = a(1 - e^2)[A\phi_g - \frac{B}{2}sen2\phi_g + \frac{C}{4}sen4\phi_g - \frac{D}{6}sen6\phi_g + \frac{E}{8}sen8\phi_g - \frac{F}{10}\phi_g + ...] \tag{13.20}$$

$$A = 1 + \frac{3}{4}e^2 + \frac{45}{64}e^4 + \frac{175}{256}e^6 + \frac{11025}{16384}e^8 + \frac{43659}{65536}e^{10} \tag{13.21}$$

$$B = \frac{3}{4}e^2 + \frac{15}{16}e^4 + \frac{525}{512}e^6 + \frac{2205}{2048}e^8 + \frac{72765}{65536}e^{10} \tag{13.22}$$

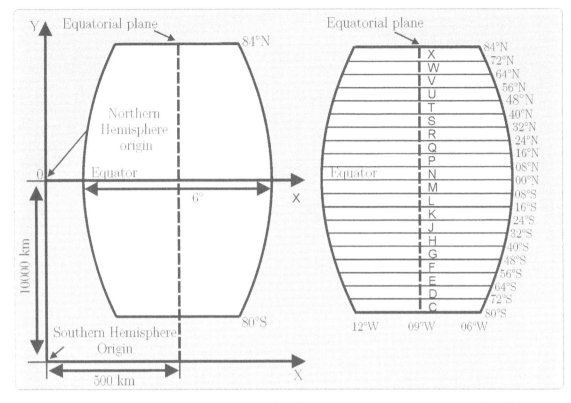

FIGURE 13.10: Coordinate determination (left) and UTM zone nomenclature (right).

$$C = \frac{15}{64}e^4 + \frac{105}{256}e^6 + \frac{2205}{4096}e^8 + \frac{10395}{16384}e^{10} \tag{13.23}$$

$$D = \frac{35}{512}e^6 + \frac{315}{2048}e^8 + \frac{31185}{131072}e^{10} \tag{13.24}$$

$$E = \frac{315}{16384}e^8 + \frac{3465}{65536}e^{10} \tag{13.25}$$

$$F = \frac{693}{131072}e^{10} \tag{13.26}$$

where, for geomatics calculations, we can perform the calculation up to equation C, disregarding equations D, E and F.

$$(II) = \frac{N \, sen\phi_g cos\phi_g (\frac{1}{\rho''})^2 k_0 10^8}{2} \tag{13.27}$$

$$(III) = (\frac{(\frac{1}{\rho''})^4 N \, sen\phi_g cos^3\phi_g}{24})(5 - tan^2\phi_g + 9e'^2 cos^2\phi_g + 4e'^4\phi_g)k_0 10^{16} \tag{13.28}$$

$$(IV) = N cos\phi_g (\frac{1}{\rho''})^2 k_0 10^4 \tag{13.29}$$

$$(V) = (\frac{(\frac{1}{\rho''})^3 N cos^3\phi_g}{6})(1 - tan^2\phi_g + e'^2 cos^2\phi_g)k_0 10^{12} \tag{13.30}$$

$$p = 0.0001\Delta\lambda" \tag{13.31}$$

$$\Delta\lambda = \lambda_g - \lambda_{MC} \tag{13.32}$$

$$A_6 = (\frac{(\frac{1}{\rho"})^6 N sen\phi_g cos^5\phi_g}{720})(61 - 58tan^2\phi_g + tan^4\phi_g + 270e'^2cos^2\phi_g - 330e'^2 sen^2\phi_g)k_0 10^{24} \tag{13.33}$$

$$B_5 = \frac{(\frac{1}{\rho"})^5 N sen\phi_g cos^5\phi_g}{120}(5 - 18tan^2\phi_g + tan^4\phi_g + 14e'^2cos^2\phi_g - 58e'^2 sen^2\phi_g)k_0 10^{20} \tag{13.34}$$

where $\rho"$ is a conversion factor from radian to arc second equal to 206264.8062", ϕ_g, the geodetic latitude of the point, λ_g, the geodetic longitude of the point, λ_{MC}, the longitude of the central meridian, k_0, 0.9996, N, the radius of the first vertical and, e, e', the first and second eccentricity, respectively.

The angular difference between geodetic north (N_G) and grid north (N_Q) is termed meridian convergence (γ). The γ is positive to the west and, negative to the east of the central meridian. The γ can be determined by (Silva and Segantine, 2015):

$$\gamma = (XII)p + (XIII)p^3 + (C_5)p^5 \tag{13.35}$$

where

$$(XII) = sen\phi_g 10^4 \tag{13.36}$$

$$(XIII) = \frac{(\frac{1}{\rho"})^4 sen\phi_g cos^2\phi_g}{3}(1 + 3e'^2cos^2\phi_g + 2e'^4cos^4\phi_g)10^{12} \tag{13.37}$$

$$C_5 = \frac{(\frac{1}{\rho"})^4 sen\phi_g cos^4\phi_g}{15}(2 - tan^2\phi_g)10^{20} \tag{13.38}$$

where e'^2 is the second quadratic eccentricity.

The γ can be determined with an approximate value by (Silva and Segantine, 2015):

$$\gamma = \Delta\lambda sen\phi_g \tag{13.39}$$

13.9.2 Reduction of geodetic distance in plane distance

The reduction of a topographic distance to the reference ellipsoid can be obtained as a function of the altimetric scale factor (k_{alt}) (Silva and Segantine, 2015):

$$K_{alt} = 1 - (\frac{H_P}{R_0 + H_P}) \tag{13.40}$$

where H_P is the orthometric altitude of the point and, R_0, the local mean radius of the Earth at the latitude of the point.

A scale factor (k_{UTM}) application is necessary to obtain the plane distance in the UTM projection (Figure 13.11) (Silva and Segantine, 2015):

$$d_{UTM} = k_{UTM}d_0 \tag{13.41}$$

where d_0 is the ellipsoidal distance.

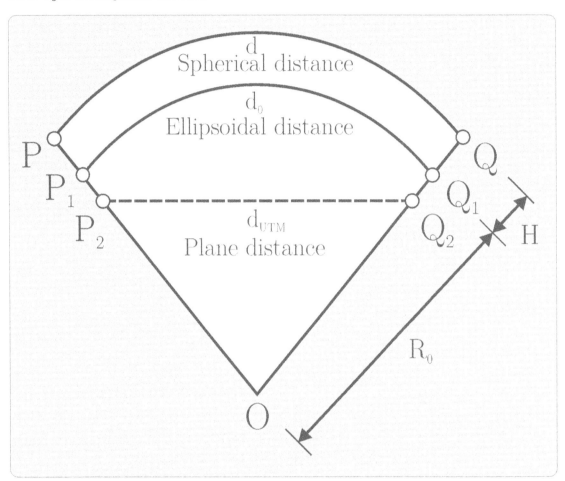

FIGURE 13.11: Geometric relations to calculate distance in the UTM projection.

A scale factor $k_0 = 0.9996$ is adopted for points on the central meridian in the UTM projection to avoid exaggerated deformations at the fuse borders. The scale factor grew westward and eastward from the central meridian, with value $k = 1.0$, at E = 320000 m, E = 680000 m and, $k = 1.00097$ at E = 166000 m and, E = 834000 m (Figure 13.12) (Silva and Segantine, 2015).

The value of the scale factor can be determined in simplified form by (Silva and Segantine, 2015):

$$k_{UTM} = k_0(1 + \frac{E'^2}{2R_0^2}) \tag{13.42}$$

where $k_0 = 0.9996$ at the central meridian, E'= E-500000 for points east of the central meridian, E'= 500000 - E for points west of the central meridian and R_0 is the local mean radius of the Earth at the point considered.

The use of the full-scale factor (k_T) is recommended for plane distance calculation from measurements with sighting instruments (Silva and Segantine, 2015):

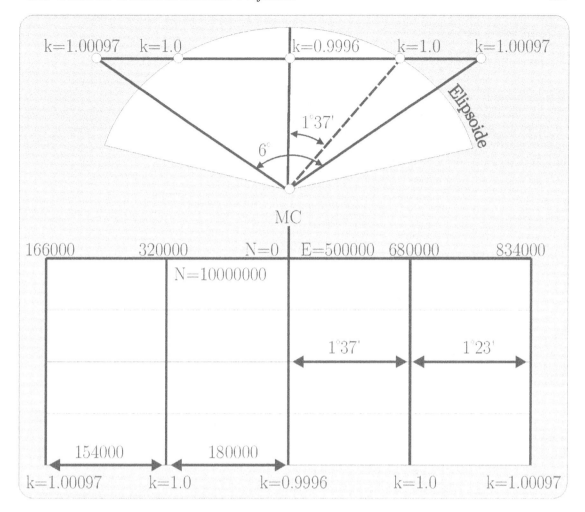

FIGURE 13.12: Scale factor of the UTM projection.

$$k_T = k_{alt}k_{UTM} \tag{13.43}$$

The plane distance (d_{UTM}) and the topographic horizontal distance (d) can be obtained by (Silva and Segantine, 2015):

$$d_{UTM} = k_T d \tag{13.44}$$

$$d = \frac{d_{UTM}}{k_T} \tag{13.45}$$

The UTM system is completed in the polar regions with the UPS (Universal Polar Stereographic) system. Regions located on the polar ice caps not covered by the UTM system can be projected by the plane stereographic projection (ellipsoidal version) on planes tangent to the north and south poles (Casaca et al., 2007; Alves and Silva, 2016).

13.10   Cartographic Projections in Computers

Map projection calculations have been more difficult to perform on calculators than on computers due to the length and magnitude of values used. Computer programs have been made available to easily perform map projection calculations. The softwares usually accept point value input or a file with the coordinate data to be transformed from one coordinate reference system (CRS) into another.

13.11   Computation

The main groups of projection types in R are conical, cylindrical and plane. In conical projection, the Earth's surface is projected in a cone along a single tangent line or two tangent lines. Distortions are minimized along the tangency lines and increased with distance from the tangency lines. This projection is most suitable for maps of mid-latitude areas. In cylindrical projection, the surface of interest is mapped in a cylinder. This projection can also be created by touching the Earth's surface along a single tangency line or two tangency lines. Cylindrical projections have been often used in mapping the Earth as a whole. In a plane projection, a plane surface touched the globe at a point or along a line of tangency to project the data, with applications, for example, in mapping polar regions (Lovelace et al., 2019b).

Projected coordinate reference systems are based on Cartesian coordinates on a plane surface. In these systems there is an origin, X and Y axes, and a linear unit of measurement, such as meters. All projected CRS are based on a geographic CRS (datum). Using map projections, 3D surface of the Earth is converted into values for East and North (X and Y) on a projected CRS. This transformation cannot be done without adding some distortion to the real surface. Some properties of the Earth's surface are distorted in this process, such as the area, direction, distance, and shape, but a projected coordinate system enabled to preserve only one or two of these properties. The projections are named based on the property preserved as equal area preserves area, azimuthal preserves direction, equidistant preserves distance, and conformal preserves local shape (Lovelace et al., 2019b).

The two main ways to describe CRS in R are either an EPSG code or a `proj4string` definition. By using an EPSG code, assigning a coordinate reference system can be more convenient, easy to remember when compared to the `proj4string` function. However, through a definition of `proj4string` there are more flexibility to specify different parameters about the coordinate system. This enables to specify different projections and modify existing ones (Lovelace et al., 2019b).

As a computational practice, we aim to check map projection definitions in the `sf` package, check available CRS codes, and check and assign a CRS to attribute vector type data. Obtain vector data with boundaries of countries on Earth and Brazil. Mollweide, equidistant cylindrical (equirectangular), Mercator and Robinson global projection transformations heve been mapped. The boundaries of Brazil are mapped and country areas are calculated in square kilometers, considering Lambert's conformal conic map projections of South America, Albers equal area projections of South America, Mercator and Equirectangular comparison results. Finally, map projection transformations are performed on data with raster geometry in the UTM system.

The boundaries of attribute vector data of Earth countries and Brazil are obtained from the packages `spData` (Bivand et al., 2020b) and `rnaturalearth` (South, 2021), respectively. Natural Earth consists of a dataset of public domain maps available at scales 1:10, 1:50 and 1:110 million.

Coordinate projection functions are used through the PROJ library of the rgdal package (Bivand et al., 2021). The area calculation of Brazil is performed with the package sf (Pebesma et al., 2021). The units package is used in the transformation of area units (Pebesma et al., 2020a). The polygon maps of Earth countries and Brazil is mapped with the tmap package (Tennekes, 2018; Tennekes et al., 2020). The transformation of the geographic projection of a SRTM digital elevation model of Lavras, Minas Gerais, Brazil (NASA, 2013) to the UTM projection is performed with the raster package (Hijmans et al., 2020).

13.11.1   Installing R packages

The install.packages function is used to install the sf, rgdal, units, spData, tmap, raster and rnaturalearth packages in the R console.

```
## install.packages("raster")
```

13.11.2   Enabling R packages

The library function is used to enable the sf, rgdal, units, spData, tmap, raster and rnaturalearth packages in the R console.

```
library(sf)
library(rgdal)
library(units)
library(spData)
library(tmap)
library(raster)
library(rnaturalearth)
```

13.11.3   Checking map projection definitions available in sf

With the sf_proj_info function, the projections, ellipses, datums and data units from the PROJ library are listed for a total of 162 projections.

13.11.4   Checking available CRS code

In the R package Spatial, a wide variety of CRS and the PROJ library are available. The EPSG codes can be searched on the Internet or by using the make_EPSG function of the R package rgdal totaling 6609 codes available.

13.11.5   Getting the CRS in sf

The function st_crs is used to obtain the CRS from a simple feature vector file with Earth country attributes. The EPSG 4326, referring to the WGS-84 datum, is observed in the case of the file "world".

13.11.6   Setting the CRS of a file

A CRS value can be assigned to a file if the coordinate system is lost or not recorded for some reason. In this case, the `st_set_crs` function is used as an example to set the CRS to a file.

13.11.7   Mapping global projection transformations

Transformations of global projections Mollweide (moll), cylindrical equidistant or Plate Carrée (eqc), Mercator (merc) and Robinson (robin) are performed on the "world" file, initially in WGS-84 datum. The projections are transformed using the `qtm` function. The projection parameter is set with the `+proj` function. The maps are arranged for mapping with the `tmap_arrange` function (Figure 13.13).

```
## tmap_arrange(w1, w2, w3, w4, widths = c(1, 1))
```

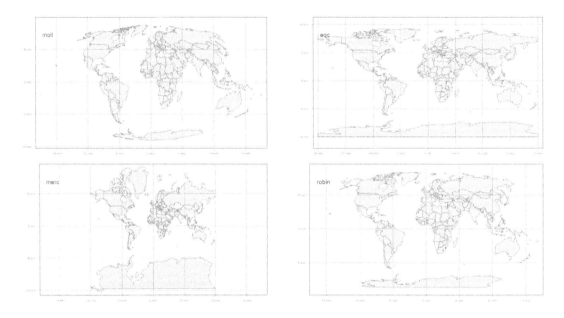

FIGURE 13.13: Mapping countries in the Mollweide (moll), Plate Carrée (eqc), Mercator (merc) and Robinson (robin) map projections.

13.11.8   Mapping Brazil in different map projections

The vector boundary of Brazil is obtained by the `rnaturalearth` package to map Brazil in different map projections.

```
brazil <- ne_countries(country = "Brazil")
```

Then the vector boundary is converted to `sf` for the subsequent mapping.

```
brazil_sf <- st_as_sf(brazil)
```

Brazil is mapped in Lambert's conformal conic for South America, Albers equal area for South America, Mercator and equirectangular map projections for the Americas (Figure 13.14).

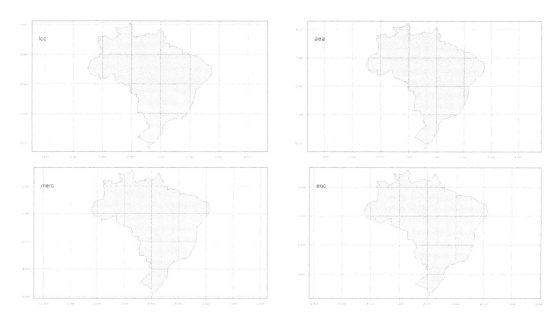

FIGURE 13.14: Mapping Brazil in the Lambert Conical Conformal projection for South America (lcc), Albers Equal Area projection for South America (aea), Mercator (merc) and Equirectangular projection (eqc) for the Americas.

13.11.9   Calculating area in different map projections

The area of Brazil in square kilometers is calculated in different map projections. The areas of Brazil in Lambert's conformal conic for South America, Albers Equal Area for South America, Mercator and Equirectangular map projections are 8235020, 8508195, 9002220, and 8800636 km^2, respectively.

```
# Lambert's Conformal Conic of South America
a1 <- st_transform(brazil_sf, "+proj=lcc +lat_1=-5 +lat_2=-42 +lat_0=-32
            +lon_0=-60 +x_0=0 +y_0=0 +ellps=aust_SA +units=m +no_defs")
st_area(a1)/1000000
```

```
## 8235020 [m^2]
```

```
units::set_units(st_area(a1), km^2)
```

```
## 8235020 [km^2]
```

```
# Albers Equal Area of South America
a2 <- st_transform(brazil_sf, "+proj=aea +lat_1=-5 +lat_2=-42 +lat_0=-32
                +lon_0=-60 +x_0=0 +y_0=0 +ellps=aust_SA +units=m +no_defs")
st_area(a2)/1000000
```

```
## 8508195 [m^2]
```

```
units::set_units(st_area(a2), km^2)
```

```
## 8508195 [km^2]
```

```
# Mercator
a3 <- st_transform(brazil_sf, "+proj=merc")
st_area(a3)/1000000
```

```
## 9002220 [m^2]
```

```
units::set_units(st_area(a3), km^2)
```

```
## 9002220 [km^2]
```

```
# Equirectangular
a4 <- st_transform(brazil_sf, "+proj=eqc")
st_area(a4)/1000000
```

```
## 8800636 [m^2]
```

```
units::set_units(st_area(a4), km^2)
```

```
## 8800636 [km^2]
```

13.11.10 Performing map projection transformations on raster data

An SRTM digital elevation model of Lavras, Brazil, used in the previous chapter, is imported to perform map projection transformations on raster data.

```
demLavras<-raster("files/demLavras_mask.tif")
```

The crs function is used to evaluate the projection of the SRTM model.

```
crs(demLavras)
```

```
## CRS arguments:
##   +proj=longlat +datum=WGS84 +no_defs +ellps=WGS84 +towgs84=0,0,0
```

Another option for evaluating the raster projection is done with the projection function.

```
projection(demLavras)
```

```
## [1] "+proj=longlat +datum=WGS84 +no_defs +ellps=WGS84 +towgs84=0,0,0"
```

The SRTM model is re-projected for the WGS-84 datum in UTM 23 S projection with the projectRaster function. For this, we configured the zone 23 South, where Lavras is located.

```
# Reproject raster
demLavrasUTM <- projectRaster(demLavras,
crs="+proj=utm +zone=23 +south +datum=WGS84
+units=m +no_defs+ellps=WGS84")
# Evaluating projection
demLavrasUTM
```

```
## class      : RasterLayer
## dimensions : 347, 389, 134983  (nrow, ncol, ncell)
## resolution : 86.5, 92.2  (x, y)
## extent     : 480788.9, 514437.4, 7632947, 7664940  (xmin, xmax, ymin, ymax)
## crs        : +proj=utm +zone=23 +south +datum=WGS84 +units=m +ellps=WGS84 +towgs84=0,0,0
## source     : memory
## names      : demLavras_mask
## values     : 750.6041, 1274.197  (min, max)
```

To assign a map projection to the raster when necessary, we use the function crs and then set the parameter +proj of the projection in which the data is transformed.

The pixels with altitude 800, 900, 1100 and 1200 m are mapped on the projected raster with the rasterToContour function, followed by setting up the altitude contour lines.

Then, the SRTM elevation model is mapped with the 800, 900, 1100 and 1200 m altitude contour values in a grey, white, black, grey90 and grey30 colors, respectively (Figure 13.15).

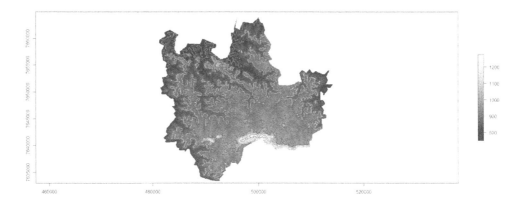

FIGURE 13.15: Mapping the relief with contour lines, in Lavras city, Minas Gerais, Brazil.

13.12   Solved Exercises

13.12.1   Calculate the value of the central meridian for UTM zone 22.

A: The value of the central meridian for UTM 22 is -51° W or +51° W.

```
# Data
fuse<-22
# MC
MC<-183-(6*fuse)
MC
```

```
## [1] 51
```

13.12.2   In a topographic survey, the UTM coordinates of points A and B were determined. Calculate the Euclidean distance between the points.

Required information: Coordinates at A: E = 471525 m; N = 6753374 m; coordinates at B: E = 475333 m; N = 6766955 m.

A: The Euclidean distance between points A and B is 14104.7660 m.

```
# Determine the distance
H<-sqrt((475333-471525)^2+(6766955-6753374)^2)
# Present the result with 4 decimal places
```

```
H<-format(H, nsmall=4)
H
```

```
## [1] "14104.7660"
```

13.12.3   Determine the spherical distance in a straight line between the cities of Arroio do Meio and Porto Alegre?

Required information: Coordinates in Arroio do Meio: $\lambda = 51°56'24"$ W; $\phi = 29°24'$ S and Porto Alegre $\lambda = 51°13'48"$ W; $\phi = 30°01'48"$ S. Earth's radius = 6378160 m.

A: The spherical distance is 98134.12 m.

```
# Enable package
library(circular)
# Data
R<-6378160
# Determine costheta
costheta<-(sin(rad(29+24/60)))*(sin(rad(30+1/60+48/3600)))+
 (cos(rad(29+24/60)))*(cos(rad(30+1/60+48/3600)))*
  (cos(rad((51+56/60+24/3600)-(51+13/60+48/3600))))
# Determine theta
theta<- deg(acos(costheta))
# Determine the distance
H<-theta*((2*pi*R)/360)
H
```

```
## [1] 98134.12
```

13.12.4   Convert the geodesic coordinates of vertex A from the WGS-84 geodetic system to UTM.

Required information: $\phi_g = -22°45'33.4523"$ S; $\lambda_g = -47°12'56.2324"$ W.

A: The UTM E and N coordinates are 272499.2525 and 7481423.8370 m, respectively.

```
# Enable package
library(rgdal)
# Enter data
LongLat <- data.frame(X = c(-(47+12/60+56.2324/3600)),
                      Y = c(-(22+45/60+33.4523/3600)))
names(LongLat) <- c("X", "Y")
# Convert to sp object
coordinates(LongLat) <- ~ X + Y
# Add coordinate system
proj4string(LongLat) <- CRS("+proj=longlat +ellps=WGS84 +datum=WGS84")
# Check input database in decimal degrees
LongLat
```

```
## class      : SpatialPoints
## features   : 1
## extent     : -47.21562, -47.21562, -22.75929, -22.75929  (xmin, xmax, ymin, ymax)
## crs        : +proj=longlat +ellps=WGS84 +datum=WGS84 +towgs84=0,0,0
```

```
# Project to UTM with the function spTransform
UTM <- spTransform(LongLat, CRS("+proj=utm +zone=23 +south +ellps=WGS84
                                 +datum=WGS84"))
# Present the results with 4 decimal places
X<-format(UTM$X, nsmall=4)
X
```

```
## [1] "272499.2525"
```

```
Y<-format(UTM$Y, nsmall=4)
Y
```

```
## [1] "7481423.8370"
```

The exercise can also be done with the equations presented in the theoretical part of the text about transformation of geodesic coordinates into UTM.

13.12.5 Determine the meridian convergence of point A in the SIRGAS-2000 geodetic system.

Required information: $\phi_g = -22°45'33.4523"$ S; $\lambda_g = -47°12'56.2324"$ W; $k_0 = 0.9996$.

A: The meridian convergence is determined in simplified form to the value of $0.8571359°$.

```
# Enable package
library(circular)
# Data
fuse<-23
long<-(-(47+12/60+56.2324/3600))
lat<-(-(22+45/60+33.4523/3600))
# Determine the central meridian
MC<-183-(6*fuse)
# Determine the lambda delta
delta<-long-(-MC)
# Determine meridian convergence
gamma<-delta*sin(rad(lat))
gamma
```

```
## [1] 0.8571359
```

13.12.6 **The slant distance (L) between vertices A and B was measured with a sighting instrument. Determine the plane distance between A and B in the SIRGAS-2000 geodetic system.**

Required information: $L_{AB} = 1724.928$ m; vertical zenith angle $= 87°27'31"$; $\phi_g = -22°45'33.4523"$ S; E $= 272499.252$ m; N $= 7481423.837$ m; H $= 854.267$ m; $\lambda_g = 47°12'56.2324"$ W; local mean radius of the Earth at point $= R_0 = 6363127.455$ m; scale factor at central meridian $= k_0 = 0.9996$.

A: The plane distance is 1723.4117 m.

```
# Enable package
library(circular)
# Data
LAB<-1724.928
z<-87+27/60+31/3600
R0<-6363127.455
H<-854.267
k0<-0.9996
E<-272499.252
El<-E-500000
# Determine the horizontal distance between A and B
d<-LAB*(sin(rad(z)))
# Determine the altimetric scale factor (kalt)
kalt<-1-(H/(R0+H))
# Determine the UTM scale factor (kUTM)
kUTM<-k0*(1+(El^2/(2*(R0)^2)))
# Determine the toal scale factor (kT)
kT<-kalt*kUTM
# Determine the plane distance between A and B
dUTM<-kT*d
# Present the results with 4 decimal places
dUTM4<-format(dUTM, nsmall=4)
dUTM4
```

```
## [1] "1723.4117"
```

13.13   Homework

Choose one exercise presented by the teacher and solve the question with different input values. Compare the results obtained. Perform a practice with a closed-path polygon traversing method in the field. Determine X, Y coordinates of the vertices using the UTM projection. Transform the UTM projection into another map projection. Calculate the area using both projections. Compare the results obtained.

13.14   Resources on the Internet

As a study guide, slides and illustrative videos are presented about the subject covered in the chapter in Table 13.1.

TABLE 13.1: Slide shows and video presentations on cartographic coordinate projection systems.

Guide	Address for Access
1	Slides on cartographic projections and plane coordinate systems in geomatics[1]
2	Animation of map projections[2]
3	Lecture on map projections[3]

13.15   Research Suggestion

The development of scientific research on geomatics is stimulated through activity proposals that can be used or adapted by the student to assess the applicability of the subject matter covered in the chapter (Table 13.2).

TABLE 13.2: Practical and research activities used or adapted by students using cartographic coordinate projection systems.

Activity	Description
1	Compare the representation of a region in different cartographic projections and choose the best projection by weighing the advantages and disadvantages of the projection adopted
2	Transform the coordinates of a closed polygon in different map projections. Compare the area values obtained and justify the choice of the best result
3	Choose map projections to represent the world and South America. Justify the choice

13.16   Learning Outcome Assessment Strategy

Perform a summary of the chapter, "Cartographic Coordinate Projection Systems with Geomatics and R", on a single A4 page in order to show the student's abilities to summarize a subject presenting key points considered of greater importance today.

---

[1] http://www.sergeo.deg.ufla.br/geomatica/book/c13/presentation.html#/
[2] https://youtu.be/gGumy-9HrSY
[3] https://youtu.be/v5fSBQRbPR0

14

GNSS Surveying

14.1 Learning Questions

The emergent learning questions answered through reading the chapter are as follows:

- What is the definition of global navigation satellite system?
- What is the difference between the Global Positioning System (GPS) and the Global Navigation Satellite System (GNSS)?
- How to measure the GNSS signal.
- What is the difference between satellite reference coordinate systems, geocentric coordinate systems, and geodetic coordinate systems?
- How to perform transformations between geocentric and geodetic coordinate systems.
- How to evaluate errors of GNSS survey observations.
- How to import, analyze and map GNSS kinematic surveying data in R software.

14.2 Learning Outcomes

The learning outcomes expected from reading the chapter are as follows:

- Understand the definition of global navigation satellite system and the difference between GPS and GNSS.
- Understand definitions and basic characteristics of satellite reference coordinate systems, geocentric coordinate systems, and geodetic coordinate systems.
- Understand fundamentals of satellite positioning and observation errors.
- Import, analyze and map kinematic topographic survey data with GNSS in R software with the `sf`, `tmap` and `mapview` R packages.

14.3 Introduction

The Global Navigation Satellite System emerged in the 1970s as a new approach to geodetic surveying. This system was funded by the military to produce a global geographic guidance and navigation system based on signals transmitted by satellites. With the success of the US Satellite Positioning System (GPS), other countries also became interested in developing their own system. Thus, the complete set of satellite systems used in positioning has been called the

DOI: 10.1201/9781003184263-14

"Global Navigation Satellite System" (GNSS). Receivers using GPS satellites and another system such as GLONASS are called "GNSS receivers". With the use of these systems, there has been synchronization of positioning information in any region of the Earth with high reliability and low cost. GNSS surveys can be performed during day, night, rain or shine, and no cleared lines of sight are required between survey stations. With this, this technology revolutionized conventional surveying procedures, which relied on observed angles and distances to determine point positions (Ghilani and Wolf, 1989; Alves and Silva, 2016). GNSS was composed of four individual systems (Silva and Segantine, 2015):

- Global Positioning System (GPS): USA;
- Global Navigation Satellite System (GLONASS): Russia;
- Global Navigation Satellite System (Galileo): Europe;
- BeiDou Navigation Satellite System (BDS): China;
- Navigation with Indian Constellation (NavIC): India.

After full GPS operation, it is possible to obtain relative positioning with millimeter accuracy with short observations of a few minutes. For distances greater than 5 km, GPS can be more accurate than electronic distance meters used in polygon traverses. The advantages of GNSS systems are (Schofield et al., 2007; Segantine, 2005; Alves and Silva, 2016):

- The results of measurements from a single line, called a "baseline", will determine not only the distance between the stations at the ends, but the direction components X, Y, Z, or E, N, h, or longitude, latitude, and altitude, respectively;
- No sighting is required. However, one must have a clear view of the sky, to observe satellites;
- Some equipment is waterproof and can be used during day and night or in foggy conditions;
- The equipment can be used by only one person, without wasting time and effort;
- No high operator skill is required;
- The position can be obtained on land, at sea, or in the air;
- Baselines of hundreds of kilometers can be determined, without the need for extensive geodetic networks, as in conventional surveys;
- Continuous measurements can be performed to improve deformation monitoring;
- Measurement stations do not need to be intervisible;
- Measurements can be performed under different weather conditions;
- Operates 24 hours a day;
- Worldwide coverage;
- High accuracy of position, velocity and time.

The disadvantages of GNSS are (Schofield et al., 2007; Alves and Silva, 2016):

- Sky visualization is required for satellite tracking, and could be a problem for surveying buildings and engineering works. Equipment does not work well inside buildings or underground;
- The equipment can be expensive. One pair of GPS can cost the same as three or four electronic total stations;
- The local projections and datum of a location must be known, as well as the satellite coordinate system, in order to obtain regionally meaningful and representative points;
- The altitude value determined by satellite may not be the same as that used in an engineering project, because the GPS coordinate system is centered on the Earth's center of mass, so any altitude of points on the Earth's surface will be relative to a datum, or the surface of an ellipsoid. Some GPS can display a geoidal model in software; however, the model can be coarse.

With the use of GNSS, it has been possible to determine specific positions with high precision and accuracy within crops in agricultural areas. As a result, various observations and measurements can be georeferenced. Decisions about the use of GNSS on farms must be based on particular needs, operational management procedures, and understanding of positioning errors. Other important issues are related to vehicle dynamics, use of pulled implements, and terrain conditions. Proper leveling and alignment, as well as installation of GNSS are necessary for effective operation in tillage. Poor quality of the steering control system, sloping terrain, or lack of implement alignment can affect the performance of the navigation system (Alves and Silva, 2016).

14.4 Global Navigation Satellite Systems

The precise distances between satellites and calculation of receiver positions are determined based on timing and signal information. In satellite surveying, satellites become the reference or control stations, and the distances to these satellites are used to calculate receiver positions. In this case, distances and angles can be observed from unknown ground station used to control known point positions (Ghilani and Wolf, 1989; Alves and Silva, 2016).

Global positioning systems have been divided into three segments (Schofield et al., 2007; Alves and Silva, 2016):

- Spatial;
- Control;
- User.

In the case of the GPS system, the spatial segment consists of 24 satellites operating in six orbital planes spaced at 60° intervals inclined at 55° from the Equator. Four additional satellites were kept in reserve as spares. With this configuration 24 h of satellite coverage is obtained between the latitudes of 80° N and 80° S (Ghilani and Wolf, 1989; Alves and Silva, 2016).

The satellites operate in nearly circular orbits with an average height of 20200 km above Earth and an orbital period of 12 sidereal hours, considering a sidereal day being approximately four minutes shorter than a solar day. On the individual satellites, there is identification of pseudo-random noise number (PRNs), satellite number or space vehicle number (SVN) and orbital positions (Ghilani and Wolf, 1989; Alves and Silva, 2016).

Precise atomic clocks are used on the satellites to control the transmission time of the signals. These clocks are extremely accurate, precise, and extremely expensive. The atomic clocks are made of cesium or rubidium. Rubidium clocks can lose one second in 30000 years, while cesium clocks can lose one second in 300000 years. Radioactive hydrogen clocks can lose one second in 30000000 years, and have been proposed for future satellites. For comparison, quartz crystal clocks used in receivers can lose one second in 30 years. If receivers uses the same clocks as satellites, their use would be prohibited, and would require users to be trained in the handling of hazardous materials. So the clocks on the receivers are controlled by oscillations of a quartz crystal that, while also accurate, are less precise than atomic clocks. However, these relatively low-cost timing devices made it possible to develop relatively inexpensive receivers (Ghilani and Wolf, 1989; Alves and Silva, 2016).

The control segment consists of monitoring stations to track the signals and follow the positions of satellites over time. The first GPS monitoring stations were set up in Colorado Springs, Hawaii, Ascension Island, Diego Garcia, and Kwajalein. The tracking information is relayed to the master control station at the Consolidated Space Operations Center (CSOC) at Schriever Air Force Base in Colorado Springs (Ghilani and Wolf, 1989; Alves and Silva, 2016).

The master control station used the received data to make accurate predictions of the satellites' orbits and correct clock parameters. This information is sent to the satellites and then transmitted by them as part of their broadcast message to be used by receivers to predict satellite positions and systematic clock errors (Figure 14.1) (Ghilani and Wolf, 1989; Alves and Silva, 2016).

FIGURE 14.1: The U.S. Air Force's eleventh launch of Boeing-built Global Positioning System IIF satellite aboard a ULA Atlas V from Space Launch Complex 41, at Cape Canaveral, Florida, October 31, 2015, at 12:13 p.m. EDT. (Courtesy of Michael Howard/SpaceFlight Insider.)

The GPS user segment consisted of two categories of receivers that are classified by their access to two services offered by the system. These services are referred to as "Standard Positioning Service" (SPS) and "Precision Positioning Service" (PPS). SPS is provided on the transmitted L1 frequency, and more recently on L2, at no cost to the user. This service is initially developed to provide accuracy of 100 and 156 m in the horizontal and vertical positions, respectively, with a confidence level of 95%. However, improvements in the system and processing software have substantially reduced the error estimates. The PPS is transmitted on frequencies L1 and L2 and available to receivers with valid cryptographic keys, reserved almost exclusively for DoD use. This service made it possible to obtain accuracy of 18 and 28 m, with a confidence level of 95%, in the horizontal and vertical, respectively (Ghilani and Wolf, 1989; Alves and Silva, 2016).

14.5   GNSS Fundamentals

The precise signal travel time is used to determine the distance, or range, to the satellite. Considering a satellite in orbit ~20200 km above the Earth, the signal travel time will be ~0.07 seconds after the receiver generates the same signal. If the time delay between the two signals is multiplied by the signal velocity (speed of light in a vacuum) c, the distance value to the satellite (r) can be determined by (Ghilani and Wolf, 1989; Alves and Silva, 2016):

$$r = ct \qquad (14.1)$$

where t is the time the wave traveled from the satellite to the receiver and c, 299792458 m.

In satellite signal receivers, the distances have been determined to the satellites by the methods of code ranging and carrier phase-shift measurements. Those that employed the code-matching method are called "series mapping receivers", and those that used phase-shift are called "series evaluation receivers". Receiver positions can be calculated based on observations made to multiple satellites. The extraordinary accuracy used to determine time intervals can be verified with the example where a time of 0.1 sec is wrong in a signal measurement. In this case, the distance will be wrong by $(0.1)(299792458) = 29979246$ km. The best GPS receivers have enabled measurement time within 1 nanosec (0.000000001 sec). An error of this magnitude will correspond to a distance error of $(0.000000001)(299792458) = 0.2997925$ m (McCormac et al., 2012; Ghilani and Wolf, 1989; Alves and Silva, 2016).

```
# Error 1
e1<-0.1*299792458
e1
```

```
## [1] 29979246
```

```
# Error 2
e2<-0.000000001*299792458
e2
```

```
## [1] 0.2997925
```

14.5.1 Code ranging

Code matching or code ranging is the method used to determine travel time from satellite signals to the receiver. With travel times, distances can be calculated by the difference between satellite and receiver coordinates. Knowing one range, the receiver would err on a sphere. If ranges are determined by two satellites, errors would occur by the interception between two spheres. Thus, the receiver would be somewhere within the circle determined by two satellites. If the interval for a third satellite is added, this interval adds a sphere producing another intersecting circle. With the addition of four satellites, the calculations at the receiver are adjusted to perform addition or subtraction of time increments for all four measurements until an intersection solution is obtained for all spheres (McCormac et al., 2012; Alves and Silva, 2016) (Figure 14.2).

In order to obtain a valid observation time, it is necessary to consider the systematic error (bias in the clocks), and the refraction of wave that passes through the Earth's atmosphere. In this example, the bias of the receiver's clock is the same for three distance intervals, because the same receiver observes each range of variation. With the introduction of a fourth satellite, the bias of the receiver's clock is determined mathematically. With this procedure, the receiver clock can be less accurate and less expensive (Ghilani and Wolf, 1989; Alves and Silva, 2016).

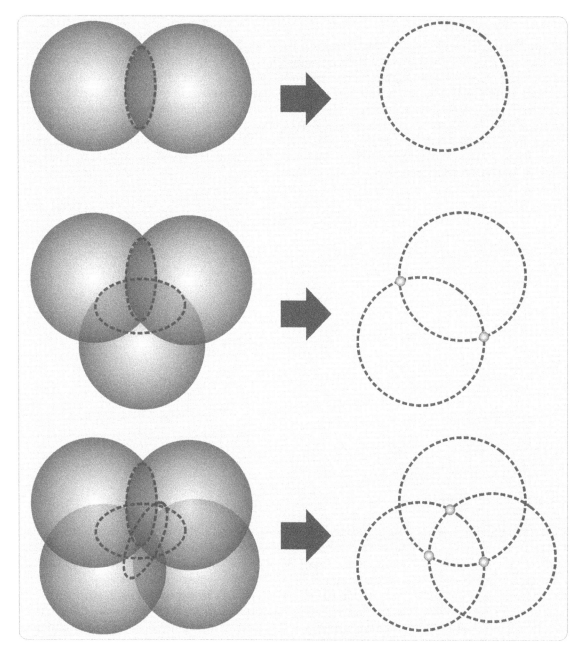

FIGURE 14.2: Intersection between two and three spheres determining the probable position of the receiver on a circle and at the intersection of two circles, respectively.

14.5.2 Carrier phase-shift measurements

Measuring distances from the ground to satellites can be made more accurate by observing the phase-shift of the signals at the time of their transmission by the satellite until they are received at the base station. This procedure is similar to that used by electronic distance measuring instruments, and enables to determine the fractional cycle of the satellite signal at the receiver. The phase-shift is measured at ~1/100 of cycle. However, the measurement does not take into account the number of full wavelengths of signal travel between the satellite and receiver. This number is the integer ambiguity or ambiguity. Unlike electronic distance measuring instruments, in satellites, a one-way form of communication is used, but since satellites are moving, their scales constantly change and the ambiguity can not be determined by simply transmitting additional frequencies. Therefore, different techniques are used to determine the ambiguity by additional observations (Ghilani and Wolf, 1989; Alves and Silva, 2016).

14.6 Global Navigation Satellite System Signal

The orbiting GPS satellites continuously transmit a single signal, on the two carrier frequencies. The carriers are transmitted in the microwave radio frequency band L identified with L1 and L2 signals at frequencies of 1575.42 and 1227.60 MHz, respectively. These frequencies derived from the fundamental frequency f_0 of 10.23 MHz. L1 and L2 bands are configured at frequencies 154 and 120 f_0, respectively (Ghilani and Wolf, 1989; Alves and Silva, 2016).

Different types of information (messages) are modulated on the carrier waves with the phase modulation technique, similar to a radio station. Some information included in the transmitted message are (Ghilani and Wolf, 1989; Alves and Silva, 2016):

- Almanac;
- Transmission ephemerides;
- Satellite clock correction coefficients;
- Ionospheric correction coefficients;
- Satellite condition (satellite health).

Developing a system for accurate measurement of travel time from satellite signal to receiver to determine basic positions of occupied stations in real time, independently, is desired for different geographic applications. In GPS, this is achieved by modulating the carriers with PRN codes. The PRN codes consist of unique sequences of binary values (zeros and ones) that look like random numbers, but are generated by a special mathematical algorithm called "return shift registers". Each satellite transmits two different PRN codes. The L1 signal is modulated with the precision code, or P code, and the coarse acquisition code, or C/A code. The L2 signal is modulated only with the P code. Each satellite transmits a single set of GOLD codes to identify the source of received signals. This identification is important when tracking on several different satellites simultaneously (Ghilani and Wolf, 1989; Alves and Silva, 2016).

C/A and P codes are used in older satellite technology. In recent satellites, new codes have been included, such as the L2 signal, called L2C. In addition, the P code is being replaced by new military codes, called code M. In 1999, the Interagency GPS Executive Board (IGEB) decided to add a third civilian signal, L5, to ensure the safety of GPS life applications. The transmission frequency of L5 is 1176.45 MHz. Both L2C and L5 are added to the IIF and III satellite block (Ghilani and Wolf, 1989; Alves and Silva, 2016).

In the C/A code, the frequency of 1.023 MHz is configured at a wavelength of ~300 m. This code is accessible to all users and consisted of a series of 1023 unique binary digits (chips) for each satellite. This chip pattern is repeated every millisecond in the C/A code. The P code has a frequency of 10.23 MHz and a wavelength of ~30 m. This code is ten times more accurate for positioning than the C/A code. The P code is repeated every 266.4 days. Each satellite assigns a single segment pattern per week that is reset at midnight every Saturday. To meet military needs, the P code is encoded with a W code to derive the Y code which can only be read by receivers with counterfeit cryptographic keys. The code consists of a sequence of +1 or -1 states, corresponding to the binary values 0 or 1. The biphasic modulation (signal shift) is accomplished by 180° inversion in the carrier phase determining change in the initial states (Ghilani and Wolf, 1989; Segantine, 2005; Alves and Silva, 2016).

The need for one-way communication has made satellite positioning systems dependent on precise timing of signal transmission. Imagine that the satellite transmitted a series of beeps, and that the beeps are transmitted in a known irregular pattern. Now imagine that this same pattern is duplicated synchronously (but not transmitted) at the receiving station. Once the transmitted signal from the satellite travels to the receiver, its reception is delayed with respect to the signal generated by the receiver. This delay can be measured and converted into a time difference. In GPS, PRN code chips replace beep signals and the exact satellite code transmission time is placed within the transmitted message, with the start time indicated by the front end of one of the chips. The receiver simultaneously generates a duplicate PRN code. The time between the emission of the satellite signal to the receiver is obtained by combining the incoming satellite signal with the signal generated by the identical receiver. This produces the delay signal, which is converted into the travel time. The distance to the satellite can be calculated knowing the signal travel time and velocity. The transmission message from each satellite is identified with a hand-over word (HOW) code to help in the code ranging, which consists of some identification bits, flags, and a number. The time of week (TOW) is determined after four repetitions of the HOW number. The TOW marks the end of the next section of the message. HOW and TOW enables detection of correspondence between the signal sent from the satellite and received by the receiver to determine signal time delay (**Figure** 14.3) (Ghilani and Wolf, 1989; Alves and Silva, 2016).

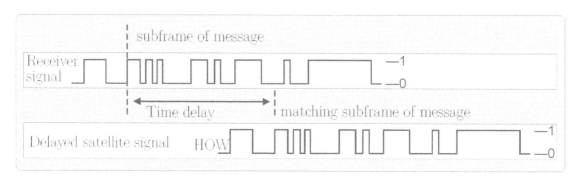

FIGURE 14.3: Determination of GPS signal travel time by code ranging.

The navigation message is transmitted relative to GPS time at the rate of 50 bits per second (bps), with a duration of 30 sec, modulated on the L1 and L2 carriers. The duration of a bit is 20 ms. The navigation message with satellite clock and orbit corrections, ephemeris information, and satellite health determines a 1500 bit message, called a "data block". The information is formatted into five sub-blocks of 6 sec duration in order to compose a block of 30 sec duration. Each sub-block is made up of 10 words with 30 bits each. The first word in each block is a telemetry word (TLM). The second word in each block is a HOW code word with the Z-counter number. The Z-counter is the integer number of 1.5 sec intervals from the beginning of the GPS week that identified the epoch at which the data is recorded. The final eight words are generated by control tracking (Segantine, 2005) (**Figure** 14.4).

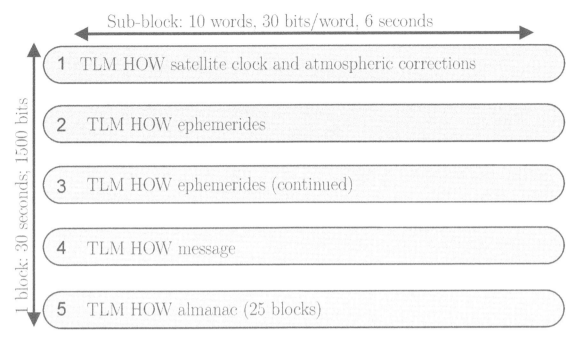

FIGURE 14.4: GPS data signal format.

In the current receivers, operation is observed with code or carrier phase reception. The satellite signal is tracked in the receiver to perform instantaneous positioning calculations.

Considering social class of use, receivers vary as:

- Civil;
- Military.

Considering applications, receivers vary as:

- Navigation;
- Aviation;
- Topography.

Considering signal frequency, receivers vary as:

- Single;
- Dual.

Considering the number of channels, receivers vary as:

- Single channel;
- Multichannel.

Considering the type of channels, receivers vary as:

- Sequential;
- Multiplexed.

Considering the type of signals, receivers vary as:

- C/A code;
- P code.

The following instruments are observed in a GNSS receiver (Segantine, 2005):

- Antenna with pre-amplifier to detect electromagnetic waves emitted by satellites;
- Radio frequency section to identify and process the signals;
- Microprocessor for control, sampling and processing of the data;
- Quartz oscillator that generates an internal wave in the receiver with characteristics similar to the wave external to the antenna;
- Interface for user control of observation sessions;
- Power supply with batteries to supply instruments;
- Receiver controller;
- Memory for data storage.

14.7 Satellite Coordinate Reference System

German astronomer Johannes Kepler (1571-1630) established three laws to define the motion of planets around the sun, which also applied to the motion of satellites around the Earth (Schofield et al., 2007; Alves and Silva, 2016):

- As satellites move around the Earth in elliptical orbits, the Earth's center of mass is located at focus points G. The other focus G' is not used. In the implication of this law, a satellite will at some point be closer or further away from the Earth's surface according to its orbit position. The orbits of GPS satellites are close to circular and therefore have very small eccentricity;
- The radius vector from the center of the Earth to the satellite determines equal areas at equal time intervals. Therefore, a satellite's velocity is not constantly the same. Satellite velocity is minimum when passing at apogee, farthest from the Earth's center, and maximum, when at perigee closest approach;
- The square of the orbital period is proportional to the cube of the semimajor axis a, $T^2 = a^3$ *constant*. The value of the constant is demonstrated by Newton:

$$\frac{\mu}{4\pi^2} \tag{14.2}$$

where μ is the Earth's gravitational constant, equal to 398601 km^3 s^{-2}.

Therefore,

$$T^2 = \frac{a^3\mu}{4\pi^2} \tag{14.3}$$

Thus, for any orbital eccentricity of the satellite, on the same semimajor axis, the period is the same.

Therefore, these laws defined the orbit geometry, speed variation of the satellite along its orbital

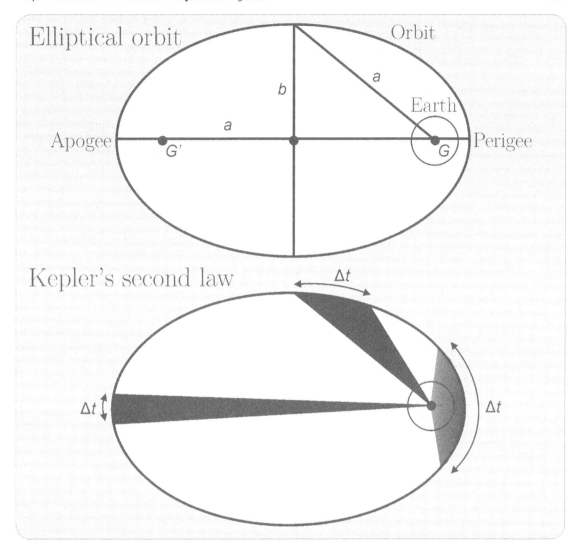

FIGURE 14.5: Elliptical orbit (top) and Kepler's second law (bottom).

path, and the time required to complete an orbit (Schofield et al., 2007; Alves and Silva, 2016) (Figure 14.5).

When determining the positions of points on the Earth based on satellite observations, three different reference coordinate systems are important (Ghilani and Wolf, 1989; Alves and Silva, 2016):

- Satellite positions, at the time of observation, specified as a satellite coordinate system;
- Transformation of satellite positions into a 3D geocentric rectangular coordinate system, physically related to the Earth;
- Determining the positions of points on the Earth by a geocentric coordinate system;
- Transformation from geocentric coordinates to geodetic coordinate system with specific datum.

14.8   Geocentric Coordinate System

Different coordinate reference systems are important when determining positions of points on Earth based on satellite observations. Satellite positions are specified by the satellite coordinate reference system at the instant they are observed. These systems are 3D rectangular systems defined by the orbits of the satellites. Satellite positions are transformed into a 3D rectangular geocentric coordinate system, physically related to the Earth. As a result of satellite positioning observations, positions of new points on Earth are determined in this coordinate system. Once a satellite is placed in orbit, its motion within the orbit is governed by the Earth's gravitational force. However, there are a number of other less important factors involved, including the gravitational forces exerted by the sun and the moon, as well as solar radiation. These forces are not uniform so that satellite motions can vary. Ignoring all forces except the gravitational pull of the Earth, the idealized orbit of a satellite is elliptical, with one of the foci at G, at the Earth's center of mass. The satellite reference coordinate system is characterized by X_S, Y_S, Z_S. Perigee and apogee are defined as the closest and farthest orbit points between the satellite and G, respectively. The apsid line joins the perigee and apogee, crossing two foci, and is the reference axis of X_S. The origin of the X_S, Y_S, Z_S coordinate system occurs at G, whose axis lies in the orbital midplane; the Y_S axis lies in the orbital midplane and, Z_S, perpendicular to that plane. The Z_S coordinate values represent the departure of the satellite from the generally small, mean orbital plane. A satellite at position S_1 will have 3D spatial Cartesian coordinates, X_{S1}, Y_{S1}, Z_{S1}. The position of the satellite in orbit can be calculated at any instant of time, based on orbital parameters that are part of the transmission ephemeris (Ghilani and Wolf, 1989; Alves and Silva, 2016).

After defining the orbit in space, the satellite is located in reference to the perigee point through the f angle. The XYZ spatial coordinate system is originated at the geocenter, G. The spatial coordinates of the satellite at time, t, are (Figure 14.6) (Schofield et al., 2007; Alves and Silva, 2016):

$$X_S = rcosf \tag{14.4}$$

$$Y_S = rsenf \tag{14.5}$$

$$Z_S = 0 \ (normal \ orbiting) \tag{14.6}$$

where r is the distance from the Earth to the satellite center.

The angular parameters required to perform the conversion from the satellite reference coordinate system to the geocentric system are (Ghilani and Wolf, 1989; Schofield et al., 2007; Alves and Silva, 2016):

- The angle i of inclination between the orbital plane and the Earth's equatorial plane;
- The i argument of the perigee, referring to the angle in the orbital plane from the Equator to the apsid line;
- The right ascension of the ascending node, Ω, referring to the angle between the Earth's equatorial plane from the vernal equinox to the line of the ascending node, in the equatorial plane;
- The hour angle between Greenwich and the vernal equinox in the equatorial plane, $GHA\gamma$.

These parameters must be known, in real time, for each satellite, based on mathematical models for predicting the orbits. Satellite coordinates in the geocentric system are determined for specific epochs based on observations at monitoring stations and distributed using precise ephemerides to achieve high accuracy (Figure 14.7).

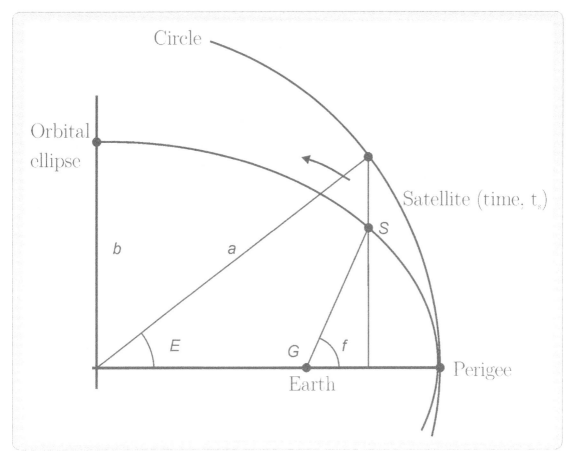

FIGURE 14.6: Orbital satellite ellipse in space.

14.9 Geodetic Coordinate System

Although the point positions of a satellite survey are calculated in the geocentric coordinate system, it is not convenient to use them by geomatics engineers because (Ghilani and Wolf, 1989; Alves and Silva, 2016):

- Geocentric coordinates show extremely large values, considering their origin at the center of the Earth;
- Axes are not related to conventional north-south or east-west directions on the Earth's surface, considering the location of the XY axis in the Equator plane;
- There is no indication of relative elevations between points.

With the advent of the GNSS positioning system, there is a need for transformation of spatial Cartesian coordinates into geodetic coordinates and vice versa. Therefore, 3D Cartesian coordinates are converted to geodetic coordinates of latitude (ϕ), longitude (λ) and altitude (h), for the purpose of greater meaning and user convenience. The reference ellipsoid used for most of the work is the 1984 World Geodetic System ellipsoid (WGS-84). Any ellipsoid can be defined by two parameters, for example, the semimajor axis (a), and the flattening ratio (f). For the WGS-84 ellipsoid, these values are $a = 6378137$ m; $f = 1/298.257223563$ (Ghilani and Wolf, 1989). Conversions from the 3D Cartesian coordinate system, (X, Y, Z), to the geodetic coordinate system

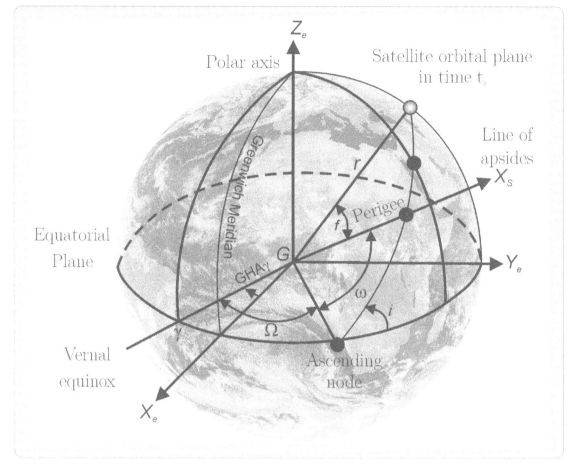

FIGURE 14.7: Parameters involved in determining the spatial orbit and transforming the satellite coordinate system to the geocentric coordinate system.

(ϕ, λ, h) can be performed (Soler and Hothem, 1988; Ghilani and Wolf, 1989; Alves and Silva, 2016) (Figure 14.8).

The Cartesian coordinates of point P can be calculated from their geodetic coordinates (Soler and Hothem, 1988; Ghilani and Wolf, 1989; Silva and Segantine, 2015; Alves and Silva, 2016):

$$X_P = (R_{NP} + h_P)cos\phi_P cos\lambda_P \tag{14.7}$$

$$Y_P = (R_{NP} + h_P)cos\phi_P sen\lambda_P \tag{14.8}$$

$$Z_P = [R_{NP}(1 - e^2) + h_P]sen\phi_P \tag{14.9}$$

$$R_{NP} = \frac{a}{\sqrt{1 - e^2 sen^2\phi_P}} \tag{14.10}$$

where XP, YP, ZP are the Cartesian coordinates of any point P, e, the eccentricity of the WGS-84 reference ellipsoid (0.08181919084), R_{NP}, the radius of the main vertical of the ellipsoid at point P, and, a, the semimajor axis of the ellipsoid. North latitudes are taken as positive, and south latitudes as negative. Similarly, east longitudes are positive, and west longitudes negative.

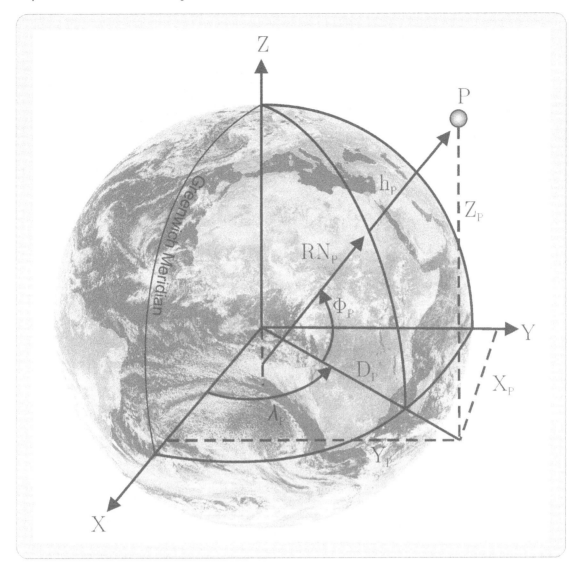

FIGURE 14.8: Parameters used in the transformation from 3D Cartesian coordinate systems to geodetic coordinate systems.

The Cartesian coordinate conversion of any point P and its geodetic value can be obtained based on the following steps (Soler and Hothem, 1988; Ghilani and Wolf, 1989; Silva and Segantine, 2015; Alves and Silva, 2016):

1. Determine the diagonal distance, D_P:

$$D_P = \sqrt{X_P^2 + Y_P^2} \tag{14.11}$$

2. Determine the longitude, λ_P:

$$\lambda_P = 2tan^{-1}(\frac{D_P - X_P}{Y_P}) \tag{14.12}$$

3. Determine the approximate value of the latitude, ϕ_0:

$$\phi_0 = tan^{-1}[\frac{Z_P}{D_P(1-e^2)}] \tag{14.13}$$

4. Determine the approximate radius of the first vertical R_N using ϕ_0:

$$R_N = \frac{a}{\sqrt{1-e^2 sen^2\phi}} \tag{14.14}$$

5. Determine a best value for latitude by:

$$\phi = tan^{-1}(\frac{Z_P + e^2 R_{NP}sen\phi_0}{D_P}) \tag{14.15}$$

Repeat the previous steps until the change in ϕ is minimal.

6. Determine the geodetic altitude of the station, where for latitudes less than 45°, adopt:

$$h_P = \frac{D_P}{cos\phi_P} - R_{NP} \tag{14.16}$$

At latitudes greater than 45°, adopt:

$$h_P = \frac{Z_P}{sen\phi_P} - R_{NP}(1-e^2) \tag{14.17}$$

Having the ellipsoidal altitudes and known elevations, we can determine geoidal altitudes with GPS (Ghilani and Wolf, 1989; Alves and Silva, 2016):

$$N_{GNSS} = h - H \tag{14.18}$$

The value obtained should be compared with that calculated in the geoidal model and the difference is determined by:

$$\Delta_N = N_{GNSS} - N_{model} \tag{14.19}$$

This procedure should be used for several landmarks dispersed in an area and the orthometric altitude (H) is corrected by an average value Δ_N of the region:

$$H = h - (N_{model} + \Delta_{Nmean}) \tag{14.20}$$

In the R package nvctr (Spinielli and EUROCONTROL, 2020) functions for non-singular position calculations and geographic position calculations for ellipsoidal and spherical models are implemented according to Gade (2010). Distance calculations are also implemented in the nvctr package. In the n vector structure, the normal vector to the Earth's ellipsoid (called n vector) is used as a non-singular position representation which is convenient for practical position calculations. With the use of the n vector, simplicity of use and exact answers are observed for all global positions and all distances, in both ellipsoidal and spherical Earth models (Spinielli and

EUROCONTROL, 2020). The application of this package is realized in the Solved Exercises of this chapter in converting Cartesian spatial coordinates to geodetic coordinates and vice versa.

Reports of accurate algorithms for transforming geocentric coordinates into geodesic coordinates are implemented computationally in **FORTRAN** in 1989 (Borkowski, 1989). A closed-form algebraic method for transforming geocentric coordinates to geodesic coordinates is proposed with a new expression excluding indeterminacy and sensitivity to error rounding around the 180° longitude discontinuity (Vermeille, 2004) compared to an earlier work (Vermeille, 2002). The effectiveness of these algorithms should be evaluated after computational implementation in later work.

14.10 GNSS Surveying

Different methods of observing GNSS data have been applied in surveying. For quality surveys, a careful pre-analysis was necessary in order to make an optimal planning of the observation session. An observation session refers to the period of time when receivers used to receive satellite signals simultaneously. Upon completion of a session, all receivers except one are taken to different observation stations. Sessions continued until all observations required by the project are completed, such as surveying a rural property or locating points in different environments (Alves and Silva, 2016).

In accurate surveying, dual-frequency receivers are preferred over single-frequency ones because (Segantine, 2005):

- Data are collected faster;
- Long distances from the base are observed with greater accuracy;
- Errors, such as ionospheric refraction, are eliminated.

Considering the available equipment and surveying error constraints, the selection of the appropriate surveying method varies according to (Ghilani and Wolf, 1989; Alves and Silva, 2016):

- Level of accuracy required in the final coordinates;
- Type of environment to be surveyed;
- Type of equipment available;
- Extent of area surveyed;
- Plant canopy and other site conditions;
- Computer programs available to analyze the data.

Surveying methods performed with satellite positioning technology are subdivided into (Silva and Segantine, 2015):

- Absolute positioning;
- Relative positioning;
- Differential positioning.

14.10.1 Absolute positioning with GNSS

In absolute positioning with GNSS, the measurement is based on pseudodistances between the receiving antenna and at least four satellites by simple positioning. Using the precise point positioning (PPP) method, precise ephemerides are used in the post-processing of data tracked at the point of interest. Accuracy of 5 to 10 cm can be obtained with single frequency receivers, and 1 to 5 cm with dual frequency receivers (Silva and Segantine, 2015). Accurate ephemeris data is obtained via the Internet (NASA, 2020). The PPP service is available for post-processing data in RINEX format via the Internet (IBGE, 2020).

14.10.2   Relative positioning with GNSS

In the GNSS relative positioning method, two or more receivers are used to track the same satellites at a given time. A vertex with known coordinates is used to install the antenna and base receiver assembly. With a remote receiver (rover), the coordinates of points are surveyed by installing another antenna and receiver set on the remote vertices. The spatial vector determined between the observation points is called the "baseline vector". The coordinates of the remote point are obtained by post-processing of points surveyed in the field (Figure 14.9) (Silva and Segantine, 2015).

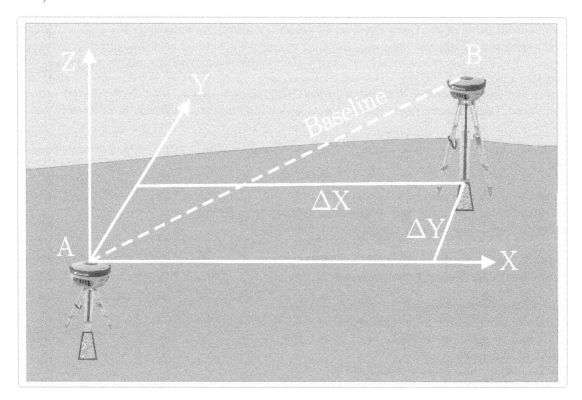

FIGURE 14.9: Baseline vector component determination in GNSS surveying.

The relative positioning between two points A and B, can be expressed by (Ghilani and Wolf, 1989; Alves and Silva, 2016):

$$X_B = X_A + \Delta X \tag{14.21}$$

$$Y_B = Y_A + YDeltaY \tag{14.22}$$

$$Z_B = Z_A + ZDelta \tag{14.23}$$

where (X_A, Y_A, Z_A) are the geocentric coordinates at base station A, (X_B, Y_B, Z_B), the geocentric coordinates at unknown station B and, $(\Delta_X, \Delta_Y, \Delta_Z)$, the basic components of calculated baseline vectors.

Relative positioning involves the use of two or more receivers simultaneously observing pseudodistances at line ends. Simultaneity is involved making observations by the receivers at the same time. It is also important that the receivers record data at the same time interval. This interval varies according to the purpose of the study and the final accuracy required; however, the most common intervals are 1", 2", 5", 10", or 15". Assuming that simultaneous observations are recorded, different linear combinations of equations can be performed in order to eliminate some single, double, and triple difference errors (Figure 14.10).

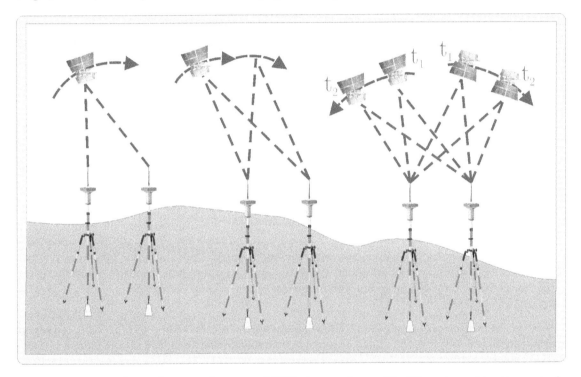

FIGURE 14.10: Recording simultaneous GNSS observations with different combinations: single difference (left), double difference (middle), and triple difference (right).

In simple difference, there is subtraction of two simultaneous observations made for one satellite from two points. This difference eliminates the satellite clock bias and some of the ionospheric and tropospheric refraction. This process would also eliminate the effects of selective availability if it was enabled. The double difference referred to the difference of two single differences obtained from two satellites j and k. This procedure eliminates the receiver clock bias by considering two single differences. With the triple difference, there is a difference between two double differences obtained in two different time periods. This difference eliminates the whole number of ambiguity, leaving only the differences between phase-shift and geometric range. The importance of using the triple difference equation in the solution is that it removed entire ambiguities without cycle loss. Cycle loss occurred in the following main situations (Ghilani and Wolf, 1989; Alves and Silva, 2016):

- Obstructions;
- Low signal-to-noise ratio (SNR);
- Incorrect signal processing.

Signal obstructions can be minimized by careful selection of tracking stations. Low SNR can be caused by undesirable ionospheric conditions, multipath, high receiver dynamics, and low satellite elevation. Malfunctioning satellite oscillators can also cause cycle loss, but this has occurred rarely. It should be noted that currently computer programs for GNSS data processing rarely use triple difference as solutions of integer ambiguities that are quickly determined by more advanced techniques (Ghilani and Wolf, 1989; Alves and Silva, 2016).

Relative surveying can be performed with static relative and kinematic relative surveying techniques. In the static relative mode, antennas of the base receiver and the remote receiver must remain static throughout the data collection period. In kinematic relative surveying, only the base station antenna is kept static over a point of known coordinates, and the other antennas are moved over the points of interest during data collection. In kinematic on-the-fly positioning, no static initialization is required as the measurement algorithm is initialized during the remote antenna displacement (Silva and Segantine, 2015).

14.10.3   Differential positioning with GNSS

In the GNSS differential positioning method, the same characteristics of relative positioning occur; however, the differential corrections between the known base station coordinates calculated by GNSS positioning are transmitted to the remote receiver by telemetric communication in order to obtain the remote antenna coordinates in real time (Figure 14.11)(Silva and Segantine, 2015).

Real-time kinematic (RTK) positioning is considered a satellite navigation technique used to enhance the precision of position data derived from GNSS, providing real-time corrections, up to centimeter-level accuracy. In practice, RTK systems use a single base-station receiver and a number of mobile units. The base station re-broadcasts the observed phase, and the mobile units compare their own phase measurements with the one received from the base. The most popular way to achieve real-time, low-cost signal transmission is to use a radio modem, typically in the UHF band. RTK provides accuracy enhancements up to about 20 km from the base station (Rietdorf et al., 2006).

Construction projects are performed by placing stakes in specific locations. The RTK surveying method enables to perform the design of projects by manual location in the field (Figure 14.12), as well as to guide vehicle during the construction process or when performing agricultural practices. This technology has been termed as machine control and enables the operator to see position of points on the construction project, cut and fill leveling in real time (Ghilani and Wolf, 1989; Alves and Silva, 2016).

The RTK surveying technique has become the most used in topographic surveys for its ease of use, velocity and accuracy obtained, with emphasis on the following applications (Silva and Segantine, 2015):

- Detail surveying;
- Cadastral surveying;
- Location of construction sites;
- Machine automation for agriculture;
- Machine automation for civil construction.

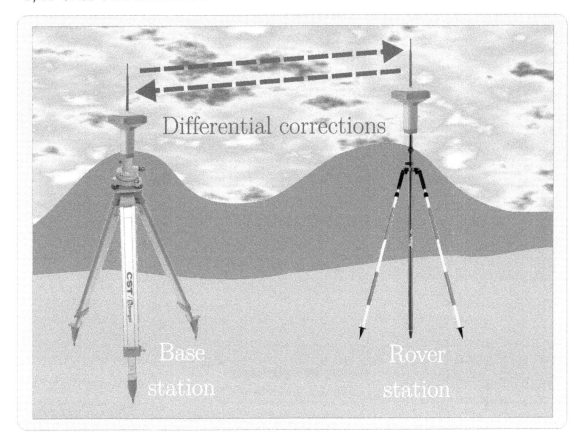

FIGURE 14.11: Differential corrections performed between base receiver and rover with compatible internal radios by surveying in RTK method.

14.11 GNSS Observation Errors

Electromagnetic waves can be affected by different sources of error during transmission. Some of the main errors that affect the quality of coordinates observed at ground stations after GNSS survey are (Ghilani and Wolf, 1989; Alves and Silva, 2016):

- Satellite and receiver clock bias;
- Ionospheric and tropospheric refraction.

Other errors in GNSS surveying are (Ghilani and Wolf, 1989; Alves and Silva, 2016):

- Satellite ephemeris errors;
- Multipath;
- Instrument leveling and centering at the point;
- Antenna height measurement;
- Satellite geometry;
- Selective availability (before May 1, 2000).

The advantage of satellites located ~ 20200 km above the Earth is the occurrence of similar

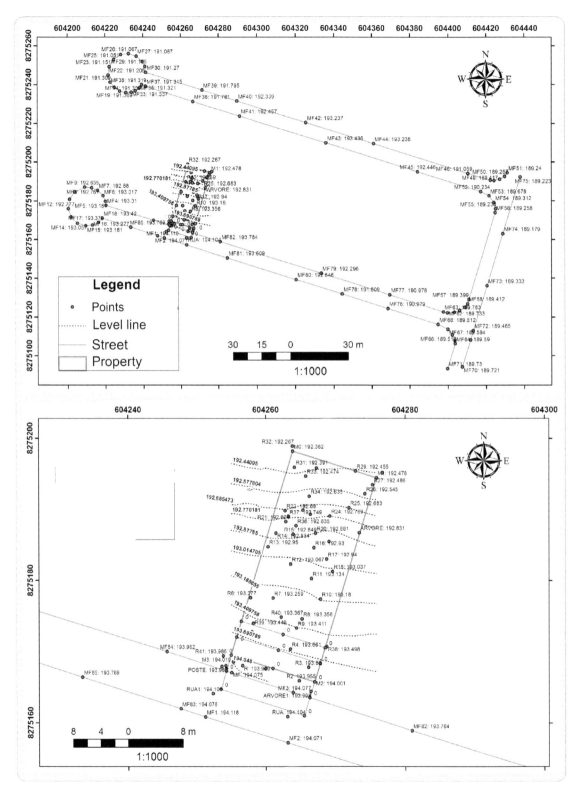

FIGURE 14.12: Base receiver and rover with compatible internal radios used in surveying the avenues of townhouse using the RTK method.

atmospheric interference for surveys performed at close time intervals. With this, in the models used, atmospheric effects on the signal are virtually eliminated. However, since signals from satellites located at the observer's horizon pass through a larger layer of atmosphere than signals above the horizon, signals from satellites below a certain threshold angle can be omitted from the observations based on an elevation mask. The specific value of this angle varies between 10° and 20°, depending on the desired accuracy (Figure 14.13) (Ghilani and Wolf, 1989; Alves and Silva, 2016).

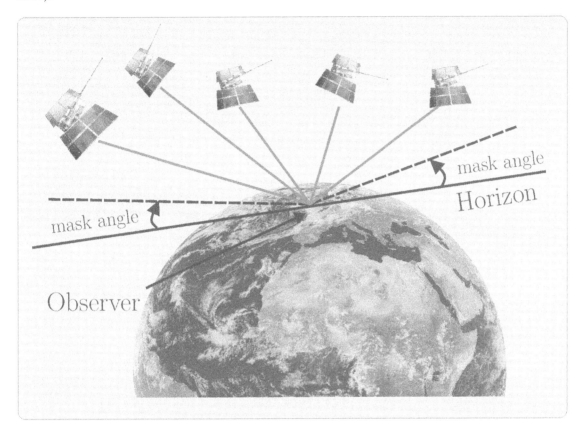

FIGURE 14.13: Satellite mask angle.

With the transmitted ephemeris, it is possible to predict satellite positions. The predicted orbital position has errors as a function of varying gravity, solar radiation, and other anomalies. In the phase-shift method, satellite position errors are translated directly into the calculated positions at ground station bases. This problem is reduced by updating orbital data based on satellite positions determined at the tracking stations. A disadvantage of this situation is the delay in obtaining updated data. The updated ephemerides in situations after surveying are (Ghilani and Wolf, 1989; Alves and Silva, 2016):

- Ultrafast ephemerides;
- Fast ephemerides;
- Precise ephemerides.

The ultrafast ephemerides are available twice a day, the fast ephemerides within two days, and the precise ephemerides after two weeks. Ultrafast and fast ephemerides are sufficient for most surveying applications (Ghilani and Wolf, 1989; Alves and Silva, 2016).

Regarding multipath, this type of error occurs when a satellite signal reflects from a surface and is directed to the receiver. This determines the arrival of multiple signals from one satellite at the receiver. Vertical structures such as buildings, fences, and poles are examples of reflective surfaces that can cause multipath errors (Figure 14.14). Mathematical techniques have been developed to eliminate these unwanted reflections, but in extreme cases, there can be loss of receiver signal due to blocking. This can be caused not only by multipath, but also by high ionospheric activity (Ghilani and Wolf, 1989; Alves and Silva, 2016).

FIGURE 14.14: Multipath caused by satellite signal reflection on building and roof of a house.

In satellite surveying, pseudodistances are observed at the receiver antennas. For precision work, antennas must be mounted on tripods, leveled and centered over a point. The improper setting of the instrument above the point is a potential source of error. Thus, it is essential to adjust the tripod with optical plummet and leveling base. The antenna height above occupied point is also another source of error (Figure 14.15) (Ghilani and Wolf, 1989; Alves and Silva, 2016).

The ellipsoidal altitude is determined at the antenna's phase center. We must use the same antenna on a receiver in an accurate survey or account for phase center offsets in GNSS data post-processing.

The geometry of observed satellites can be weak or strong. Small angles between satellite signals received by the receiver determine weak geometry and larger errors. On the other hand, in the strong geometry, larger angles between received satellite signals and the receiver determines a better solution (Figure 14.16). The effect of geometry on position accuracy is determined by least squares (Ghilani and Wolf, 1989; Alves and Silva, 2016).

The different types of errors that have occurred in satellite positioning, if no corrections or compensations are made, can be of the order of ±7.5 m, as a result of ionospheric refraction. In

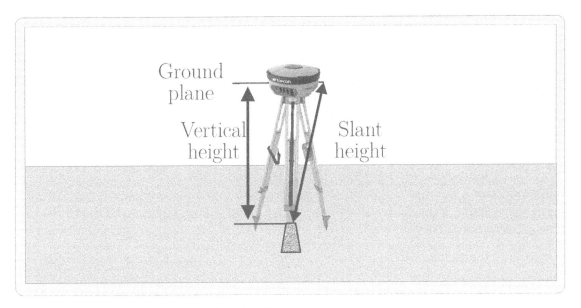

FIGURE 14.15: GNSS antenna slant height measurement.

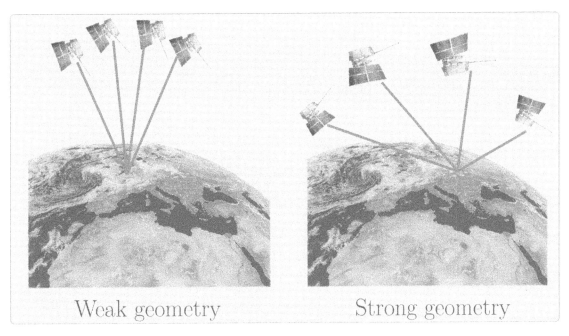

FIGURE 14.16: Weak (left) and strong (right) geometry in satellite observation.

this case, satellite geometry errors are not included. The error is minimized including L2C and L5 signals. The total user equivalent range error (UERE) drops to approximately ±2.8 m with inclusion of the L2C and L5 signals, by reducing the ionospheric refraction error. The UERE is determined by the square root of the summation of each error source squared (Ghilani and Wolf, 1989; Alves and Silva, 2016):

$$UERE = \sqrt{sumerror_i^2} \tag{14.24}$$

Based on the number and position of satellites visible at a given time and location, the least squares method can determine an estimated accuracy of the point coordinate, called "dilution of precision" (DOP). The DOP is calculated by error propagation. The DOP, when multiplied by the expected GPS errors, results in the geometry errors of the observed satellite constellation. Thus, the smaller DOP value, the better the expected accuracy in calculated ground station positions. The DOP factors of most interest in the survey are position dilution of precision (PDOP), horizontal dilution of precision (HDOP), and vertical dilution of precision (VDOP). The best possible satellite constellation has the mean value for HDOP less than two and PDOP less than five. Other factors, such as geometry dilution of precision (GDOP) and time dilution of precision (TDOP) can also be evaluated, but are of less importance in the survey (Table 14.1) (Ghilani and Wolf, 1989; Alves and Silva, 2016).

TABLE 14.1: Important categories of dilution of precision (DOP) with meanings in terms of standard deviation and equations.

Type of DOP	Standard Deviation	Equation	Max. Accept. Value
PDOP (Positional DOP)	σ in geocentric coordinates, X, Y, Z	$\sqrt{\sigma_X^2 + \sigma_Y^2 + \sigma_Z^2}$	6
HDOP (Horizontal DOP)	σ in local coordinates, $x,$ y	$\sqrt{\sigma_X^2 + \sigma_Y^2}$	3
VDOP (Vertical DOP)	σ in altitude, h	σh	5
TDOP (Time DOP)	σ in time, t	σt	
GDOP (Geometric DOP)	σ in position and time	$\sqrt{\sigma_X^2 + \sigma_Y^2 + \sigma_Z^2 + +\sigma_t^2}$	

Multiplying the DOP by the probable error coefficient and UERE, the probable positional error by precision dilution (EP) at 95% probability is determined by:

$$EP_{95} = \pm \sigma DOP\ 1.96\ UERE \tag{14.25}$$

Even though the satellite constellations of global positioning systems are not complete, manufacturers of satellite receiver technology are already building receivers that use all available GNSS. The advantage of using multiple systems is that more satellites will be available for tracking by receivers. By combining these systems, the surveyor can expect improvements in velocity and accuracy. In addition, the combination of systems will provide a viable method of achieving satisfactory satellite positioning in difficult areas such as canyons, deep mines, natural areas, and urban areas with obstacles such as buildings and trees (Ghilani and Wolf, 1989; Alves and Silva, 2016).

14.12 Computation

GNSS devices are present in a variety of applications, from watches to cars to cell phones. GNSS datasets can be used to solve transportation situations (Lovelace et al., 2019b), farm machinery, determine animal walking patterns, and collect biotic and abiotic samples along a traveled path. In this computational practice the use of Garmin Forerunner 235 GPS data is emphasized in a walk performed in rural and urban areas of Lavras city, Minas Gerais, Brazil. Data are processed and mapped in order to extract information about the walk performed and map the path traveled with auxiliary databases from the Internet.

The R package `sf` (Pebesma et al., 2021) is used to process the GPS data in order to perform statistical calculations and map the path traveled. The R packages `tmap` (Tennekes, 2018; Tennekes et al., 2020) and `mapview` (Appelhans et al., 2020) are used to perform interactive web region mapping with algorithms used in the transportation field (Lovelace et al., 2019a,b).

14.12.1 Installing R packages

The `install.packages` function is used to install the `sf`, `tmap` and `mapview` packages in the R console.

14.12.2 Enabling R packages

The `library` function is used to enable the `sf`, `tmap` and `mapview` packages in the R console.

```
library(sf)
library(tmap)
library(mapview)
```

14.12.3 Obtaining and mapping GPS garmin forerunner 235 data

The gpx data of hiking performed with GPS garmin forerunner 235 used in the computational practice is obtained using the `download.file` function.

The data is unzipped with the `unzip` function.

The working directory into which the data is imported can be observed using the `getwd` function.

The data is imported into R with the `st_read` function.

```
rota<- st_read("files/activity_4871799754.gpx", layer = "tracks")
```

```
## Reading layer `tracks' from data source
##   `C:\bookdown\surveying-with-geomatics-and-r_R1_03102021\files\activity_4871799754.gpx'
##   using driver `GPX'
## Simple feature collection with 1 feature and 12 fields
## Geometry type: MULTILINESTRING
```

```
## Dimension:     XY
## Bounding box:  xmin: -44.97712 ymin: -21.21822 xmax: -44.96131 ymax: -21.21166
## Geodetic CRS:  WGS 84
```

The imported data is mapped with the plot function (Figure 14.17).

```
plot(rota[1], axes=T, col="black")
```

FIGURE 14.17: Mapping a GPS registered route with sf package functions.

The route can also be mapped using the qtm function from the R package tmap.

```
qtm(rota)
```

14.12.4   Mapping the route with Internet database spatial infrastructure

The interactive function of tmap is enabled with the ttm function to obtain an interactive visualization of the map on the Internet. Next, route vectors are mapped with the qtm function. In this case, it was possible to better exploit the map zoom, as well as background themes with existing vector and remote sensing image bases.

```
# Enable interactive display of the tmap
ttm()
# Map with qtm
qtm(route)
```

An interactive web visualization, with the same database superimposed on a very high spatial resolution satellite image of the area as well as the digital elevation model and public utilities cadastral information, is performed using the `mapview` function from the `mapview` package (Figure 14.18).

```
mapview::mapview(rota)
```

FIGURE 14.18: Mapping a GPS registered route over satellite imagery with `mapview`.

14.12.5 Importing route data by points and performing time and speed calculations

Data can be imported by the `sf` package by different ways of reading simple features from the file as needed:

- `sf::st_read()`, returned a `sf-data.frame`, an object of class `c("sf", "data.frame")`;
- `sf::read_sf()`, returned an `sf-tibble`, an object of class `c("sf", "tbl_df", "tbl", "data.frame")`;
- `sf::st_as_sf()`, returned just an `sf`, from class `data.frame`.

In this case, the data is imported with the `read_sf` function.

```
pRote = read_sf("files/activity_4871799754.gpx",
                layer = "track_points")
```

The `plot` function is used for a preview of the imported vector attribute point data (Figure 14.19).

```
plot(pRote, max.plot=5)
```

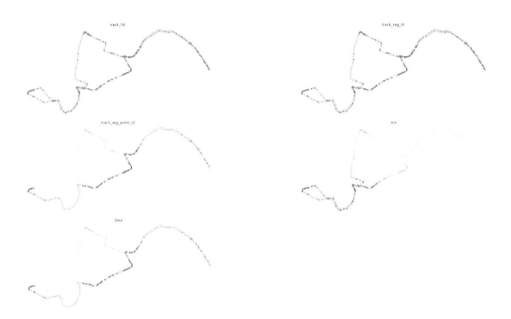

FIGURE 14.19: Route mapping of attribute points recorded by GPS with `sf` package functions.

The statistics of the time spent on the route are obtained with the `summary` function.

```
summary(pRote$time)
```

```
##                     Min.                1st Qu.                  Median
## "2020-05-02 16:57:44" "2020-05-02 17:09:20" "2020-05-02 17:18:47"
##                     Mean                3rd Qu.                    Max.
## "2020-05-02 17:20:09" "2020-05-02 17:33:00" "2020-05-02 17:44:16"
```

Afterward, a graph of the time measurements is made using the `plot` function (Figure 14.20).

```
plot(pRote$time, 1:nrow(pRote))
```

The time resolution of measurements can be determined with the `difftime` function.

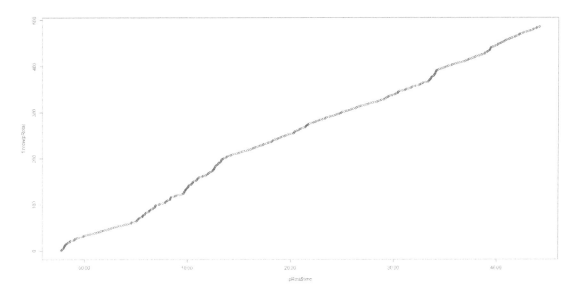

FIGURE 14.20: Time mapping as a function of route traveled with GPS.

```
difftime(pRote$time[11], pRote$time[10])
```

```
## Time difference of 1 secs
```

A function is defined to convert points to lines in order to calculate the speed, v, along the trajectory.

```
points2line_trajectory = function(pRote) {
  c <- st_coordinates(pRote)
  i <- seq(nrow(pRote) - 2)
  l <- purrr::map(i, ~ sf::st_linestring(c[.x:(.x + 1), ]))
  v <- purrr::map_dbl(i, function(x) {
    geosphere::distHaversine(c[x, ], c[(x + 1), ]) /
      as.numeric(pRote$time[x + 1] - pRote$time[x])
  }
  )
  lfc <- sf::st_sfc(l)
  a <- seq(length(lfc)) + 1
  p_data <- cbind(sf::st_set_geometry(pRote[a, ], NULL), v)
  sf::st_sf(p_data, geometry = lfc)
}
```

The conversion function from points to lines is used to map the speed along the trajectory.

```
l <- points2line_trajectory(pRote)
```

The speed map along the trajectory is obtained using the `plot` function (Figure 14.21).

```
plot(l["v"], lwd = l$v, axes=T)
```

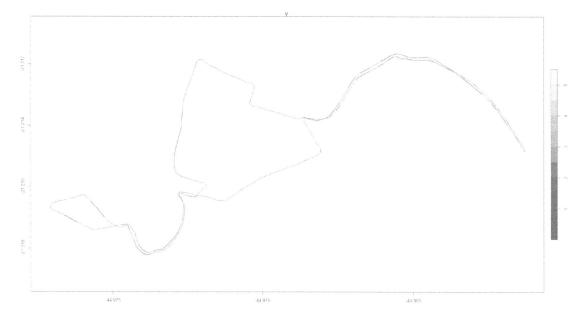

FIGURE 14.21: Mapping the speed at which the route was traveled with GPS.

14.13   Solved Exercises

14.13.1   Name the main components of a GNSS receiver.

A: Antenna, pre-amplifier, signal processor, oscillator, microprocessor and memory.

14.13.2   What are the parts (segments) of the global positioning satellite system?

A: User segment, control segment, and space segment.

14.13.3   Explain the difference between GPS and GNSS.

A: The Global Positioning System (GPS) was developed by the US Department of Defense (DoD) and, the Global Navigation Satellite System (GNSS) refers to the complete set of satellite systems used in positioning, including positioning systems developed in countries, such as those in continental Europe, the US, Russia, and China.

14.13.4 Name one advantage of GNSS surveying over conventional sighted instrument surveying.

A: In GNSS surveying, baselines of hundreds of kilometers can be determined, without the need for extensive geodetic networks, as in conventional surveying.

14.13.5 Calculate the orthometric altitude of a station as a function of the geodetic altitude.

Required information: Geodetic altitude = 59.1 m; geoidal swell = -21.3 m.

A: The orthometric altitude is 80.4m.

```
# Data
h<-59.1
N<--21.3
# Orthometric altitude
H<-h-N
H
```

```
## [1] 80.4
```

14.13.6 Determine the geodesic coordinates of a point with known spatial Cartesian coordinates.

Required information: The spatial Cartesian coordinates X, Y, Z are 1241581.343, -4638917.074, and 4183965.568 m, respectively.

A: The exercise is solved using functions from the nvctr package. The longitude, latitude and altitude coordinates are -75.01628°, 41.25506°, and 312.3907 m, respectively.

```
library(nvctr) # Enable Package
```

```
# Data
x<-1241581.343
y<--4638917.074
z<-4183965.568
# Create a numerical vector of coordinates
c<-c(x, y, z)
# Convert cartersian coordinates to n-vector
n_EB_E <- p_EB_E2n_EB_E(c)
# Convert n-vector to latitude and longitude in radians
g<-n_E2lat_lon(n_EB_E$n_EB_E)
# Convert latitude and longitude to degrees
(deg(g))
```

```
## [1]   41.25506 -75.01628
```

```
# Note that altitude z_EB = -height
-1*n_EB_E$z_EB
```

```
## [1] 312.3907
```

14.13.7 Determine spatial Cartesian coordinates of the point A using the WGS-84 ellipsoid.

Required information: Latitude, longitude and geodetic altitude of point A are 41°15′18.2106″ N, 75°00′58.6127″ W, 312.391 m, respectively.

A: The exercise is solved using functions from the nvctr package. The geocentric coordinates X, Y, Z of A are 1241581.3427; -4638917.0743, and 4183965.5682 m, respectively.

```
# Data
lon<--(75+58.6127/3600)
lat<-(41+15/60+18.2106/3600)
alt<-312.391
# Create a numeric vector of coordinates
m<-c(lon, lat, alt)
# Convert coordinates into n-vector
n<-lat_lon2n_E(rad(m[2]), rad(m[1]))
# Convert n-vector to Cartesian coordinates
c<-n_EB_E2p_EB_E(n_EB_E=n, z_EB = -m[3], a = 6378137,
                 f = 1/298.257223563)
# Evaluate the results to 4 decimal places
c4<-format(c, nsmall=4)
c4
```

```
## [1] " 1241581.3427" "-4638917.0743" " 4183965.5682"
```

14.14 Homework

Use the set of theory equations in this chapter to check the results obtained with the nvctr package in the Solved Exercises section. Perform a field practice with GNSS and plot the results with the sf package.

14.15 Resources on the Internet

As a study guide, slides and illustrative videos are presented about the subject covered in the chapter in Table 14.2.

TABLE 14.2: Slide shows and video presentations on topographic surveying with GNSS.

Guide	Address for Access
1	Slides on global positioning satellite systems in geomatics[1]
2	RTK receiver with centimeter precision[2]
3	Practices for reducing GNSS surveying errors[3]
4	GNSS surveying in RTK mode[4]
5	Surveying indirect measurements with GNSS[5]

14.16 Research Suggestion

The development of scientific research on geomatics is stimulated by the activity proposals that can be used or adapted by the student to assess the applicability of the subject matter covered in the chapter (Table 14.3).

TABLE 14.3: Practical and research activities used or adapted by students using topographic surveying with GNSS.

Activity	Description
1	In the content about satellite positioning system, interest may arise to do the work based on measurements and determinations of geodetic coordinates over time
2	Carry out static measurement with GNSS in the field for a few hours. Evaluate the post-processing options to correct the obtained coordinates. Evaluate the quality of the results based on the sigma error
3	Install a GNSS base in the field and take measurements with the rover at different points with RTK mode surveying
4	Take measurements with a garmin forerunner 235 GPS on a given route and evaluate the speed variation along the route

14.17 Learning Outcome Assessment Strategy

Perform a summary of the chapter, "GNSS Surveying with Geomatics and R", on a single A4 page in order to show the student's abilities to summarize a subject presenting key points considered of greater importance today.

---

[1] http://www.sergeo.deg.ufla.br/geomatica/book/c14/presentation.html#/
[2] https://youtu.be/UgRGAB8Kyvg
[3] https://youtu.be/KLCDQ8yafY0
[4] https://youtu.be/pOuYdxBWVyU
[5] https://youtu.be/A4PuDVh1hsY

Bibliography

ABNT (1994). NBR-13133 Execução de levantamento topográfico. Technical report, Associação Brasileira de Normas Técnicas - ABNT, Rio de Janeiro.

Adrian Raymond Legault (1956). *Surveying: An introduction to enginering measurements (Civil engineering and engineering mechanics series)*. Bailey & Swinfen.

Ali, A. E. (1995). Accuracy of stadia tacheometry with optical theodolites and levels. *Journal of King Saud University - Engineering Sciences*, 7(2):175–183. https://doi.org/10.1016/S1018-3639(18)30625-1.

Ali, A. E. (2001). Stadia tacheometry with electronic theodolites. *Journal of King Saud University - Engineering Sciences*, 13(1):25–36. https://doi.org/10.1016/S1018-3639(18)30723-2.

Alves, M. C. and Silva, F. M. (2016). *Geomática para levantamento de ambientes: base para aplicações em topografia, georreferenciamento e agricultura de precisão*. Editora UFLA, Lavras, 1st edition. ISBN 978-85-8127-047-0.

Appelhans, T., Detsch, F., Woellauer, C. R. S., Forteva, S., Nauss, T., Pebesma, E., Russell, K., Sumner, M., Darley, J., Roudier, P., Schratz, P., Marburg, E. I., and Busetto, L. (2020). *mapview: Interactive Viewing of Spatial Data in R*. R package version 2.9.0.

Aragonés, L., Pagán, J., López, M., and Serra, J. (2019). Cross-shore sediment transport quantification on depth of closure calculation from profile surveys. *Coastal Engineering*, 151:64–77. https://doi.org/10.1016/j.coastaleng.2019.04.002.

Arvanitou, E.-M., Ampatzoglou, A., Chatzigeorgiou, A., and Carver, J. C. (2021). Software engineering practices for scientific software development: A systematic mapping study. *Journal of Systems and Software*, 172:110848. https://doi.org/10.1016/j.jss.2020.110848.

Barcellos, T. C. (2003). Distâncias. In Erba, D. A., Thum, A. B., Silva, C. A. U., de Souza, G. C., Veronez, M. R., Leandro, R. F., and Maia, T. C. B., editors, *Topografia para estudantes de arquitetura, engenharia e geologia*, chapter II, pages 1–34. Editora Unisinos, São Leopoldo, 1st edition. ISBN 85-7431-191-X.

Barrell, H. and Sears, J. E. (1939). The refraction and dispersion of air for the visible spectrum. *Philosophical Transactions of the Royal Society of London. Series A. Mathematical and Physical Sciences*, 238(786):1–64. https://doi.org/10.1098/rsta.1939.0004.

Bhattacharjee, Y. and Clery, D. (2013). A gallery of planet hunters. *Science*, 340(6132):566–569. https://doi.org/10.1126/science.340.6132.566.

Bird, R. G. (1970). Least squares adjustment of a traverse. *Survey Review*, 20(155):218–230. https://doi.org/10.1179/sre.1970.20.155.218.

Birk, M. A. (2019). *measurements: Tools for Units of Measurement*. R package version 1.4.0.

Bivand, R., Keitt, T., and Rowlingson, B. (2019). *rgdal: Bindings for the Geospatial Data Abstraction Library*. R package version 1.4-8.

Bivand, R., Keitt, T., Rowlingson, B., Pebesma, E., Sumner, M., Hijmans, R., Baston, D., Rouault, E., Warmerdam, F., Ooms, J., and Rundel, C. (2021). *rgdal: Bindings for the 'Geospatial' Data Abstraction Library*. R package version 1.5-19.

Bivand, R., Lewin-Koh, N., Pebesma, E., Archer, E., Baddeley, A., Bearman, N., Bibiko, H.-J., Brey, S., Callahan, J., Carrillo, G., Dray, S., Forrest, D., Friendly, M., Giraudoux, P., Golicher, D., Rubio, V. G., Hausmann, P., Hufthammer, Ove, K., Jagger, T., Johnson, K., Lewis, M., Luque, S., MacQueen, D., Niccolai, A., Pebesma, E., Lam, O. P., and Turner, R. (2020a). *maptools: Tools for Handling Spatial Objects*. R package version 1.0-2.

Bivand, R., Nowosad, J., Lovelace, R., Monmonier, M., and Snow, G. (2020b). *spData: Datasets for Spatial Analysis*. R package version 0.3.8.

Bivand, R., Rundel, C., Pebesma, E., Stuetz, R., Hufthammer, K. O., Giraudoux, P., Davis, M., and Santilli, S. (2018). *rgeos: Interface to Geometry Engine - Open Source ('GEOS')*. R package version 0.5-5.

Björk, B.-C., Kanto-Karvonen, S., and Harviainen, J. T. (2020). How frequently are articles in predatory open access journals cited. *Publications*, 8(2):17. https://doi.org/10.3390/publications8020017.

Borges, A. C. (2013). *Topografia aplicada à Engenharia Civil*. Blucher, São Paulo, 3rd edition. ISBN 978-85-212-0762-7.

Borkowski, K. M. (1989). Accurate algorithms to transform geocentric to geodetic coordinates. *Bulletin Géodésique*, 63(1):50–56. https://doi.org/10.1007/BF02520228.

Briz-Redon, A. and Serrano-Aroca, A. (2020). *LearnGeom: Learning Plane Geometry*. R package version 1.5.

Brownrigg, R. (2018). *maps: Draw Geographical Maps*. R package version 3.3.0.

Bullo, D., Villela, A., and Bonomo, N. (2016). Azimuth calculation for buried pipelines using a synthetic array of emitters, a single survey line and scattering matrix formalism. *Journal of Applied Geophysics*, 134:253–266. https://doi.org/10.1016/j.jappgeo.2016.09.016.

Burrough, P. A. and McDonnell, R. (1998). *Principles of geographical information systems*. Oxford University Press, Oxford. ISBN 0-19823366-3.

Cáceres, A. M., Gândara, J. P., and Puglisi, M. L. (2011). Scientific writing and the quality of papers: Towards a higher impact. *Jornal da Sociedade Brasileira de Fonoaudiologia*, 23(4):401–406. https://doi.org/10.1590/S2179-64912011000400019.

CAPES (2020). Web of science - coleção principal. https://www-periodicos-capes-gov-br.ezl.periodicos.capes.gov.br/.

Casaca, J., Matos, J., and Baio, M. (2007). *Topografia geral*. LTC Editora, Rio de Janeiro, 4th edition. ISBN 978-85-216-1561-3.

Chernetsov, N., Pakhomov, A., Kobylkov, D., Kishkinev, D., Holland, R. A., and Mouritsen, H. (2017). Migratory eurasian reed warblers can use magnetic declination to solve the longitude problem. *Current Biology*, 27(17):2647–2651.e2. https://doi.org/10.1016/j.cub.2017.07.024.

Chinchusak, W. and Tipsuwanporn, V. (2018). Investigation of yaw errors in measuring tape calibration system. *Measurement*, 125:142–150. https://doi.org/10.1016/j.measurement.2018.04.053.

Chuerubim, M. L. (2013). Use of the software Mapgeo2010 as a teaching resource in the study of the reference surfaces and geodetic reference in geodesy. *Revista Geográfica Acadêmica*, 7(2):31. https://doi.org/10.18227/1678-7226rga.v7i2.2990.

Comastri, J. A. and Junior, J. G. (1998). *Topografia aplicada - Medição, divisão e demarcação*. Editora UFV, Viçosa. ISBN 85-7269-036-0.

Danielsen, J. (1989). The area under the geodesic. *Survey Review*, 30(232):61–66. https://doi.org/10.1179/sre.1989.30.232.61.

Dunnington, D., Pebesma, E., Rubak, E., Ooms, J., and Google, I. (2021). *s2: Spherical Geometry Operators Using the S2 Geometry Library*. R package version 1.0.4.

Fitz, P. R. (2008). *Cartografia básica*. Editora Oficina de Textos, São Paulo, 1st edition. ISBN 978-8586238765.

Gade, K. (2010). A non-singular horizontal position representation. *Journal of Navigation*, 63(3):395–417. https://doi.org/10.1017/S0373463309990415.

Garcia, G. J. and Piedade, G. C. R. (1987). *Topografia aplicada às ciências agrárias*. Editora Nobel, São Paulo, 5th edition. ISBN 85-213-0133-2.

Ghilani, C. D. (2017). *Adjustment computations: Spatial data analysis*. John Wiley & Sons, New Jersey, 6th edition. ISBN 978-1119385981.

Ghilani, C. D. and Wolf, P. R. (1989). *Elementary surveying: An introduction to geomatics*. John Wiley & Sons, Upper Saddle River, 13th edition. ISBN 2010032525.

Gőbel, E., Mills, I., and Wallard, A. (2006). The international system of units (si).

González-Bustamante, B. (2021). Evolution and early government responses to COVID-19 in South America. *World Development*, 137:105180. https://doi.org/10.1016/j.worlddev.2020.105180.

Grassi, D. and Memoli, V. (2016). Political determinants of state capacity in Latin America. *World Development*, 88:94–106. https://doi.org/10.1016/j.worlddev.2016.07.010.

Graves, S., Dorai-Raj, S., and Francois, R. (2020). *sos: Search contributed R packages, sort by package*. R package version 2.0-2.

Guidotti, E. (2021). *COVID19: R Interface to COVID-19 Data Hub*. R package version 2.3.2.

Hager, J. W., Behensky, J. F., and Drew, B. W. (1989). The universal grids: Universal Transverse Mercator (UTM) and Universal Polar Stereographic (UPS). Technical report, Washington, D.C.

Hartmann, K., Krois, J., and Waske, B. (2018). E-Learning Project SOGA: Statistics and geospatial data analysis. https://www.geo.fu-berlin.de/en/v/soga/index.html.

Helmert, F. R. (1964). *Mathematical and physical theories of higher geodesy, Part 1, Preface and the Mathematical Theories*. Zenodo, St Louis. https://doi.org/10.5281/zenodo.32050.

Hengl, T., Roudier, P., Beaudette, D., Pebesma, E., and Blaschek, M. (2020). *plotKML: Visualization of Spatial and Spatio-Temporal Objects in Google Earth*. R package version 0.6-1.

Hernández-Quintero, E., Goguitchaichvili, A., Cejudo, R., Cifuentes, G., García, R., and Cervantes, M. (2020). Spatial distribution of historical geomagnetic measurements in Mexico. *Journal of South American Earth Sciences*, 100:102556. https://doi.org/10.1016/j.jsames.2020.102556.

Hijmans, R. J., van Etten, J., Sumner, M., Cheng, J., Baston, D., Bevan, A., Bivand, R., Busetto, L., Canty, M., Fasoli, B., Forrest, D., Ghosh, A., Goliche, D., Gray, J., Greenberg, J. A., Hiemstra, P., Hingee, K., Geosciences, I. f. M. A., Karney, C., Mattiuzzi, M., Mosher, S., Nowosad, J., Pebesma, E., Shortridge, O. P. L., Ashton, E. B. R., Rowlingson, B., Venables, B., and Wueest, R. (2020). *raster: Geographic Data Analysis and Modeling*. R package version 3.4.5.

Hijmans, R. J., Williams, E., and Vennes, C. (2019). *geosphere: Spherical Trigonometry*. R package version 1.5-10.

Holt, P. (2003). An assessment of quality in underwater archaeological surveys using tape measurements. *The International Journal of Nautical Archaeology*, 32(2):246–251. https://doi.org/10.1016/j.ijna.2003.04.002.

IBGE (1995). *Tabelas para Cálculos no Sistema de Projeção Universal Transverso de Mercator (UTM)*. Instituto Brasileiro de Geografia e Estatística - IBGE, Rio de Janeiro, 2nd edition. ISBN 8524004509.

IBGE (1999). *Noções básicas de cartografia*. Instituto Brasileiro de Geografia e Estatística - IBGE, Rio de Janeiro. ISBN 85-240-0751-6.

IBGE (2020). Serviço online para pós-processamento de dados GNSS - IBGE-PPP. https://www.ibge.gov.br/geociencias/informacoes-sobre-posicionamento-geodesico/servicos-para-posicionamento-geodesico/16334-servico-online-para-pos-processamento-de-dados-gnss-ibge-ppp.html?=&t=o-que-e.

Iliffe, J. and Lott, R. (2008). *Datums and map projections: For remote sensing, GIS and surveying*. Whittles Publishing, 2nd edition. ISBN 978-1420070415.

INMETRO (2012). *Sistema Internacional de Unidades: SI*. Instituto Nacional de Metrologia, Qualidade e Tecnologia - INMETRO, Duque de Caxias, 1st edition. ISBN 978-85-86920-11-0.

IOGP (2009). Geomatics: Epsg geodetic parameter relational data base – developers guid. Technical report, International Association of Oil & Gas Producers - IOGP, London.

IOGP (2012). Surveying and positioning: Using the epsg geodetic parameter dataset. Technical report, International Association of Oil & Gas Producers - IOGP, London.

IOGP (2021). Geomatics: Coordinate conversions and transformations including formulas. Technical report, International Association of Oil & Gas Producers - IOGP, London.

Jammalamadaka, S. R. and SenGupta, A. (2001). *Topics in circular statistics*, volume 5 of *Series on multivariate analysis*. World Scientific.

Jensen, J. R. and Jensen, R. R. (2012). *Introductory geographic information systems*. Prentice Hall, Upper Saddle River, 1st edition. ISBN 978-0136147763.

Kahmen, H. and Faig, W. (1988). *Surveying*. De Gruyter, Berlin, Boston. ISBN 978-3110845716.

Karney, C. F. (2013). Algorithms for geodesics. *Journal of Geodesy*, 87(1):43–55. https://doi.org/10.1007/s00190-012-0578-z.

Kavanagh, B. and Slattery, D. K. (2015). *Surveying with construction applications*. Pearson Education Limited, Harlow, 8th edition. ISBN 978-0-13-276698-2.

Kelland, N. (1994). Developments in integrated underwater acoustic positioning. *Hydrographic Journal*, 71:19–27.

Kelley, D., Richards, C., Layton, C., and British Geological Survey (2020). *oce: Analysis of Oceanographic Data*. R package version 1.2-0.

Kuchmister, J., Gołuch, P., Ćmielewski, K., Rzepka, J., and Budzyń, G. (2020). A functional-precision analysis of the vertical comparator for the calibration of geodetic levelling systems. *Measurement*, 163:107951. https://doi.org/10.1016/j.measurement.2020.107951.

Legendre, L. and Legendre, P. (2012). *Numerical Ecology*. Elsevier, 3rd edition. ISBN 0167-8892.

Lindsay, D. (2011). *Scientific Writing = Thinking in Words*. CSIRO Publishing, Collingwood. ISBN 9780643101579.

Lo, C. P. and Yeung, A. K. W. (2007). *Concepts and techniques of geographic information systems*. Pearson Prentice Hall, Upper Saddle River, 2nd edition. ISBN 0-13-149502-x.

Longley, P. A., Goodchild, M. F., Maguire, D. J., and Rhind, D. W. (2001). *Geographic information systems and science*. John Wiley & Sons, Inc, Chichester. ISBN 0-471-49521-2.

Lovelace, R., Nowosad, J., and Muenchow, J. (2019a). *geocompkg / metapackage for the book Gecomputation with R*.

Lovelace, R., Nowosad, J., and Muenchow, J. (2019b). *Geocomputation with R*. CRC Press, Boca Raton, 1st edition. ISBN 978-1-138-30451-2.

Lund, U., Agostinelli, C., Arai, H., Gagliardi, A., Portugues, E. G., Giunchi, D., Irisson, J.-O., and Agostinelli, C. (2017a). *circular: Circular Statistics*. R package version 0.4-93.

Lund, U., Agostinelli, C., Arai, H., Gagliardi, A., Portugues, E. G., Giunchi, D., Irisson, J.-O., Pocernich, M., and Rotolo, F. (2017b). *circular: Circular Statistics*. R package version 0.4-93.

Marsella, M., Nardinocchi, C., Paoli, A., Tini, M. A., Vittuari, L., and Zanutta, A. (2020). Geodetic measurements to control a large research infrastructure: The Virgo detector at the European Gravitational Observatory. *Measurement*, 151:107154. https://doi.org/10.1016/j.measurement.2019.107154.

McCormac, J. C. (2007). *Topografia*. LTC, Rio de Janeiro, 5th edition. ISBN 85-216-1523-X.

McCormac, J. C., Sarasua, W. A., and Davis, W. J. (2012). *Surveying*. John Wiley & Sons, Inc, 6th edition. ISBN 978-0-470-49661-9.

MTE (2015). Nr 18 condições e meio ambiente de trabalho na indústria da construção. Technical report, Ministério do Trabalho e Emprego - MTE, Brasília.

Mullen, L. A. and Bratt, J. (2018). USAboundaries: Historical and contemporary boundaries of the United States of America. *Journal of Open Source Software*, 3(23):314. https://doi.org/10.21105/joss.00314.

NASA (2013). NASA Shuttle Radar Topography Mission Global 1 arc second [Data set]. https://doi.org/10.5067/MEaSUREs/SRTM/SRTMGL1.003.

NASA (2020). GNSS Orbit Products. https://cddis.nasa.gov/.

Neves, M. C., Spadotto, C. A., Luiz, A. J. B., and Quirino, T. R. (1998). Caracterização espaço-temporal do uso de agrotóxicos para o estado de São Paulo. In Assad, E. D. and Sano, E. E., editors, *Sistema de informações geográficas. Aplicações na agricultura*, chapter 12, pages 233–240. Embrapa-SPI/Embrapa-CPAC, Brasília. ISBN 85-7383-045-X.

NOAA (2020a). Magnetic field calculators. https://www.ngdc.noaa.gov/geomag/.

NOAA (2020b). State Plane Coordinate System - SPCS. https://geodesy.noaa.gov/SPCS/.

ON (2020). ASTRO Um conjunto de ferramentas de astronomia. https://daed.on.br/astro/.

Pebesma, E. (2018). Simple Features for R: Standardized Support for Spatial Vector Data. *The R Journal*, 10(1):439–446.

Pebesma, E. (2021). *sf: Simple Features for R*. R package version 1.0-2.

Pebesma, E. and Bivand, R. (2021). Spatial Data Science with applications in R. https://keen-swartz-3146c4.netlify.app/.

Pebesma, E., Bivand, R., Racine, E., Sumner, M., Cook, I., Keitt, T., Lovelac, R., Wickham, H., Ooms, J., Müller, K., Pedersen, T. L., and Baston, D. (2021). *sf: Simple Features for R*. R package version 0.9-7.

Pebesma, E. and Dunnington, D. (2020). In r-spatial, the Earth is no longer flat. https://www.r-spatial.org/r/2020/06/17/s2.html.

Pebesma, E., Mailund, T., and Hiebert, J. (2016). Measurement Units in R. *The R Journal*, 8(2):486–494. https://doi.org/10.32614/RJ-2016-061.

Pebesma, E., Mailund, T., and Kalinowski, T. (2019). *units: Measurement Units for R Vectors*. R package version 0.6-4.

Pebesma, E., Mailund, T., Kalinowski, T., Hiebert, J., and Ucar, I. (2020a). *units: Measurement Units for R Vectors*. R package version 0.6-7.

Pebesma, E., Rundel, C., Teucher, A., and Developers, L. (2020b). *lwgeom: Bindings to Selected 'liblwgeom' Functions for Simple Features*. R package version 0.2-5.

Pedzich, P. and Kuźma, M. (2012). Application of methods for area calculation of geodesic polygons on Polish administrative units. *Geodesy and Cartography*, 61(2):105–115. https://doi.org/10.2478/v10277-012-0025-6.

Penedo, J. and Borges, N. (2017). The challenge of choosing where to publish: The Predatory Journals! *Revista Portuguesa de Cirurgia*, II(42):5–6.

Price, W. F. and Uren, J. (1988). *Laser surveying*. Spon Press, Londres, 1st edition. ISBN 978-0747600237.

R Core Team (2021). *The R Stats Package*. R package version 4.1.0.

Rietdorf, A., Daub, C., and Loef, P. (2006). Precise positioning in real-time using navigation satellites and telecommunication. https://citeseerx.ist.psu.edu/viewdoc/summary?doi=10.1.1.581.2400.

Rule, N. (1989). The Direct Survey Method (DSM) of underwater survey, and its application underwater. *The International Journal of Nautical Archaeology and Underwater Exploration*, 18(2):157–162.

Saadati, S., Abbasi, M., Abbasy, S., and Amiri-Simkooei, A. (2019). Geodetic calibration network for total stations and GNSS receivers in sub-kilometer distances with sub-millimeter precision. *Measurement*, 141:258–266. https://doi.org/10.1016/j.measurement.2019.04.044.

Sallis, P. and Benwell, G. (1993). Geomatics: the influence of informatics on spatial information processing. In *Software engineering education*, pages 199–208. Elsevier. https://linkinghub.elsevier.com/retrieve/pii/B9780444815972500279.

Savje, F. (2019). *distances: Tools for Distance Metrics*. R package version 0.1.8.

Schofield, W., Mark, and Breach (2007). *Engineering surveying*. CRC Press, Oxon, 6th edition. ISBN 0-7506-6949-7.

Scott, D. N., Brogan, D. J., Lininger, K. B., Schook, D. M., Daugherty, E. E., Sparacino, M. S., and Patton, A. I. (2016). Evaluating survey instruments and methods in a steep channel. *Geomorphology*, 273:236–243. https://doi.org/10.1016/j.geomorph.2016.08.020.

Segantine, P. C. L. (2005). *GPS Sistema de Posicionamento Global*. EESC/USP, São Paulo, 1st edition. ISBN 8585205628.

Silva, C. A. U. (2003). Ângulos. In Erba, D. A., Thum, A. B., Silva, C. A. U., de Souza, G. C., Veronez, M. R., Leandro, R. F., and Maia, T. C. B., editors, *Topografia para estudantes de arquitetura, engenharia e geologia*, chapter III, pages 1–15. Editora Unisinos, São Leopoldo, 1st edition. ISBN 85-7431-191-X.

Silva, F. A. M. and Assad, E. D. (1998). Análise espaço-temporal do potencial hídrico climático do estado de Goiás. In Assad, E. D. and Sano, E. E., editors, *Sistema de informações geográficas. Aplicações na agricultura*, chapter 15, pages 273–309. Embrapa-SPI/Embrapa-CPAC, Brasília. ISBN 85-7383-045-X.

Silva, I. and Segantine, P. (2015). *Topografia para engenharia: Teoria e prática de geomática*. Elsevier, Rio de Janeiro, 1st edition. ISBN 978-85-352-7748-7.

Sinnott, R. W. (1984). Virtues of the Haversine. *Sky and Telescope*, 68(2):159.

Slocum, T. A., McMaster, R. B., Kessler, F. C., and Howard, H. H. (2014). *Thematic cartography and geovisualization*. Pearson Prentice Hall, Essex, 3rd edition. ISBN 978-1292040677.

Snyder, J. P. (1993). *Flattening the Earth: Two thousand years of map projections*. University of Chicago Press. ISBN 978-0226767475.

Soler, T. and Hothem, L. D. (1988). Coordinate systems used in geodesy: Basic definitions and concepts. *Journal of Surveying Engineering*, 114(2):84–97. https://doi.org/10.1061/(ASCE)0733-9453(1988)114:2(84).

South, A. (2021). *rnaturalearth: World Map Data from Natural Earth*. R package version 0.1.0.

Souza, G. C. (2003). Levantamentos planimétricos. In Erba, D. A., Thum, A. B., Silva, C. A. U., de Souza, G. C., Veronez, M. R., Leandro, R. F., and Maia, T. C. B., editors, *Topografia para estudantes de arquitetura, engenharia e geologia*, chapter V, pages 1–35. Editora Unisinos, São Leopoldo. ISBN 85-7431-191-X.

Spinielli, E. (2020). *nvctr: The n-vector Approach to Geographical Position Calculations using an Ellipsoidal Model of Earth*. R package version 0.1.4.

Spinielli, E. and EUROCONTROL (2020). *nvctr: The n-vector Approach to Geographical Position Calculations using an Ellipsoidal Model of Earth*. R package version 0.1.4.

Sun, K., Chen, J., and Viboud, C. (2020). Early epidemiological analysis of the coronavirus disease 2019 outbreak based on crowdsourced data: A population-level observational study. *The Lancet Digital Health*, 2(4):e201–e208. https://doi.org/10.1016/S2589-7500(20)30026-1.

Tennekes, M. (2018). tmap : Thematic maps in R. *Journal of Statistical Software*, 84(6). https://doi.org/10.18637/jss.v084.i06.

Tennekes, M. (2021). *tmap: Thematic Maps*. R package version 3.3-2.

Tennekes, M., Nowosad, J., Gombin, J., Jeworutzki, S., Russell, K., Zijdeman, R., Clouse, J., Lovelace, R., and Muenchow, J. (2020). *tmap: Thematic Maps*. R package version 3.2.

Tippmann, S. (2015). Programming tools: Adventures with R. *Nature*, 517(7532):109–110. https://doi.org/10.1038/517109a.

Turbek, S. P., Chock, T. M., Donahue, K., Havrilla, C. A., Oliverio, A. M., Polutchko, S. K., Shoemaker, L. G., and Vimercati, L. (2016). Scientific writing made easy: A step-by-step guide to undergraduate writing in the biological sciences. *The Bulletin of the Ecological Society of America*, 97(4):417–426. https://doi.org/10.1002/bes2.1258.

UKOOA (1994). The use of differential GPS in offshore surveying. Technical report, International Association of Oil & Gas Producers - IOGP, London.

Vermeille, H. (2002). Direct transformation from geocentric coordinates to geodetic coordinates. *Journal of Geodesy*, 76(8):451–454. https://doi.org/10.1007/s00190-002-0273-6.

Vermeille, H. (2004). Computing geodetic coordinates from geocentric coordinates. *Journal of Geodesy*, 78:94–95. https://doi.org/10.1007/s00190-004-0375-4.

Vincenty, T. (1975). Direct and inverse solutions of geodesics on the ellipsoid with application of nested equations. *Survey Review*, 23(176):88–93. https://doi.org/10.1179/sre.1975.23.176.88.

Wheaton, J. M., Garrard, C., Whitehead, K., and Volk, C. J. (2012). A simple, interactive GIS tool for transforming assumed total station surveys to real world coordinates – the CHaMP transformation tool. *Computers & Geosciences*, 42:28–36. https://doi.org/10.1016/j.cageo.2012.02.003.

Wickham, H. (2021). *tidyr: Tidy Messy Data*. R package version 1.1.3.

Wickham, H., Chang, W., Henry, L., Pedersen, T. L., Takahashi, K., Wilke, C., Woo, K., Yutani, H., Dunnington, D., and RStudio (2020). *ggplot2: Create Elegant Data Visualisations Using the Grammar of Graphics*. R package version 3.3.3.

Wickham, H., François, R., Henry, L., and Müller, K. (2021). *dplyr: A Grammar of Data Manipulation*. R package version 1.0.7.

Yoder, C. (1995). Astrometric and geodetic properties of Earth and the solar system. In *American Geophysical Union*, pages 1–31. American Geophysical Union. http://hdl.handle.net/2014/32032.

Zeisk, K. (1999). TPS1100 Professional series professional series, A new generation of total station from Leica Geosystem. Technical Report May, Leica Geosystems AG, Heerbrugg.

Zhou, J., Xiao, H., Jiang, W., Bai, W., and Liu, G. (2020). Automatic subway tunnel displacement monitoring using robotic total station. *Measurement*, 151:107251. https://doi.org/10.1016/j.measurement.2019.107251.

Index

For Product Safety Concerns and Information please contact our EU
representative GPSR@taylorandfrancis.com
Taylor & Francis Verlag GmbH, Kaufingerstraße 24, 80331 München, Germany

www.ingramcontent.com/pod-product-compliance
Ingram Content Group UK Ltd.
Pitfield, Milton Keynes, MK11 3LW, UK
UKHW051828180425
457613UK00007B/253